Carl Diener

Himálayan Fossils

Vol. 1, Part 1

Carl Diener

Himálayan Fossils
Vol. 1, Part 1

ISBN/EAN: 9783337299484

Printed in Europe, USA, Canada, Australia, Japan

Cover: Foto ©berggeist007 / pixelio.de

More available books at **www.hansebooks.com**

MEMOIRS

OF

THE GEOLOGICAL SURVEY OF INDIA.

Palæontologia Indica,

BEING

FIGURES AND DESCRIPTIONS OF THE ORGANIC REMAINS PROCURED DURING THE
PROGRESS OF THE GEOLOGICAL SURVEY OF INDIA.

PUBLISHED BY ORDER OF THE GOVERNMENT OF INDIA.

Series XV.

Vol. I, Part 1.

UPPER-TRIASSIC AND LIASSIC FAUNÆ OF THE EXOTIC BLOCKS OF MALLA JOHAR IN THE BHOT MAHALS OF KUMAON.

PLATES I to XVI.

By

CARL DIENER, Ph.D.,

Professor of Palæontology at the University of Vienna.

CALCUTTA:
SOLD AT THE OFFICE OF THE GEOLOGICAL SURVEY, 27, CHOWRINGHEE ROAD.
LONDON: MESSRS. KEGAN PAUL, TRENCH, TRÜBNER & CO.
BERLIN: MESSRS. FRIEDLÄNDER UND SOHN.

1908.

MEMOIRS OF THE GEOLOGICAL SURVEY OF INDIA.

𝔓𝔞𝔩𝔞𝔬𝔫𝔱𝔬𝔩𝔬𝔤𝔦𝔞 𝔍𝔫𝔡𝔦𝔠𝔞.

SERIES XV.

Vol. I, Part 1.

UPPER-TRIASSIC AND LIASSIC FAUNÆ OF THE EXOTIC BLOCKS OF
MALLA JOHAR IN THE BHOT MAHALS OF KUMAON.

MEMOIRS

OF

THE GEOLOGICAL SURVEY OF INDIA.

Palæontologia Indica,

BEING

FIGURES AND DESCRIPTIONS OF THE ORGANIC REMAINS PROCURED DURING THE
PROGRESS OF THE GEOLOGICAL SURVEY OF INDIA.

PUBLISHED BY ORDER OF THE GOVERNMENT OF INDIA.

Series XV.

Vol. I, Part 1.

UPPER-TRIASSIC AND LIASSIC FAUNÆ OF THE EXOTIC BLOCKS OF MALLA JOHAR IN THE BHOT MAHALS OF KUMAON.

PLATES I to XVI.

CALCUTTA :

SOLD AT THE OFFICE OF THE GEOLOGICAL SURVEY, 27, CHOWRINGHEE ROAD.
LONDON : MESSRS. KEGAN PAUL, TRENCH, TRÜBNER & CO.
BERLIN: MESSRS. FRIEDLÄNDER UND SOHN.

1 ***.

CALCUTTA
SUPERINTENDENT GOVERNMENT PRINTING, INDIA
8, HASTINGS STREET

HIMALAYAN FOSSILS.

VOLUME I, PART I.

UPPER-TRIASSIC AND LIASSIC FAUNÆ OF THE EXOTIC BLOCKS OF MALLA JOHAR IN THE BHOT MAHALS OF KUMAON.

(COLLECTIONS MADE BY THE GEOLOGICAL SURVEY OF INDIA IN THE YEAR 1900.)

BY

CARL DIENER, Ph.D.

PROFESSOR OF PALÆONTOLOGY AT THE UNIVERSITY OF VIENNA.

INTRODUCTION.

The following descriptions of fossils are based upon the rich collections gathered by the late Dr. A. von Krafft during his survey of the frontier district between Hundes and Malla Johar in the year 1900. They represent, for the most part, completely new material.

During the expedition of 1892, in which Griesbach, Middlemiss, and myself took part, a single fossiliferous block of red limestone was discovered near Sangcha Talla encamping-ground, at the head of the Kiogarh river. It yielded some fragments of *Jovites nov. sp. ex aff. J. bosnensis*, which, according to E. v. Mojsisovics (Palæont. Ind. ser. XV, Himálayan Fossils, Vol. III, Pt. 1, p. 18), proved the block to belong to the carole stage of the upper trias. In 1900 Dr. A. v. Krafft made an exhaustive study of the exotic blocks in the neighbourhood of the Balchdhura. He found the upper flysch and the basic igneous rocks of the district to abound with exotic blocks, some of which yielded a large number of well-preserved fossils.

The geological results of this survey have been summarized by A. v. Krafft in a very interesting paper, forming Vol. XXXII, Pt. 3, of the Memoirs of the Geological Survey of India (Calcutta, 1902). The majority of fossiliferous blocks discovered were of permian age. They have been marked on the map accompanying Dr. v. Krafft's memoir as B. B. Nos. 9, 11, 12, 13, 15, 18, 19. The rich fauna of block No. 0 has been described in my memoir on the permian fossils from the Central Himálayas (Palæont. Ind. ser. XV, Vol. I, Pt. 5, pp. 62-100).

B

Besides this considerable number of exotic blocks of permian age, some fossiliferous blocks containing triassic and liassic faunæ have been recorded by A. v. Krafft. Evidence has been obtained of the representation of the following stratigraphical horizons :—

1. *Lower Trias* (E. B. No. 20 on map). A large block of a dark red, earthy limestone, thin bedded, with *Danubites nivalis* Dien., *Flemingites sp.*, *Meekoceras sp.*, probably homotaxial with the Hedenstroemia beds of the main region of the Himálayas.

2. *Lower Muschelkalk*, doubtful, some badly-preserved Ammonites in a loose block (*Procladiscites cf. Yasoda* Dien. ?) pointing to the fauna of the Middlemiss crag in the Chitichun area.

3. *Ladinic or lower carnic stage* (E. B. No. 1). A dark red, very ferruginous limestone yielded a few specimens of *Daonella indica* Bittn.

4. *Upper carnic stage* (E. B. No. 2). From a bright red marine marble like that of the Hallstatt beds, about 1 mile to the north-west of the Balcholhura pass, a very large collection of Ammonites was obtained, most of them closely related to species characteristic of the zone of *Tropites subbullatus*. The following list of fossils, based on a cur-ory examination of the materials from this block, is given by A. v. Krafft on p. 143 of his memoir :—

> *Cladiscites sharetos* Mojs.
> *Phylloceras Ebneri* Mojs.
> *Juvavites (Griesbachites) Mojsvaans* Stol.
> *Arcestes dic. sp.*
> *Jovites sp. ex aff. Isaeusis* Mojs.
> *Placites sp. ind.*
> *Tropites sp. aff. subbullatus* Hauer.
>   ,,   *sp. aff. acutangulus* Mojs.
>   ,,   *sp. aff. Barthi* Mojs.
>   ,,   *sp. aff. episcus* Mojs.
> *Nautilus sp. ind.*

5. *Dachsteinkalk* (E. B. No. 8). No fossils.

6. *Lower Lias* (E. B. Nos. 4, 6, 7, 16, 17). In a very earthy, brick red, thinbedded, nodular limestone several species of *Arietites* and *Phylloceras* have been found. In preservation they are exactly ident cal with the Lias of Adneth near Salsburg in the Eastern Alps.

The faunæ representing the three horizons, 3, 4, 6, have been submitted to me for examination. They are of particular interest. The Subbullatus fauna from exotic block No. 2 is the first upper-triassic fauna obtained from a facies identical with the famous Hallstatt limestone of Austria, and differing considerably from what is seen in the main region of the sedimentary belt of the Central Himálayas. The liassic fossils from exotic blocks Nos. 16 and 17 are the first as yet recorded from India.

I.—FOSSILS FROM EXOTIC BLOCK NO. 1 (BALCHDHURA HEIGHTS).

The fossil materials from this block available for examination are extremely scanty, consisting of two fairly well-preserved casts of *Daonella indica* Bittn. and of the fragment of a *Halobia*.

DAONELLA INDICA Bittner. Pl. VIII, fig. 1.

1899. *Daonella indica* Bittner, Himálayan Foss., Palæontologia Ind. ser. XV, Vol. III, Pt. 7, p. 39, Pl. VII, figs. 4-11.

1901. *Daonella indica* A. v. Krafft, General Report, Geol. Surv. of India, for 1899-1900, p. 206.

1902. *Daonella indica* A. v. Krafft, Mem., Geol. Surv. of India, XXXII, Pt. 3, p. 148.

1907. *Daonella indica* Diener, Himál. Foss. l. c. Vol. V, Pt. 3, Pt. III, figs. 6, 7, 10.

The specimen illustrated—a right valve with fairly well-preserved sculpture—agrees in all its characters with the types of this species, which have been figured by A. Bittner. It is of very large size and scarcely inferior in its dimensions to any of the examples known from the Central Himálayas of Painkhánda and Spiti. Its exact measurements cannot be given, the outlines of the shell being but partly accessible to examination.

There is no anterior ear developed, but the radial ribbing reaches the hinge-margin on either side of the apex almost in full strength. In accordance with the large size of the specimen the majority of secondary ribs are subdivided in the vicinity of the margins, especially so in the middle of the shell.

The ribbing is very regular, being characterized by the almost uniformly bipartite structure of ribs and furrows.

HALOBIA sp, ind. Pl. VIII, fig. 3.

Together with the illustrated specimen of *Daonella indica* a fragmentary cast has been noticed, showing an unmistakable anterior ear, which has probably been flat and narrow. The ribs are very delicate and undulating, but are curved backward in the middle of the height of the valve. They begin near the umbo, but are rather distant and exhibit a bundle-like arrangement, which is due to a very regular intercalation of new ribs of a second and third order. The wrinkles of growth are closer set and stronger than the majority of ribs.

Among European species it is probably the group of *Halobia raretristriata* Mojs. to which the present form is most nearly allied, but its preservation is not sufficiently perfect to justify a specific determination.

CONCLUSIONS.

As has been remarked by A. v. Krafft, no definite age can be assigned to the limestone of exotic block No. 1 with *Daonella indica*. In the Himálayas th species is not confined to a distinct stratigraphical horizon, as had been suggested by Bittner, but is of comparatively wide stratigraphical distribution, ranging throughout the ladinic into the lower carnic stage. In Painkhánda it has its main layer in the bed with *Norella Kingi* and *Norella tibetica*, immediately overlying the Traumatocrinus limestone of lower carnic age, but it has also been discovered by A. v. Krafft in the thin-bedded limestone series separating the Traumatocrinus beds from the main layer of *Ptychites rugifer*. In the Cephalopoda-bearing limestone near the Ralphu glacier (Lissar valley) it is associated with a fauna of Ammonites which is certainly older than that of the Traumatocrinus limestone. In Spiti it occurs both in the Daonella limestone and in the Daonella shales, where it is associated with *Daonella Lommeli*, *Protrachyceras Archelaus*, and numerous species characteristic of the ladinic stage.

Exotic block No. 1 may therefore represent either the ladinic or lower carnic stage. The presence of a true Halobia is rather in favour of the latter alternative.

II.—FAUNA OF EXOTIC BLOCK NO. 2 (BALCHDHURA HEIGHTS, ABOUT 18,000 FEET).

Among all mesozoic faunæ from the exotic blocks of Malla Johar this is by far the richest and best preserved. The specimens are in preservation exactly identical with those from the Hallstatt marble of the Roethelstein near Aussee and they are often provided with their test.

LAMELLIBRANCHIATA.

The materials collected by A. v. Krafft are so scanty and so imperfectly preserved that it is impossible to venture on a specific identification of the single valves, none of them allowing an observation of any characters of generic importance. Another reason which renders the fragmentary examples lying before me unfit for determination is the incomplete state of our knowledge of the Lamellibranchiata of the Hallstatt limestone, *Daonella*, *Halobia*, and perhaps *Monotis*, being the only genera of the class which have been treated thoroughly by previous authors. There is only one single specimen admitting of a special description, although a more minute investigation into its relationship is impracticable on account of its incomplete state of preservation.

CASSIANELLA sp. ind. Pl. VIII, fig. 6.

A cast of a left valve of large size deserves mentioning on account of its similarity with Alpine representatives of the genus *Cassianella*.

It is considerably inflated, of nearly equal length and height. The central inflation of the valve exhibits a saddle-like depression in the middle, which is, however, but faintly marked and extends from the umbonal region to the ventral margin. The strongly vaulted posterior wing is separated from the central portion of the cast by a sharp incision. Otherwise there is no interruption to the regularity of the posterior slope.

The anterior wing has, unfortunately, not been preserved. But two slight keel-like ribs are recognized along the anterior slope of the umbo. These are the only radial ribs, otherwise the sculpture of the shell, which has been partly preserved, consists of numerous and delicate concentric lines of growth. The beak is strongly incurved and with its apex shifted anteriorly. It projects above a flat area of moderate width, whose lower borders are not distinctly known.

That this specimen belongs really to the genus *Cassianella*, is very probable but not certain. The shape of the beak, the presence of an area, the plug-shaped incision in the posterior margin, and the anterior radial ribs are all structures characteristic of the genus *Cassianella*. But the entire absence of an anterior wing and of any characters of the hinge leaves a shade of doubt about our determination.

GASTEROPODA.

Gen. : LOXONEMA Phill.

LOXONEMA (POLYGIRINA) cf. ELEGANS Hoernes. Pl. I, fig. 6.

1855. *Loxonema elegans* Hoernes. Gastropoden und Acephalen der Hallstaetter Schichten, Denkschr. Kais. Akad. d. Wissensch. Wien, LX, p. 36, Taf. I, fig. 3.

1897. *Loxonema (Polygirina) elegans* Koken, Gastropoden der Trias am Hallstatt, Abhandl. K. K. Geol. Reichsanst. XVII, p. 94, Taf. XV, figs. 6, 7, 18.

Only one fragmentary specimen, consisting of the last whorl and two coils of the spire, are contained in A. v. Krafft's collections. Notwithstanding the absence of its apical portion it is more complete than the majority of the Alpine examples described by Hoernes and Koken, the typical shape of this species being only known from a combination of several fragmentary specimens.

My specimen agrees entirely with equal-sized fragments of *Loxonema elegans* from the Hallstatt limestone of Aussee. It is turreted and provided with high whorls, which are slightly convex and distinctly impressed along the suture. Their greatest transverse diameter is situated considerably below the middle of their height.

The aperture has not been entirely preserved, but the presence of a small canal at the posterior border is distinctly marked.

The shell, which has been partly preserved, is covered with delicate, falciform transverse striæ of growth. Spiral wrinkles are faintly developed.

Remarks.—*Loxonema elegans* is of little stratigraphical value, ranging from the carnic into the noric stage.

Gen. : SAGANA Koken.

SAGANA cf. GEOMETRICA Koken.   Pl. II, fig. 3.

1887. *Sagana geometrica* Koken, Gastropoden der Trias von Hallstatt, Abhandl. K. K. Geol. Reichsanst. Bd. XVII, p. 39, Taf. VI, fig. 10.

This species is represented in A. v. Krafft's collections by a single but beautifully preserved and almost complete specimen, which differs from the typical shape of *Sagana geometrica* only by such details of ornamentation as are noticed in some of the intermediate forms between *Sagana geometrica* and *S. Hoernesi* Stur (Jahrb. K. K. Geol. Reichsanst. 1899, p. 285).

The turbinate shell consists of five whorls, which are strongly convex and inflated, especially the last one, which is separated from the umbilicus by a distinct spiral edge.

The slit band corresponds to the greatest convexity of the whorls. It is flatly concave, not broader than the rest of the bands enclosed between the numerous spiral ridges. The slit of the aperture has been entirely preserved. It is short and rounded. The numerous and sharp, closely set lunulæ, which cover the slit band, run nearly parallel to the outlines of the apertural slit. Two spiral keels are counted between the keel bordering the slit band and the suture, and eight between the slit band and the umbilical edge. Thus the number of spiral keels is smaller than in the typical *Sagana geometrica*, but larger than in *S. Hoernesi*. My specimen moreover recalls the intermediate shapes connecting the two species by the height of its last volution, which is smaller than in *Sagana geometrica*.

Of the shelly substance small fragments only are accessible to examination, showing the beautiful lattice-shaped ornamentation characteristic of the genus *Sagana*. The transverse striæ are very close-set and sharp, but inferior in strength to the spiral keels. The meshes, which are formed by the crossing of transverse ribs and spiral edges, are considerably higher than broad. The points, where the spiral ridges are intersected by the transverse ribs, are not marked by any tubercles as is the case in the typical *S. Hoernesi*.

Dimensions.

The measurements of this specimen are as follows :—

Entire length of the shell	. . . .	36·5 mm.
Greatest breadth of the shell	. . . .	23 „
Height of the last volution at the aperture	. . . . . .	13·5 „
Apical angle	. . . .	84°

Remarks.—*Sagana geometrica* is not rare in the carnic Hallstatt limestone of Aussee. *S. Hoernesi* and the intermediate shapes connecting the two species are known both from the carnic and noric stages of the Salzkammergut.

Gen. : CAPULUS Montf.

CAPULUS (PHRYX) JOHARENSIS nov. sp. Pl. II, fig. 4.

The distinction of fossil *Patellidæ* and *Capulidæ* is attended with great difficulty, since the characters of generic importance are but exceptionally preserved. The cup-shaped shells exhibit very little variation in form, and their systematic position is nearly always doubtful in the fossil state.

A single cast of a cup-shaped shell, in its external shape, more closely recalls the subgenus *Phryx* Blaschke (Gastropodenfauna der Pachycardientuffe der Seiseralpe in Suedtirol, Beiträge zur Geol. und Paläont. Oesterreich-Ungarns, etc., XVII, p. 174) than any genus of *Patellidæ*. Montfort (Conchyl. système 1810, II, p. 54) gives the following diagnosis of the genus *Capulus*: " Coquille libre, univalve, en bonnet phrygien ; à sommet plus ou moins aigu ou roulé, bouche entière ; intérieur marqué de deux musculaires. " K. v. Zittel (Handbuch der Palæozoologie II, p. 216) adds to this diagnosis, that the apex is always directed backwards and that the muscular scars take the shape of a horseshoe. Such forms of *Capulus*, which are provided with a symmetrical apex, which is neither enrolled spirally nor shifted laterally, have been distinguished from *Capulus s. s.* as *Phryx* by Blaschke, and have been elevated by this author to the rank of a proper subgenus. The only species of this subgenus hitherto known was *Phryx bilateralis* Blaschke (l. c. p. 172, Taf. XIX, fig. 9). Another triassic species, *Capulus Apollinis* Boehm (Die Gastropoden des Marmolatakalkes, Palæontographica XLII, p. 201), differs from *Phryx* by its spirally enrolled apex, which has been shifted considerably to the right.

The present shell agrees in its characters of subgeneric value with *Phryx bilateralis*, not with *Capulus Apollinis*. Its aperture is regularly elliptical, rather wide. The shell recalls by its shape a low Phrygian cap with its apex greatly removed posteriorly. The apex is neither enrolled nor twisted but contracted into a sharp and slightly incurved beak of symmetrical position. Muscular scars have not been noticed.

From *Phryx bilateralis* the present species differs by the following characters: The summit of the shell is not situated centrally, but nearly coincides with the apical beak. The apex is not elevated above the posterior margin of the aperture but removed anteriorly. The distance between the apex and the posterior shell margin is considerably larger and less strongly concave.

From all the triassic representatives of *Patellidæ*, which have been described by Boehm, Kittl, S. von Woehrmann, Koken, and Blaschke, my Himálayan species is distinguished by the curved shape of its apical beak. I therefore deemed it preferable to unite it with the subgenus *Phryx* among the family of *Capulidæ*.

Gen. : Naticopsis McCoy.

Naticopsis sp. ind. ex aff. obvallatæ Koken. Pl. I, fig. 5.

In making use of the generic name *Naticopsis* M'Coy for a triassic species I am not in accordance with E. Kittl, who in his monograph of the Gasteropoda of the Esino limestone (Annalen des K. K. Naturhistorischen Hofmuseums XIV, 1899, p. 23) reserves this denomination for *Natica ampliata* Phill, and its allies. But my materials are so scanty, consisting of two shells only, one of them rather incomplete, that I cannot detect such characters of distinction as might enable me to establish their generic determination with any certainty. The only internal feature which I have been able to discover is the presence of a median tooth projecting from the inner lip, which is curved and flattened. This character clearly shows that our species has nothing to do with *Natica* and its allies, but is not sufficient for its grouping among the family of *Naticopsidæ*. The most important characters for a generic distinction, namely, the reabsorption of inner walls in the earlier whorls, are not accessible to examination in my examples. As it has been customary to group under the general head of *Naticopsis* such fossil shells as externally recall *Natica*, and are not specially distinguished by some other characters, I shall keep provisionally to this denomination.

Among Alpine species the present one bears a close resemblance to *Naticopsis obvallata* (Gastropoden der Trias um Hallstatt, Abhandl. K. K. Geol. Reichsanst. XVII, p. 70, Taf. XII, fig. 5), from which it differs, however, by the insignificant depression of its sutures.

Shell globose. Spire with a pointed apex, whorls increasing rapidly. Last whorl inflated, with a regular aperture whose outline is somewhat flattened along the upper portion of the outer lip. Striæ of growth describing a curve, which is turned backwards considerably.

CEPHALOPODA.

AMMONOIDEA.

ARCESTOIDEA.

Fam. : CLADISCITIDÆ.

Gen. : Cladiscites v. Mojsisovics.

The genus *Cladiscites* is rather richly represented in the fauna of exotic block No. 2. The majority of the specimens belong to representatives of *Cladiscites s. s*, distinguished by the serial arrangement of their sutural elements. Two species belong to the subgenus *Hypocladiscites* (group of *Cladiscites subtornati*).

Of the subgenus *Paracladiscites* no representative is known to me, although from the triassic belt of the main region of the Himálayas a species of this subgenus (*Paracladiscites indicus*) has been described by E. v. Mojsisovics.

1. CLADISCITES CRASSESTRIATUS v. Mojsisovics, Pl. II, figs. 5-8; Pl. IV, figs. 1, 2, 8.

1873. *Arcestes crassestriatus* E. v. Mojsisovics, Ceph'dopoden der Hallstaetter Kalke, Abhandl. K. Geol. Reichsanst. VI-1, p 79, Taf. XXX, fig. 4.

1902. *Cladiscites crassestriatus* E. v. Mojsisovics, ibidem, Supplementbd., p 281.

This is the most common species of the genus and, indeed, one of the leading fossils of the red limestone of exotic block No. 2. It attains considerable dimensions. My largest fragment, which has been illustrated on Pl. II, fig. 7, is yet entirely chambered. As one more volution at least must be reckoned for the body-chamber, the diameter of the adult individual cannot have been less than 150 mm. Thus the Indian types of the species are scarcely inferior in size to Alpine specimens from the Rorthelstein, among which some examples, consisting of air-chambers only, have a diameter of 90 mm.

I should not have ventured on a direct identification of my Himálayan examples with the Alpine species without a personal examination of large materials from the Hallstatt limestone of julic and tuvalic age. The most prominent features of the species have not been reproduced satisfactorily in the illustration on Pl. XXX, fig. 4, of E. v. Mojsisovics' memoir, as will be seen from a comparison with the illustration of his type-specimen on Pl. IV, fig. 9, of the present monograph. It must, however, be remarked that this type-specimen is of rather small size and that some of the specific characters of *Cladiscites crassestriatus*, especially the angular shape of the cross-section and the insignificant development of striæ on the siphonal area are less prominent than in the majority of specimens from the Hallstatt limestone. An illustration of an average-sized specimen from the *Subbullatus* beds of Aussee is given on Pl. V, fig. 5, of this memoir.

The most important distinctive feature between *Cl. crassestriatus* and other congeneric forms in the carnic stage is the presence of distinctly marked angular margins, bordering the siphonal area. The siphonal area is gently arched and separated sharply from the flat lateral parts. The width of the siphonal area is but slightly inferior to the greatest transverse diameter of the cross-section. It is only in very young stages of growth (Pl. II, fig. 8) that the marginal angles are rounded off. In later stages of development the rectangular shape of the cross-section is as distinctly marked as in *Cl. morosus* v. Mojsisovics (l. c. p. 76, Pl. XXIX, fig. 3) or in *Paracladiscites diuturnus* v. Mojsisovics (l. c. p. 89, Taf. XXXI, fig. 2).

The volutions are comparatively high, the width of the shell being as a rule more than three-quarters of its length, but invariably inferior to the latter.

Another feature of specific importance is the remarkable difference in the strength of ornamentation on the ventral area and lateral parts. The striation of the lateral parts is rather coarse, especially so in specimens of large dimensions. The ventral area is either perfectly smooth or covered with very delicate longitudinal striae only. Examples with a siphonal striation, as distinctly marked as in the type-specimen from the Roethelstein (Pl. IV, fig. 9), are quite an exception.

Dimensions.

		Pl. IV, fig. 1.	Pl. II, fig. 5.
Diameter of the shell		53 mm.	67 mm.
" " " umbilicus . . .		0 "	0 "
Height of the last } above the umbilical suture		31 "	43 "
volution } " " preceding whorl		21 "	27 "
Thickness of the last volution		25 "	34 "

Sutures.—The sutural line of this species was not known to E. v. Mojsisovics in 1873. I have succeeded in developing the sutures of a specimen from the Hallstatt limestone of the Roethelstein, although not in its details. The illustration represented in Pl. IV, fig. 8, is taken from one of my largest Himálayan examples. The complication of the sutural line renders its preparation extremely difficult.

In the genus *Cladiscites s.s.* special importance is attributed by E. v. Mojsisovics to the shape of the siphonal saddle. Three external branches are developed, besides the bipartite culminating outer branch of the dimeroidic siphonal saddle. The arrangement of the lobes is typically serial, the lateral and auxiliary lobes standing at an equal level. The number of auxiliary lobes is not exactly known to me.

Remarks.—Among the species of *Cladiscites* from the triassic beds of Sicily which have been described and figured by Gemmellaro (I cefalopodi del trias superiore della parte occidentale della Sicilia, Palermo, 1904), *Cl Ferdinandi* Gemmellaro (l. c. p. 275, Tav. XXIX, figs. 43, 44) is probably identical with the present species. The species illustrated by Gemmellaro is of small size, attaining a diameter of 26 mm. only, but it is provided with a part of its body-chamber. It recalls very strongly *Cladiscites crassestriatus*. The features of distinction enumerated by Gemmellaro, namely, the more delicate striation and the compressed shape of the lateral parts, are either insignificant or due to a misinterpretation of the unsatisfactory illustration given by E. v. Mojsisovics.

2. CLADISCITES cf. GORGIÆ Gemmellaro. Pl. IV, fig. 5.

1904. *Cladiscites Gorgiæ Gemmellaro.* I cefalop di del Trias superiore della parte occidentale della Sicilia, p. 276, Tav. III, figs 19, 20; XXIII, figs. 3-5.

This species shows in its external characters a great resemblance to *Cladiscites Gorgiæ* Gemm. from the carnic limestone of Sicily. My Himálayan examples differing from the European type in some subordinate details, the species is recorded here as *Cl. cf. Gorgiæ.*

All my Himálayan examples are entirely chambered but do not reach the dimensions of the largest types from Sicily. They agree best in size and shape with the specimen illustrated on Pl. III, figs. 19, 20, by Gemmellaro. The width of the shell is either equal or inferior to the height. The specimen figured is as strongly compressed as the Sicilian types, whose dimensions have been chosen for standard by Gemmellaro. The flattened siphonal area is not marked off from the gently and regularly arched flanks by sharp margins, but passes into them more gradually than in *Cl. crassestriatus*. The umbilicus is closed, as in the majority of congeneric species.

The ornamentation recalls *Cl. crassestriatus* by the faint development of spiral striæ in the siphonal region. Transverse striæ of growth have not been noticed in any of my specimens, whereas such have been mentioned in the diagnosis of the Mediterranean species by Gemmellaro.

Dimensions.

Diameter of the shell	40 mm.
,, ,, ,, umbilicus	0 ,,
Height of the last whorl above the umbilical suture	20 ,,
,, ,, ,, preceding whorl	14 ,,
Thickness of the last volution	20 ,,

Sutures.—I have succeeded in chiseling out the inner nucleus of a large specimen, which shows the sutural line fairly well. It agrees very closely with the sutures of *Cl. Gorgiæ*, as illustrated by Gemmellaro on Pl. XXIII, fig. 5.

There are from seven to eight saddles counted from the umbilical suture to the middle of the siphonal area. Their stems and branches are thinner than in the majority of congeneric species, perhaps, even a little more robust than in the Sicilian type-specimen of *Cl. Gorgiæ*. All saddles are distinctly dimeroidic. In the main saddle each portion of the dimeroidic top is again subdivided into two smaller secondary phylla. The arrangement of lobes and saddles is typically serial. The siphonal lobe is the deepest. The lateral lobes show a tripartite arrangement of their basal branches.

3. CLADISCITES cf. PUSILLUS v. Mojsisovics. Pl. IV, figs. 6, 7.

1873. *Arcestes pusillus* E. v. Mojsisovics, Cephalopoden der Hallstätter Kalke, Abhandl. K. K. Geol. Reichsanst. VI-1, p. 77, Taf. XXVIII, fig. 4.
1902. *Cladiscites pusillus* E. v. Mojsisovics, ibidem, Supplementbd, p. 281.

Having E. v. Mojsisovics' type-specimens of *Cladiscites crassestriatus*, *Cl. pusillus* and *Cl. strialissimus* at hand for comparison, I find their distinction from each other and from *Cl. Gorgiæ* Gemm. a difficult matter. All of them are certainly most nearly allied and differ by very subordinate details only.

The type-specimen of *Cladiscites pusillus* from the *Elipticus* beds of the Roethelstein is a small Ammonite of 24 mm. in diameter (not 28, as has been stated by E. v. Mojsisovics). Its cross-section is of nearly equal height and width. The lateral parts are very gently arched. The demarcation between them and the flat siphonal area is not so sharp as in *Cl. crassestriatus*, but a little sharper than in *Cl. Gorgia*. The only distinction which I can find in the ornamentation is confined to the very delicate striation of the lateral and siphonal parts. The latter is not smooth as in the majority of examples of *Cl. crassestriatus*. An equally delicate striation is found in *Cl. striatissimus* r. Mojsisovics (L. c. p. 77, Taf. XXX, fig. 1), but this species is more strongly inflated, and the ventral area is considerably less broad than the largest transverse diameter of the shell.

With the type-specimen of *Cl. pusillus* some Himálayan representatives of *Cladiscites* agree very well in their external shape and sculpture. Both the flanks and siphonal part are covered with numerous and thin concentric striæ. Their transverse section is less distinctly angular than in *Cl. crassestriatus*, but in this respect transitional shapes between the two species are known to me.

Dimensions.

	E. r. Mojsisovics' type-specimen (Pl. XXVIII, fig. 4).	Himálayan specimen (Pl. IV, fig. 7).
Diameter of the shell	24 mm.	27 mm.
„ „ „ umbilicus	0 „	0 „
Height of the last ⎰ above the umbilical suture	14 „	15 „
volution ⎱ „ „ preceding whorl	7 „	7 „
Thickness of the last volution . . .	16 „	16 „

Sutures.—Not known in detail.

4. CLADISCITES sp. ind. cf. CORACIS, Gemm.   Pl. V, fig. 4.

A single, imperfectly preserved specimen of a large *Cladiscites* is mentioned here on account of its resemblance to *Cl. Coracis* Gemmellaro (I cefalopodi del Trias superiore della regione occidentale della Sicilia, p. 273, Tav. XXII, fig. 7; XXIII, figs. 1, 2) from the carnic limestone of Sicily. It is strongly compressed, high-mouthed and considerably higher than broad. The lateral parts are regularly although gently arched. Their greatest transverse diameter is situated below the middle of the height. They unite with the flattened siphonal area without forming a sharp angle. The concentric striation is coarse and not restricted to the lateral parts, as in *Cl. crassestriatus*.

The affinity with *Cl. Coracis* shows itself in the shape of the cross-section and in the strong compression of the whorls. From *Hypocladiscites subtornatus* and its allies, which are also provided with strongly compressed volutions, our species is distinguished by its flat siphonal area and by its more distinctly defined lateral margins.

Dimensions.

Diameter of the shell	59 mm.
„ „ „ umbilicus	0 „
Height of the last ⎰ above the umbilical suture	53 „
volution ⎱ „ „ preceding whorl	34 „
Thickness of the last volution	20 „

Remarks.—The specimen has been referred provisionally to *Cladiscites Coracis* as *sp. ind.*, but its possible connection with the subgenus *Hypocladiscites* cannot be excluded, the sutural line being entirely unknown.

Subgen. : *HYPOCLADISCITES* v. Mojs.

5. HYPOCLADISCITES SUBCARINATUS Gemmellaro. Pl. IV, fig. 4.

1904 *Hypocladiscites subcarinatus* Gemmellaro, I cefalopodi del Trias superiore della regione occidentale della Sicilia, p. 279, Tav. XXII, figs. 8-11; XXV, fig. 21; XXVI, figs. 1, 2.

This species of the upper carnic limestone of Sicily is represented in A. v. Krafft's collections by numerous and well-preserved specimens which permit of an accurate determination. The largest of my Himálayan examples, which consists of air-chambers only, has been chosen for illustration.

In its general shape this species recalls *Cladiscites crassestriatus* v. Mojsisovics, especially in the rectangular outlines of its cross-section, but in the character of its siphonal area it is at a glance distinguished from all congeneric forms. The siphonal area is slightly excavated, forming a shallow depression, which is interrupted in the middle by a low and rounded keel. The two marginal elevations separating the depressed siphonal area from the lateral parts are higher than the median keel and more broadly rounded. The lateral parts are very gently arched. Their greatest transverse diameter is situated in the umbilical region.

Both the median keel and the rounded marginal elevations are developed at very early stages of growth. I have seen them clearly marked in a chambered nucleus attaining a diameter of 10 mm. only. Thus a determination of this species is even possible if inner nuclei of small size only are available for examination.

The lateral parts are covered with numerous and delicate longitudinal striæ. On the siphonal area those striæ are still more delicate and are gradually obliterated in the direction towards the median keel, which remains free from any sculpture. Transverse striæ or folds, as noticed by Gemmellaro in a minority of his Sicilian specimens, are entirely absent.

The strongly inflated variety which has been mentioned as *Hypocladiscites nov. form. ind. prox. H. subcarinato* by Gemmellaro (l. o. p. 280, Tav. XXII, figs. 12, 13) is not represented among my Himálayan materials.

Dimensions.

Diameter of the shell	.	37 m.m.
„ „ „ umbilicus	.	0 „
Height of the last { above the umbilical suture	.	27·5 „
volution { „ „ preceding whorl	.	11 „
Thickness of the last volution	.	16 „

Sutures.—Not known in detail, but in their general arrangement agreeing with the illustration on Tav. XXV, fig. 21, of Gemmellaro's memoir. The second lateral lobe is considerably deeper than the following lobes.

6. HYPOCLADISCITES SUBARATUS v. Mojsisovics.   Pl. IV, fig. 3.

1896. *Cladiscites cf. subtornatus* E. v. Mojsisovics in Diener, Ergebnisse einer geologischen Expedition in den Central-Himalaya, etc.. Denkschr. K.als. Akad. d. Wissensch. LXII, p. 544.

1895. *Cladiscites (Hypocladiscites) subtornatus* E. v. Mojsisovics, Obertriadische Cephalopodenfaunen des Himalaya, Denkschr. kals. Akad. d. Wiss. LXIII, p. 657, Taf. XX, fig. 3.

1899. *Cladiscites (Hypocladiscites) subaratus* E. v. Mojsisovics, Palaeont. Ind. ser. XV Himálayan Foss., Vol. III. Pt. 1, p. 102, Pl. XX, fig. 2.

A species of *Hypocladiscites* from the carnic Daonella beds of Lauka, which in its outward shape agrees very closely with *H. subtornatus*, has been separated from this Alpine form by E. v. Mojsisovics, on account of some differences in its sutural line.

Some fragments of high-mouthed and strongly compressed *Cladiscitidæ* in A. v. Krafft's collection must be united with *Hypocladiscites subaratus*. The close agreement of the sutures will be recognised by comparing the illustrations. In my specimen the principal lateral saddle is still higher and broader, and is provided with more richly serrated lateral branches than in the type illustrated by E. v. Mojsisovics. Thus the features of distinction between *H. subtornatus* and *H. subaratus* are still more strongly marked in the present example than in the specimen from the Daonella beds of Lauka.

Fam. : *ARCESTIDÆ.*

Gen. : ARCESTES, SUESS.

7. ARCESTES cf. PERIOLCUS, v. Mojs.   Pl. III, fig. 1.

1873. *Arcestes periolcus* E. v. Mojsisovics, Cephalopoden der Hallstätter Kalke, Abhandl. K. K. Geol. Reichsanst. VI-1, p. 109, Taf. L, figs. 1-3 ; LII, figs. 4, 5 ; LIII, fig. 27.

My type-specimen is of large size, nearly complete, and has its peristome entirely preserved. It recalls most closely the Alpine *Arcestes periolcus*, both in its general outlines and in the development of a spiral sulcus surrounding the widely open umbilicus.

Of the penultimate whorl a small portion only is accessible to examination. It is globose, regularly rounded, smooth and without any constrictions or varices. The last whorl changes its shape in a manner agreeing with that in *A. periolcus*. The greatest width of the flattened siphonal area corresponds to the peristome, whereas it is most strongly compressed in the opposite middle portion of the whorl. The lateral parts are flattened. The umbilical wall is vertical and separated from the adjoining parts of the flanks by a raised marginal band, which is bordered externally by a shallow spiral sulcus. This sulcus is lower than in the majority of Alpine specimens, which have been united in this species by E. v. Mojsisovics.

The apertural margin is strongly contracted in the middle, and protracted into two lateral lappets in the siphonal region. This is the normal shape of the peristome in the group of *Arcestes coloni*.

The shell is nearly smooth. It is only in the vicinity of the peristome that delicate striæ of growth are developed, following in their direction the outlines of the apertural margin.

Traces of transitorial paulostomes, as have been noticed by E. v. Mojsisovics in some of his European examples, are altogether absent.

Dimensions.

Diameter of the shell	.	.	.	.	.	.	.	.	.	8: mm.
„ „ „ umbilicus	.	.	.	.	.	.	.	.	8 „
Height } of the last volution	.	.	.	.	.	.	42 „
Thickness									32 „

Sutures.—Not known.

2. ARCESTES of. RICHTHOFENI v. Mojsisovics. Pl. III, fig. 2.

1873. *Arcestes Richthofeni* E. v. Mojsisovics. Die Cephalopoden der Hallstatter Kalke, Abhandl. K. K. Geol. Reichsanst, VI-1, p 132. Taf. XLIX, figs. 4, 5 ; LIII, fig 18.

The only specimen available for investigation is a nearly complete cast, with its peristome almost entirely preserved and with large portions of the shelly substance adhering. It recalls very strongly the European *Arcestes Richthofeni* from the julic stage of the carnic Hallstatt limestone.

The apertural margin, although partly injured, is sufficiently well preserved to indicate the absence of any emargination along the broadly flattened siphonal area as is noticed in the majority of *Arcestidæ.* In this character it agrees exactly with the apertural margin of *A. Richthofeni* and *A. agnatus.* Peristome contracted laterally ; umbilicus narrow, but open, not closed by any callosity Siphonal area flattened in the vicinity of the peristome, but sharply rounded and narrow in the opposite quadrant of the last volution.

E. v. Mojsisovics noticed in one of his specimens an internal siphonal ridge, which was, however, restricted to the quadrant of the volution bordering the aperture. This internal siphonal ridge is clearly developed in my specimen, as is obvious from the illustration. Where the shelly substance has been preserved, its surface is perfectly smooth, but in the cast a deep impression of rectangular cross-section follows the median line of the siphonal part, marking the presence of an internal ridge. This ridge was probably intended to protect the siphuncle, which is situated more internally. Similar internal ridges have been described in *Phylloceras wermæsense* Herb. by Wæhner (Beiträege zur Palæontologie Oesterreich-Ungarns und des Orients, Bd. XI, p. 176). Casts of this species, in which a deep impression marks the place of the original shelly ridge, have been described as *Phylloceras aulomotum* by Herbich (Das Szeklerland. Mittell. aus dem Jahrb. der Kgl. ungar. geol. Anstalt. Bd. V, p. 115).

Dimensions.

Diameter of the shell	. . . .	85 mm.
„ „ „ umbilicus	. . . .	2·5 „
Height of the last { above the umbilical suture	. . .	20·5 „
volution { „ „ preceding whorl	. . .	9 „
Thickness of the last volution	. . . . . .	21·5 „

Sutures. — Not known.

3. Arcestes sp. ind. aff. decipiens v. Mojs.   Pl. VIII, fig. 3.

A single nucleus, consisting of air-chambers only, most probably is related to *Arcestes decipiens* from the tuvalic substage of the Hallstatt limestone.

It agrees with this species in the presence of a very narrow umbilicus and remarkably strong constrictions. The greatest transverse diameter coincides with the rounded-off siphonal margin, from which the lateral parts converge very gradually towards the umbilicus.

The only difference between the Alpine and Himálayan forms consists in the direction of the strong constrictions, which in my specimen are not turned back-wards in crossing the siphonal area, as they are in *Arcestes decipiens* E. v. Moj-sisovics (Cephalopoden der Hallstaetter Kalke, l. c. VI-1, p. 133, Taf. LIV, figs. 2, 3).

Dimensions.

Diameter of the shell	. . . .	74 mm.
„ „ „ umbilicus	. . . .	ca. 1 „
Height } of the last volution .	. . . . .	{ 13 „
Thickness }		{ 12 „

Sutures. — Not known.

4. Arcestes cf. placenta v. Mojs.   Pl. IX, fig. 1.

1873. *Arcestes placenta* E. v. Mojsisovics. Cephalopoden der Hallstaetter Kalke, Abhandl. K. K. Geol. Reichsanst. VI-1, p. 116, Taf. LV, figs. 2—7.

A single but nearly complete specimen of *Arcestes* recalls very closely *A. placenta* v. Mojs. from the tuvalic Hallstatt limestone of Aussee. It is of larger size than full-grown European examples, but agrees with them in all characters of specific importance as far as such are accessible to examination.

The whorls are moderately compressed and provided with flatly arched lateral parts. The umbilicus is small but not closed. Siphonal part broadly rounded in the vicinity of the aperture and in the corresponding quadrant of the last volution, but the difference in width between this and the opposite quadrant is not consider-able. Peristome slightly emarginated along the siphonal area but not contracted along the lateral parts.

There are also some Alpine specimens in the collection of the K. K. Geologische Reichsanstalt in Vienna, in which this lateral contraction of the peristome is almost insignificant.

Dimensions.

Diameter of the shell	. . . .	65 mm.
„ „ umbilicus	. . . .	1·6 „
Height of the last { above the umbilical suture	. . . .	34 „
volution { „ „ preceding whorl	. . . .	11 „
Thickness of the last volution	. . . . . .	31 „

Sutures.—Not known.

Subgen. : PROARCESTES v. Mojs.

1. PROARCESTES GAYTANI, v. Klipstein.   Pl. III, figs. 3, 4, 5, 7, 8.

1845. *Ammonites Gaytani* v. Klipstein. Beitrage zur geologischen Kenntnis der oestlichen Alpen, p. 110, Taf. V. fig. 4.

1847. *Ammonites Gaytani* F. v. Hauer, Neue Cephalopoden von Aussee, Haidingers Naturwissenschaftl. Abhandl. I, p. 347.

1849. *Ammonites Gaytani* F. v. Hauer, Neue Cephalopoden von Hallstatt und Aussee, ibid, III, p. 17, Taf. IV, figs. 13, 14.

1869. *Arcestes Gaytani* Laube, Fauna der Schichten von St. Cassian, Denkschr. Kais. Akad. d. Wissensch. math. nat. Cl. XXX, 161, p. 89, Taf. XLIII, fig. 5.

1873. *Arcestes Gaytani* E. v. Mojsisovics, Die Cephalopoden der Hallstaetter Kalke, Abhandl. K. K. Geol. Reichsanst. VI-1, p. 105, Taf. LVIII, figs. 1—3.

1892. *Proarcestes Gaytani* E. v. Mojsisovics, ibidem, Supplem-nibl., p. 239.

1896. *Proarcestes Gaytani* O. v. Arthaber, Die alpine Trias des Mediterrangebietes, Lethaea mesozoica I. Taf. XLIV, fig. 2.

1908. *Proarcestes cf. Gaytani* Diener, Fauna of the Tropites limestone of Byans, Palaeont. Ind. ser. XV, Vol. V, Pt. I, p. 177, Pl. XII, figs. 10, 11.

It is this well-known species of the julic substage to which a considerable number of *Arcestidæ* in A. v. Krafft's collections can be safely assigned. The character of the sutural line is that which is peculiar to the group of *Arcestes bicarinati*, and which is distinguished by the presence of a moderately high and richly serrated median prominence. From *Proarcestes bicarinatus* Münst. and from *P. Aussenmus* v. Hauer our species differs remarkably by its less globose shape and by the faint development of varices. On the other hand the majority of specimens agrees very closely with *P. Gaytani*.

In some of them, it is true, the flattening of the lateral parts is less strongly marked than in typical shapes of *P. Gaytani*. This is especially the case with inner volutions, exactly as in the examples from the Tropites limestone of Byans, which have been referred to the present species as *cf.* in Himál. Foss., Vol. V, Pt. 1. The specimen illustrated in fig. 4 affords a good instance of this character. In the penultimate whorl, which consists of air-chambers only, the lateral parts are almost regularly rounded, describing a graceful and uninterrupted curve from the siphonal convexity to the rounded-off umbilical margin. In the last volution, however, which has been partly preserved, the lateral parts are high, compressed and flattened as distinctly as in full-grown individuals from the cærnic Hallstatt limestone of the Salzkammergut. This specimen is scarcely inferior in size to the largest European examples which have been noticed by E. v. Mojsisovics.

The umbilicus is widely open, both in chambered nuclei and in specimens provided with their body-chambers. Constrictions on varices are but faintly marked, confined to inner nuclei and often entirely absent. No constriction has been noticed in the example illustrated in fig. 5, whereas in the two nuclei, illustrated in figs. 7 and 8, two varices are counted within the circumference of the last whorl. Their direction is nearly straight with a convexity slightly turned forward in the siphonal area.

Dimensions.—The measurements of my largest specimen (fig. 4), with a portion of its body-chamber preserved, are as follows :—

Diameter of the shell	.	.	.	.	.	.	ab.	146 mm.
„ „ umbilicus	.	.	.	.	.	.		8 „
Height of the { above the umbilical suture	.	.	.	.		46 „
last volution { „ „ preceding whorl	.	.	.	.		15 „
Thickness of the last volution	.	.	.	.	.	.		44 „
Height } of the penultimate whorl	.	.	.	.		22 „
Thickness }								38 „

Sutures.—Agreeing entirely with those of the Alpine types of *P. Gaytani*.

2. PROARCESTES sp. cf. AUSSEANUS v. Hauer. PL III, fig. 6.

The comparatively high and serrated median prominence of the siphonal lobe proves this species to be a representative of the group of *Arcestes bicarinati*.

The only specimen available for examination is an inner nucleus with part of the last volution adhering, which is yet entirely chambered. It shows three constrictions, which are narrow and faintly developed, as in inner nuclei of *Arcestes Ausseanus* v. Hauer (Haidingers Naturwissenschaftl. Abhandl. I. p. 20⁵, Taf. VIII, figs. 6-8). From typical shapes of *Proarcestes Ausseanus*, as illustrated by E. v. Mojsisovics (Cephalopoden der Hallstaetter Kalke, Abhandl. K. K. Geol. Reichsanst. VI-1, Taf. LI, figs 1, 4) my specimen differs by its less globose shape, although its height is yet considerably inferior to its thickness.

Dimensions (of the nucleus).

Diameter of the shell	.	.	.	.	.	.	.	39 mm.
„ „ umbilicus	.	.	.	.	.	.	.	3 „
Height } of the last volution	.	.	.	.		23 „
Thickness }								19 „

Sutures.—Agreeing entirely with those of *Proarcestes Ausseanus*. The principal lateral lobe is situated on the convexity, by which the siphonal area passes into the lateral parts. Two auxiliary lobes outside the umbilical margin.

Remarks.—My scanty materials do not allow one to decide the question of specific identity or close affinity of the Himálayan shell with the European *P. Ausseanus*.

PROARCESTES (?) sp. ind. ex aff. BARRANDEI Laube. Pl. VIII. fig. 5.

The systematic position of the only specimen available for examination is somewhat doubtful. It is only on account of its external similarity with *Arcestes Barrandei* Laube (Die Fauna der Schichten von St. Cassian, Denkschr. Kais. Akad. d. Wissensch. Bd. XXX, 1869, 5. Abt. p. 90, Taf. XLIII, fig. 2) that I have united it with *Proarcestes*, not with *Arcestes s. s.* It must, however, be borne in mind, that not even the connection of *Arcestes Barrandei* itself with the group of *Arcestes extralabiati* (*Proarcestes*) has been ascertained by E. v. Mojsisovics (Cephalopoden der Hallstætter Kalke l. c. VI-1, p. 91), although the sutures exhibit some characters peculiar to this group.

In its general shape my specimen agrees with *Proarcestes Barrandei*, especially so in the outlines of its transverse section, but it differs remarkably by the presence of a large and widely open, funnel-shaped umbilicus. The inner volutions are globose and strongly inflated, the place of greatest inflation corresponding to the rounded umbilical margin. In the last whorl, which apparently belongs to the body-chamber, the regularly rounded siphonal part becomes considerably narrower, the transverse section thus assuming a cordiform shape. No traces of furrows are noticed on the body-chamber volution preceding the aperture, but a shallow contraction is marked on the cast of the penultimate whorl.

Dimensions.

Diameter of the shell	.	.	.	.	.	.	.	.	.	57 mm.
„ „ umbilicus	.	.	.	.	.	.	.	.	.	9 „
Height } of the last volution	.	.	.	.	.	.	.	.	.	25 „
Thickness }										21 „

Sutures.—As far as known, agreeing with those of *P. Barrandei*. Median prominence moderately high and laced at its base by converging digitations of the siphonal lobe. Siphonal saddle not perfectly symmetrical, but with larger external branches. Number of auxiliary lobes not known exactly.

PROARCESTES sp. ind. (GROUP OF EXTRALABIATI). Pl. VIII. fig. 4.

Among the *Arcestidæ* from exotic block No. 2 a species belonging to the group of *extralabiati* is rather richly represented, although by incomplete specimens only. In none of them has the body-chamber been entirely preserved, nor have I succeeded in developing the sutural line. The relation of this species is therefore based on its external features only, exactly as in the case of *P. Danai* v. Mojsisovics (Cephalopoden der Hallstætter Kalke, l. c. VI-1, p. 93, Taf. LVII, fig. 4), to which species it appears to be most nearly allied.

In the specimen illustrated, the inner nucleus is globose with siphonal and lateral parts regularly rounded. In the last volution the lateral parts are flatly arched and marked off indistinctly from the broad and flattened siphonal area. No constrictions or varices have been noticed on the inner volutions as far as they

have been exposed, but, near the aperture, the shallow furrows and folds characteristic of the group of *Arcestes extralabiati* make their appearance. In a second
specimen these furrows are even more strongly marked and deeper than in the
present one, but their exact number is not known to me, the body-chamber being
incomplete in all the examples available for examination.

From *Proarcestes Danai* our species is distinguished by its more slowly
increasing whorls and by its narrower umbilicus.

Dimensions.

Diameter of the shell	. . . .	crs. 67 mm.
„ „ umbilicus .	. . . .	. 3 „
Height of the ʃ above the umbilical suture	. . . . . .	. 29 „
last rolation ｛ „ „ preceding whorl	. . . . .	. 7 „
Thickness of the last rolation	. . . .	. 31 „

Sutures. — Not known.

PINACOCERATITOIDEA.

Fam. : *LYTOCERATIDÆ.*

Gen. : PHYLLOCERAS Suess.

Subgen. : *DISCOPHYLLITES* Hyatt.

There is much discrepancy of opinion among different authors as to the range
and interpretation of the subgenera of the genus *Phylloceras* Suess. The majority
of triassic species have hitherto been grouped with the subgenus *Rhacophyllites* v.
Zittel, but there are strong reasons against the correctness of this grouping, which
seems to be at variance with the circumscription of *Rhacophyllites*, introduced in
the memoirs on liassic cephalopod faunae.

K. v. Zittel (Handbuch der Palæontologie II, p. 439) proposed the new subgenus (or genus) *Rhacophyllites* for the accommodation of such species of *Phylloceras* as are distinguished by wide umbilici, by a steep umbilical slope, and by a
smaller number of auxiliary lobes than are noticed in typical shapes of *Phylloceras*.
In this interpretation, *Rhacophyllites* comprises species of triassic, liassic and even
jurassic age (*Phyll. fortisulcatum* d'Orb.) and is not at all identical with a group of
Phylloceras for which E. v. Mojsisovics (Cephalopoden der Mediterranen Triaspro-
vinz, Abhandl. K. K. Geol. Reichsanst. X, p. 151) had claimed a special systematic
position, chiefly on account of its body-chamber differing materially from the
chambered portion of the shell, and on account of a special arrangement of the
auxiliary series, which is united into a sloping suspensive lobe.

Geyer in his valuable memoir on the liassic Cephalopoda of the Hierlatz near
Hallstatt (Abhandl. K. K. Geol. Reichsanst. XII, p. 223) agrees with E. v. Mojsisovics in attributing a paramount importance to the character of the body-chamber

and sutural line, and not to the widely umbilicated shape of the whorls. According to his view special stress ought to be laid on the difference in the terminal phylla of the saddles, which are regularly oval in *Phylloceras*, conically elongated or club-shaped in *Rhacophyllites*, and on the position of the branches which in *Rhacophyllites* never produce an angular geniculation of the stems of the main saddles, as is commonly noticed in typical species of *Phylloceras*.

This interpretation of *Rhacophyllites* would make v. Zittel's subgenus include *Phylloceras psilomorphum* Neum. or *Ph. planispira* Reyn. but would peremptorily exclude triassic species of the groups of *Ammonites debilis* v. Hauer or *Ammonites neojurensis* Quenst.

In his memoir on the liassic Cephalopoda of the Schafberg (Abhandl. K. K. Geol. Reichsanst. XV, p. 74) Geyer does not any longer insist on differences in the shape of the phylla, but considers the presence of an abnormal body-chamber and of a suspensive lobe as the only distinctive features on which a generic separation of *Rhacophyllites* and of widely umbilicated species of *Phylloceras s. s.* ought to be based.

Waehner in his "Beitraege zur Kenntnis der tieferen Zonen des unteren Lias in den nordoestlichen Alpen" (Beitraege zur Palæontologie Oesterreich-Ungarns, etc., XI, Bd. p. 173) is altogether averse to a generic separation of *Phylloceras* and *Rhacophyllites*. According to his observations the widely umbilicated shapes allied to *Phylloceras stella*, which might he considered as descendants from triassic *Phylloceratidæ*, are most intimately connected in the lower lias with narrowly umbilicated species agreeing with the typical forms of *Phylloceras*, no important characters of difference affording any clue for their generic distinction.

In the palæontological works of E. v. Mojsisovics three different views have been taken regarding the classification of triassic species of *Phylloceras*. In 1882 (l. c. p. 151) this learned author advocated the generic separation of *Ammonites eximius* v. Hauer and of *A. lariensis* Menegh. from *Phylloceras* on account of their abnormally shaped body-chambers. In this new genus *A. rakoscusis* Herbich, *A. transsylvanicus* Herbich, and *A. mimatensis* d'Orb. should be included, with *A. eximius* as prototype, but the majority of triassic species ought to be left with *Phylloceras s. s.* In the description of the upper triassic Cephalopoda of the Himálayas (Pal. Ind. ser. XV, Himál. Foss., Vol., III, Pt. 1, p. 114) the triassic forms which are grouped round *Phylloceras neojurense* and had been assigned to *Rhacophyllites* by Zittel and Steinmann are explicitly considered as direct ancestors of the liassic *Phylloceratidæ* and as true representatives of the genus *Phylloceras*.

"*Phylloceras neojurense* and its contemporaries from the same group "—E. v. Mojsisovics remarks—"are distinguished from the typical representatives of the genus *Phylloceras*, as, for instance, from *Ph. heterophyllum*, only by the wider umbilicus and by the smaller number of auxiliary saddles, connected with the lesser degree of involution. Out of the evolute species of *Phylloceras* are developed on the one hand the strongly involute typical species of *Phylloceras* of the Jura, and on the other hand the subgenus *Rhacophyllites* Zittel, which is distinguished by inclined auxiliary lobes and a variable body-chamber and is confined to the lias."

From this diagnosis it is evident that E. v. Mojsisovics has adopted the subgenus *Rhacophyllites* in a much narrower circumscription than it had been established originally by K. von Zittel and that in 1899 none of the triassic species of *Phylloceras* were by himself considered referable to this subgenus. It must, however, be remarked that one triassic species at least, *Phylloceras occultum* v. Mojs. (Abhandl. K. K. Geol. Reichsanst. VI-1, p. 38, Taf. XVI, figs. 3-6), must be included in the subgenus *Rhacophyllites*, even if the latter is taken in the narrower circumscription, as is evident from the shape of its body-chamber, differing from the chambered whorls by the development of broad transverse plications.

In 1902 E. v. Mojsisovics entirely abandoned his former view and adopted *Rhacophyllites* in the original circumscription proposed by v. Zittel. In the supplement to the Cephalopoda of the Hallstatt limestone (Abhandl. K. K. Geol. Reichsanst. VI-1. Supplem. p. 317) he states the practical advantages of retaining the name of *Rhacophyllites* for all the widely umbilicated *Phylloceratidæ* of triassic age, which K. von Zittel himself had enumerated among the leading types of his new subgenus.

It is to these triassic types of the genus *Phylloceras*, as *Ph. neojurense* Quenst., not to the liassic species of *Rhacophyllites*, that the Himálayan form which I am going to describe, is most nearly allied. If we accept the subgeneric designation of *Rhacophyllites* for this species, it would mean a contradiction of the majority of authors dealing with liassic ammonites, who in their memoirs have accepted *Rhacophyllites* in the narrow interpretation proposed by Geyer. If we wish to choose a proper subgeneric name for the widely umbilicated *Phyllocerata* of triassic age, we must make use of the name *Discophyllites*, which has been proposed by Hyatt for the group of *Phylloceras patens* (Zittel's Text-book of Palæontology, Vol. II, p. 566).

No diagnosis of *Discophyllites* had been given by Hyatt, who was content with fixing the prototype of his new subgenus. E. v. Mojsisovics (Cephalopoden Hallstaetter Kalke l. c. VI-1, p. 318) restricted the name to such types as he considered to be transitional shapes between *Monophyllites* and *Phylloceras* (or *Rhacophyllites* in the wider interpretation). But in including *Phylloceras Ebneri* v. Mojsisovics (Himálayan Foss. l. c. Vol. III, Pt. 1, p. 116, Pl. XIX, fig. 6) among them, he was misled by an erroneous reconstruction of the sutures of the strongly weathered type-specimen of *Phylloceras Ebneri* from the Daonella beds of Lauka. As I have shown in my memoir on the fauna of the Tropites limestone of Byans (Himál. Foss. l. c. Vol. V, Pt. 1, p. 173), the sutural line of excellently preserved examples from Kalapani and Libathi exhibits a distinctly diphyllic development of the siphonal and a rich foliation of the principal lateral saddles. It is only in the second lateral saddle that the monophyllic plan is still clearly marked. Thus this species is, indeed, a transitional shape connecting *Discophyllites patens* most intimately with the rest of the widely umbilicated *Phylloceratidæ* of triassic age.

We should not be in contradiction with Hyatt's views in uniting all triassic *Phyllocerata* with *Ph. patens* in the subgenus *Discophyllites*, all these forms being linked together very closely and differing among each other by features of specific value only. In this narrow interpretation both *Discophyllites* and *Rhacophyllites* would comprise well-defined groups of subgeneric rank. A third group of widely umbilicated *Phyllocerata* of liassic age corresponds to Hyatt's subgenus *Schistophylloceras*.

DISCOPHYLLITES FLOWERI nov. sp. Pl. VIII, fig. 2; IX, fig. 2.

An examination of extensive materials of this species has convinced me that it must be separated from the European *Discophyllites neojurensis* Quenstedt (Cephalopoden, p. 255, Taf. XIX, fig. 8) on account of some subordinate but constant features.

In its external shape one of the most remarkable characters of *Discophyllites Floweri* is the considerable height of the whorls. In *D. neojurensis* the height and width of the volutions are nearly equal. Among my materials, consisting of more than twenty individuals, there is not a single one in which the proportion of the height and thickness were not conspicuously in favour of the first dimension. This remark applies equally to smaller and larger examples. In this character our Himàlayan species more strongly recalls *Discophyllites Ebneri* Mojs. than *D. neojurensis*, with which it otherwise agrees in all the rest of the external features.

Subordinate differences are also exhibited in the arrangement of the sutural line. The general structure of the sutures is the same in both species, and a close examination of their details is needed for observing the points of difference.

All the main saddles are distinctly diphyllic, exactly as in *D. neojurensis*. The illustrations of the sutural line of the latter species, as reproduced by E. v. Mojsisovics (Cephalopoden der Hallstaetter Kalke VI, Supplementd. Taf. XXIII, figs. 2, 3) after Quenstedt and F. v. Hauer, show the constancy of all the characters of importance, notwithstanding a certain individual variability in some minor details. From a comparison of these illustrations with the sutures of *D. Floweri* the following differences are evident:—

In *D. neojurensis* the saddles are more richly serrated, the terminal phylla of the saddles are more slender and elongated, the stems are very narrow, especially near the base. In the main saddles of *D. Floweri* the base is broad and the terminal leaves are very large, recalling in this respect *D. Zitteli* v. Mojsisovics (Cephalopoden der Hallstaetter Kalke, l. c. VI-1, p. 318, Taf. XVII, figs. 3, 4). Of the two terminal leaves in the siphonal saddle the internal one is decidedly the lower in *D. neojurensis*, whereas it is the higher one in *D. Floweri*. There are three auxiliary lobes present in the Himálayan species, the third one being divided by the umbilical suture. They slope obliquely towards the umbilical suture, but are not united into a suspensive lobe, as in typical species of the subgenus *Rhacophyllites* Zittel.

Dimensions.

	Pl. VIII, fig. 2.	Pl. IX, fig. 7.
Diameter of the shell	76 mm.	ab. 32 mm.
„ „ umbilicus . . .	34 „	16 „
Height of the {above the umbilical suture	36 „	22 „
last volution { „ „ preceding whorl	ab. 30 „	14 „
Thickness of the last volution . . .	25 „	17 „

Remarks.—With the exception only of the higher and more strongly compressed whorls and of the bulky shape of the phylloid saddles in the sutural line, I could not detect any characters among the congeneric forms of the Alpine trias in which this species differs from *Discophyllites neojurensis*.

Discophyllites Zitteli v. Mojs. from the carnic Hallstatt limestone of Aussee is also among its nearest allies. In its external shape, especially in the proportion of height and width, it agrees even better with this species than with *D. neojurensis*. A closer comparison of the sutural lines is, however, rendered difficult by the great difference in the dimensions of the European and Indian type-specimens. The small example illustrated by E. v. Mojsisovics has also diphyllic main saddles with bulky terminal leaves and comparatively broad stems. In the siphonal saddle the internal phyllum is the higher one, exactly as in *D. Floweri*. There are only two auxiliary lobes present, but this character is evidently connected with the smaller size of the Alpine type-specimen. A more important feature of difference is the larger size of the monophyllic auxiliary saddle. This saddle is united with the preceding lateral one, thus forming a kind of suspensive lobe, and is not distinctly separated from it by the first auxiliary lobe. In *D. Floweri*, as in *D. neojurensis* and in *D. debilis*, the auxiliary saddles are considerably smaller, dwindling down to rather insignificant elements of the sutural line, but they are not united into a sloping suspensive lobe.

There are no other European triassic species of the genus *Discophyllites* to which the present one might advantageously be compared. *D. pumilus* v. Mojs., *D. despectus* v. Mojs., *D. insolidus* v. Mojs. are all dwarf species. So are the three triassic species from Sicily, which have been described by Gemmellaro as *D. billimiensis*, *D. Laubei* and *D. Jacquoti* (I cefalopodi del Trias superiore, della regione occidentale della Sicilia, Palermo, 1904, pp. 294-297). *D. debilis* v. Hauer has triphyllic saddles. *D. occultus* v. Mojs. agrees with *D. Floweri* in the general arrangement of the sutures, but is provided with an abnormal body-chamber.

An Indian species to which the present one bears a remarkable similarity in its external shape is *Discophyllites Ebneri* v. Mojs. (Himálayan Foss. Palæont. Ind. ser. XV, Vol. III, Pt. 1, p. 116, Pl. XIX, fig. 6). I should, indeed, be at a loss how to distinguish them without an examination of their sutural lines, in whose details considerable differences are noticed (*vide* Himálayan Foss. l. c. Vol. V, Pt. 1, Pl. V, fig. 5).

Fam. : *PINACOCERATIDÆ.*

Gen. : PINACOCERAS v. Mojsisovics.

PINACOCERAS sp. ind. aff. REX Mojs.

In the fauna from exotic block No. 2 the genus *Pinacoceras s. s.* is represented by several specimens, all of them too imperfectly preserved to permit a specific determination. They are distinguished by slowly increasing volutions, with wide umbilici, thus recalling the group of *P. rex* v. Mojsisovics (Cephalopoden der Hallstaetter Kalke, Abhandl. K. K. Geol. Reichsanst. VI-1, p. 65, Taf. XXIII, figs. 8, 9 ; XXIV, fig. 8) from the Alpine trias. Several forms belonging to this group have been described from the triassic rocks of Sicily by Gemmellaro, especially *P. Zitteli* Gemm. (I cefalopodi del trias superiore della parte occidentale della Sicilia, p. 283, Pl IX, fig. 20 ; X, figs. 3, 4) and *P. Haueri* Gemm. (ibidem, p. 289, Pl. XIX, figs. 1, 2 ; IX, figs. 17, 18, 19).

That our Indian species does really belong to the group of *Pinacoceras rex* is evident from the character of the sutural line, which I have been able to trace in a fragment which is otherwise too poorly preserved to allow a reconstruction of the original shell. In common with *P. rex* this fragment has the circular arrangement of lobes and saddles, the apex of the curve coinciding with the innermost adventitious saddle. There are six adventitious saddles present, all of which are distinctly diphyllic, in sharp contrast to the pyramid-shaped main saddles, which occur to the number of two, exactly as in *P. rex*.

Gen. : PLACITES v. Mojsisovics.

PLACITES cf. PERAUCTUS v. Mojsisovics. Pl. IX, figs. 3, 4.

The majority of the species of *Placites,* belonging to the group of *Pl. platyphyllus* v. Mojs, are distinguished by the presence of two adventitious lobes and saddles. There is only one Alpine species, *Pl. perauctus* v Mojsisovics (Cephalopoden der Hallstaetter Kalke, Abhandl K. K. Geol. Reichsanst. VI-1, p. 53, Taf. XXI, figs. 7-8), in whose sutural line a larger number of adventitious lobes and saddles has been developed. The typical form of *Placites perauctus* has been collected in the noric Hallstatt limestone of the Sommeraukogel, but types closely related are already found in the julic deposits of the Salzkammergut. They are only imperfectly known and have been described cursorily as *Pl. cf. perauctus* by E. v. Mojsisovics (l. c. p. 53).

Types very closely allied to *Pl. perauctus* have also been met with in the Himálayan trias. A chambered fragment measuring one half volution, from the upper Daonella beds of the Bambanag section, has been mentioned by E. v. Mojsisovics (Upper triassic cephalopod faunæ of the Himálayas, Palæont. Ind. ser. XV, Himál. Foss. Vol. III, Pt. 1, p. 111, Pl. XVIII, fig. 9) as *Placites sp. ind. aff. peraucti*. A similar fragment has been discovered by myself among the materials collected by Smith in the Tropites limestone of Lilinthi (Himál. Foss. l. c. Vol. V, Pt. 1, p. 167, Pl. XXV, fig. 6). To these fragmentary examples several better preserved specimens have been added recently from the fauna of exotic block No. 2. They undoubtedly are related to *Placites perauctus*, not to *Pl. Oldhami* v. Mojsisovics, as is shown by the occurrence of three adventitious lobes and saddles. The lobes are as deeply incised and the saddles nearly as richly serrated as in the typical form of *Pl. perauctus*. The number of auxiliary lobes is not exactly known to me.

The measurements of the two specimens figured are as follows :—

										Fig. 3.	Fig. 4.
Diameter of the shell	.	.	.	.	.	.	.	.	.	67 mm.	95 mm.
„ „ „ umbilicus	.	.	.	.	.	.	.	.	.	0 „	0 „
Height } of the last volution	.	.	.	.	.	40 „	18 „
Thickness }	.	.	.	.	.	14 „	6 „

I dare not identify my specimens with the typical *Placites perauctus*. The type-specimens from the Sommeraukogel are all of large size, showing a diameter of the last volution from 100 to 125 mm. The only Himálayan example of similar dimensions is a body-chamber fragment, in which a height of 55 mm. corresponds to a thickness of 17 mm. It is, however, not the more slender and compressed shape of the whorl but rather the difference in the shape of the cross-section on which I am inclined to lay a special stress.

In the typical *Placites perauctus* the flanks run nearly parallel from the siphonal shoulder to the umbilical region, whereas in my Himálayan fragment they converge strongly towards the latter. Whether this character is sufficient for a specific distinction can only be decided after an examination of larger materials. For the moment it will be preferable to mention the Indian species as *Placites cf. perauctus*, but without insisting on an identification with the Alpine form from the carnic Hallstatt limestone of the Roethelstein.

TROPITOIDEA.

Fam. : *TROPITIDÆ.*

Gen. : Discotropites Hyatt and Smith.

1905. *Discotropites Hyatt and Smith, The triassic Cephalopod genera of America, U. S. Geol. Survey Professional Papers No. 40, p. 89.*

In 1877 A. Hyatt (in Meek, U. S. Geol. Exploration 40th Parall., Vol. IV, p. 126) introduced the generic name of *Eutomoceras* for a triassic Ammonite from Nevada, which E. v. Mojsisovics considered as the nearest ally to the Alpine

Ammonites sandlingensis v. Hauer. Thus the generic name of *Entomoceras* was assigned by him to the European group of *Ammonites sandlingensis* and was, on his authority, accepted unanimously in this interpretation.

A re-examination by Hyatt and Smith of the type-specimen of *Entomoceras Laubei* Meek, the American prototype of the genus, led to the surprising result that the American Ammonite was a member of the section of *Ceratitoidea*, showing close affinities to *Hungarites* v. Mojs., but differing widely from *Ammonites sandlingensis*, which is a typical representative of the longidome section of *Tropitoidea*. Thus the grouping of the latter species with *Entomoceras* Hyatt having been based on a misinterpretation of the true affinities of the American genus by E. v. Mojsisovics, a new generic designation, *Discotropites*, was proposed for *Ammonites sandlingensis* by Hyatt and Smith.

It must be conceded that from Meek's memoir, in which the name *Entomoceras* is used for the first time, a satisfactory decision on the distinguishing characters of the new genus is not at all easy. E. v. Mojsisovics therefore did not meet with any opposition in applying the name to the European group of *Ammonites sandlingensis*, which ever since has been considered as the true prototype of the genus *Entomoceras* in substitution of the almost forgotten American species.

The rules of priority being decidedly and indubitably in favour of the first species described under the generic designation of *Entomoceras*, the inconvenience of changing the generic name, usurped hitherto by *Ammonites sandlingensis*, cannot be avoided. I feel consequently obliged to follow P. Smith in adopting the generic name *Discotropites* for the European form from the Hallstatt limestone, notwithstanding the confusion in the nomenclature of triassic Ammonites which will undoubtedly result from the misinterpretation of Hyatt's genus.

DISCOTROPITES cf. SANDLINGENSIS v. Hauer. Pl. VI, fig. 4

1849. *Ammonites sandlingensis* F. v. Hauer, Ueber neue Cephalopoden aus den Marmorschichten von. Hallstatt und Aussee, Haidingers Naturwiss. Abhandl. III, p 10, Taf. III, figs. 10-12.

1866. *Ammonites sandlingensis* A. v. Dittmar, Zur Fauna der Hallstaetter Kalke, Geognost. Palaeont. Beitr. von Benecke, etc., I, p. 370.

1893. *Entomoceras sandlingense* E. v. Mojsisovics, Die Cephalopoden der Hallstaetter Kalke, Abhandl. K. K. Geol. Reichsanst. VI-2, p 285, Pl. CXXX, figs. 11-13 ; CXXXI, figs. 1-11.

1896. *Entomoceras cf. sandlingense* E. v. Mojsisovics, Himál. Foss. Pal. Ind. ser. XV, Vol. III, Pt. I, p. 49.

1904. *E. sandlingense* Gemmellaro, I cefalopodi del Trias superiore della parte occidentale della Sicilia, p. 77, Tav. VIII, figs. 8-10.

1904. *E. sandlingense* I. P. Smith, The comparative stratigraphy of the marine trias of Western America. Proceed. Californian Acad. of sciences 3rd ser.Vol. I, p. 397, Pl. XLVI, fig. 10, XLVIII, figs. 5, 6.

1906. *Discotropites sandlingensis* Hyatt and Smith, Triassic cephalopod genera of America, U. S. Geol. Surv. Ser. C. Profess. Papers No. 40, p. 63, Pl. XXXV, figs. 1-12.

1906. *Entomoceras cf. sandlingense* Diener, Fauna of the Tropites Limestone of Byans, Himál. Foss. Vol. V, Pt. I, p. 138.

It is a rather astonishing fact that this wide-spread species of carnic age is numerously represented in the upper trias of the Himálayas, but in fragments only, and that no complete specimen has as yet been noticed.

The only specimen known to me from exotic block No. 2 is less fragmentary and in a better state of preservation than any of the examples from the Tropites limestone of Byans. Its identification with F. v. Hauer's species is, indeed, pretty certain. It is only my aversion against an unrestricted determination of incomplete specimens which has induced me to add the designation *cf.* to the specific name.

My specimen is of moderate size. One half volution and a large part of the umbilicus have been preserved. The siphonal edge is surmounted by a low and distinct keel. The steep umbilical wall is separated from the flattened lateral parts by a sharply rounded edge.

The ornamentation agrees almost perfectly with the delicate sculpture of the American example illustrated by Hyatt and Smith on Pl. XXXV, fig. 3, and of the Alpine specimen illustrated by E. v. Mojsisovics on Pl. CXXXI, fig. 8. The majority of the sickle-shaped ribs are dichotomous. There is no angular geniculation in their gentle backward curve along the middle of the flanks. Of umbilical tubercles faint traces only have been noticed. The spiral striation is very delicate.

Dimensions.

Height } of the last whorl	. . . .	27 mm.
Thickness		11 „
Diameter of the umbilicus	. . . . .	7 5 „

Sutures.—Not known in detail.

Remarks.—*Discotropites sandlingensis* is one of the leading fossils of the zone of *Tropites subbullatus* in the Eastern Alps and in California, but it has also been met with, although very rarely, in the geologically older (julic) *Ellipticus* beds of Aussee.

Gen : TROPITES v. Mojsisovics

TROPITES cf. SUBBULLATUS v. Hauer.   Pl. VI, fig. 9.

1849. *Ammonites subbullatus* F. v. Hauer, Ueber neue Cephalopoden aus den Marmorschichten von. Hallstadt und Aussee, Haidingers Naturwiss Abhandl. III, p. 19, Taf. IV, figs. 1-4 (non 5-7).

1893. *Tropites subbullatus* F. v. Mojsisovics, Die Cephalopoden der Hallstaetter kalke, Abhandl, K. K. Geol. Reichsanst, VI-2, p. 1-7, Taf CVI, figs. 1-3, 5, 7 ; CVII, figs. 1-5 ; CVIII, figs. 1-6 ; CX, fig. 6

1905. *Tropites subbullatus,* Hyatt and Smith, Triassic cephalopod genera of America, U. S. Geol. Surv. ser. C, Prof. Pap. No. 40, p. 67, Pl. XXXIII, figs. 1-7 ; XXXIV, figs. 1-14 ; LXXIX, figs. 1-10.

1906. *Tropites subbullatus* Diener, Fauna of the Tropites limestone of Byans, Himal. Foss. Vol. V, Pt. 1, p. 165, Pl. IV, figs. 6-7

1907. *Tropites cf. subbullatus* Diener, Ladinic, carnic and noric fauna of Spiti, ibid., Vol. V, Pt. 3, Pl. XIV, figs. 2, 3.

There are only three fragmentary casts of inner nuclei available for examination. Although more or less deformed and damaged they show a remarkable resemblance in shape and sculpture to *Tropites subbullatus.* The whorls are considerably thicker than high. In one of my casts the siphonal part has been preserved satisfactorily enough for illustration. It exhibits the low median keel towards which numerous straight ribs converge at steep angles.

Although it is not possible to establish the identity of the Indian fossil with the European species, the presence of the group of *T. subbullatus* in the fauna of exotic block No. 2 has been ascertained.

TROPITES sp. ind. aff. ACUTANGULO v. Moja. Pl. X, fig. 3.

1902. *Tropites aff. acutangulus* A. v. Krafft, Mem. Geol. Surv. of India, XXXII, Pt 3, p. 143.

The figured specimen belongs to all appearance to a species nearly allied to *Tropites acutangulus* v. Mojsisovics (Cephalopoden der Hallstaetter Kalke, l. c. VI-2, p. 203, Taf. CXII, figs. 1, 2). It is not sufficient for an exact determination, the outer whorls only having been preserved.

The cast, which has been partly injured by weathering, agrees in its size and shape with the type-specimen from the Subbullatus beds of the Salzkammergut and belongs probably to an adult specimen, in which a small part of the body-chamber only is wanting. It differs from *Tropites acutangulus* by the regularity of its involution, the umbilical suture not leaving the normal spiral. Height and thickness of the last volution increase slowly, but quite regularly, from the beginning of the whorl to the very aperture. The cross-section agrees exactly with that in *T. acutangulus*, being only a little wider than high.

In its sculpture our specimen shows a remarkable resemblance to *T. acutangulus*. The ribs are numerous and sharp and exhibit an angular geniculation outside the umbilical margin. In the anterior half of the last volution all the ribs are simple and undivided. In the posterior half of this whorl the sculpture has, unfortunately, suffered considerably from weathering along the umbilical margin. It cannot be decided, therefore, whether or not sharply edged umbilical tubercles have been present, but from the direction of the ribs it is obvious that the majority of them dichotomised in the umbilical region. No secondary bifurcations have been noticed in the marginal region.

Although the siphonal part has been severely damaged, I have been able to state the presence of a keel, accompanied by a distinctly marked keel-furrow.

Dimensions.

Diameter of the shell	. . . .	69 mm.
„ „ „ umbilicus	. . . .	27 „
Height of the last { above the umbilical suture	. . . ab.	28 „
volution { „ „ preceding whorl	. . . ab.	21 „
Thickness of the last volution	. . . . . .	22 „

Sutures.—Not known.

Remarks.—Species very closely allied to *Tropites acutangulus* have been mentioned from the Tropites limestone of Byans by E. v. Mojsisovics (Himálayan Foss., Vol. III, Pt. 1, p. 46, Pl. XI, fig. 4) and by myself (*ibid.*, Vol. V, Pt. 1, p. 51). As they are represented by specimens still more fragmentary than the

present one, their specific identity with the form here described cannot be
ascertained.

TROPITES sp. ind. aff. WODANI v. Mojs.   Pl. VI, fig. 11.

A fragmentary cast of a small species of *Tropites* recalls in its external
characters *T. Wodani* v. Mojsisovics (Cephalopoden der Hallstaetter Kalke,
Abhandl. K. K. Geol. Reichsanst. VI-2, p. 221, Taf. CXVI, fig. 6). It agrees with
the Alpine type-specimen in size and in the shape of the transverse section. The
volutions are of equal height and thickness, the flattened lateral parts being
separated from the high and perpendicular umbilical wall by a sharp edge and
passing gradually into the rounded external part in a very regular curve.

The delicate ribs originate in pairs from faintly developed umbilical nodes.
They are nearly straight and turned forward very strongly on the siphonal area,
thus meeting the median keel at acute angles.

The presence of *Tropites Wodani* in the Tropites limestone of Byans (Himál.
Foss., Vol. V, Pt. 1, p. 152, Pl. V, fig. 6) is in favour of a closer comparison of this
undeterminable cast with the Alpine species from Aussee.

Subgen.: ANATROPITES v. Mojsisovics.

ANATROPITES cf. SPINOSUS v. Mojsisovics.   Pl. VI, fig. 10.

1893. *Tropites (Anatropites) spinosus* K. v. Mojsisovics, Cephalopoden der Hallstaetter Kalke, Abhandl.
     K. K. Geol. Reichsanst. VI-2, p. 223, Taf. CX, fig. 3.
1902. *Tropites aff. spinosus* A. v. Krafft, Mem. Geol. Surv. of India, Vol. XXIII, Pt. 2, p. 143.

The figured specimen, tolerably well preserved, is a typical representative of
the group of *Tropites spinosi* (*Anatropites*), as is shown by the large thorns
adorning the umbilical edge of its inner volutions. In the last volution, which
forms part of the body-chamber, the umbilical thorns are replaced by tubercles,
from which faint radial ribs arise, which become obsolete before reaching the siphonal
part. The median keel is comparatively high and sharp, but not accompanied by
lateral furrows.

My specimen agrees so closely with the Alpine type of *Anatropites spinosus*
that I should have ventured on a direct identification, if the inner whorls of the
Himálayan example had been preserved more completely. Although the latter
is slightly inferior in size, its proportions and sculpture agree almost exactly. The
thickness of the last volution is considerably greater than the height, but is equal
to the width of the umbilicus.

Seventeen umbilical thorns or nodes are counted within the circumference of
the last volution.

From *Tropites (Anatropites) Adalgi* v. Mojsisovics (ibid., p. 225, Taf. CX,
fig. 3) our specimen is distinguished by the greater height of its whorls and by
its less strongly depressed external shoulders.

Dimensions.

Diameter of the shell		23·6 mm.
„ „ „ umbilicus		10 „
Height ⎫ of the last volution		7·5 „
Thickness ⎭		10 „

Sutures.—Not known.

Remarks.—*Anatropites spinosus* is a very rare species in the julic stage of the Salzkammergut. One single specimen only has been quoted by E. v. Mojsisovics from the *Ellipticus* beds of the Roethelstein.

AnatropiTES PilgrimII nov. sp.   Pl. VI, figs. 6, 7; IX, fig. 5.

That this species belongs to the genus *Tropites*, not to *Tropiceltites*, is proved by the character of its sutural line, which, although not entirely known, is decidedly dolichophyllic, not clydonitic. It is, however, less easy to fix its systematic position among the five groups which have been distinguished among the genus *Tropites* by E. v. Mojsisovics, and to which even a subgeneric rank has been attributed by that learned author.

The inner volutions are not adorned with proper thorns along their umbilical edges, but rather with very broad and stout ribs. A similar type, *Tropites Geyeri*, has been grouped with the subgenus *Anatropites* by E. v. Mojsisovics. Our species, indeed, agrees with typical representatives of this subgenus in the majority of its leading features, especially in the external shape, which recalls *Tropiceltites*, and in the gradual weakening of the ornamentation in the outer whorls.

Three fairly well-preserved examples, consisting both of inner nuclei and body-chambers, are available to me for examination. They are provided with numerous, slowly increasing whorls, which leave a wide umbilicus open. The transverse section is more strongly compressed than in *A. Geyeri* and higher than thick.

In the inner volutions the sculpture consists of strong umbilical ribs, which are but exceptionally elevated into obtuse tubercles near the umbilical edge and die out gradually towards the external shoulders. Some of them, although thinning out into delicate folds, reach across the siphonal area to the median keel, which is high and sharp. The body-chamber whorl is distinguished by its more delicate ornamentation. The umbilical ribs are reduced to obtuse nodes, and the delicate folds are arranged in crescents with their convexity turned backward.

Dimensions.

Diameter of the shell		21 mm.
„ „ „ umbilicus		11 „
Height ⎫ of the last volution		10 „
Thickness ⎭		8·4 „

Sutures.—Not known in detail. Siphonal lobe narrow, with two diverging points at its base, which are separated by a low median prominence. Siphonal saddle large and dolichophyllic, as in typical species of *Tropites*.

Remarks.—Among Alpine species of *Anatropites A. Geyeri* v. Moja. is certainly most nearly allied to the present one. It is less strongly compressed, provided with more slowly increasing whorls, and exhibits a large umbilicus. But otherwise the two species agree pretty well in their general shape and ornamentation, especially in the reduction of umbilical thorns, which are replaced by stout ribs.

From the triassic rocks of Sicily two species with a similar ornamentation have been described by Gemmellaro, namely, *Anatropites Mojsisovicsi* Gemmellaro (I cefalopodi del trias superiore della regione occidentale della Sicilia, p. 123, Tav. VII, figs. 40-42) and *A. Frechi* Gemm. (*ibid.*, p. 120, Tav. XXV, figs. 13-15). But both of them are easily distinguished from the Himalayan form by their compressed and slender shape, especially *A. Mojsisovicsi*, which has a very large umbilicus and numerous radiating costæ.

Gen.: MARGARITES v. Mojsisovics.

MARGARITES IRREGULARICOSTATUS nov. sp. Pl. VI, fig. 8.

This species represents a type of the group of *Margarites unispinosi*, which is distinguished from the most closely related forms of the Alpine trias by the irregular character of its sculpture.

My type-specimen, which is fairly well preserved and provided with a part of its body-chamber, shows slowly increasing and low whorls of a nearly quadrangular section, thus recalling *Marg. subunctus* v. Mojsisovics (Cephalopoden der Hallstaetter Kalke, l. c. VI-2, p. 305, Taf. CXCV, fig. 17). The external keel is low and not crenulated.

The ornamentation of the inner whorls is distinguished by the lack of umbilical tubercles, and by the alternation of stout primary and delicate secondary ribs. The stouter ribs are disposed at considerable distances and elevated into marginal spines. Their direction is not exactly radial but slightly turned backward. The delicate secondary ribs, which are of irregular strength and number, are directed backward more strongly, thus including an acute angle with the primary costæ. On the last volution the disposition of primary and secondary ribs is much more irregular. The development of marginal spines or tubercles is no longer confined to the primary ribs, some of which occasionally bifurcate. This tendency towards bifurcation is also exhibited in some of the secondary ribs, which are of very different strength and number and occasionally of an undulating direction.

There is no species of *Margarites* known to me, either in the carnic Hallstatt limestone of the Salzkammergut, or among the triassic fauna of Sicily, which for irregularity of ornamentation can be compared with the present one.

Dimensions.

Diameter of the shell	.	.	.	.	.	.	.	.	.	.	23·5	mm.
„ „ „ umbilicus	.	.	.	.	.	.	.	5	.	.	10	„
Height } of the last volution	.	.	.	.	.	.	.	.	.	.	7	„
Thickness }											4·5	„

Sutures—Not known.

Fam.: *HALORITIDÆ.*

Gen.: JOVITES v. Mojsisovics.

JOVITES cf. SPECTABILIS Dien. Pl. V, fig. 1.

1906. *Jovites spectabilis*, Diener, Fauna of the Tropites limestone of Byans, Himal. Foss. Palæont. Ind. ser. XV, Vol. V, Pt. 1, p. 183, Pl. XVI, fig. 1; Pl. IX, fig. 10.
1907. *Jovites spectabilis*, Diener, Ladinic, carnic, and noric faunæ of Spiti, *ibid*, Vol. V, Pt. 3, Pl. XV, fig. 3.

The specimen figured is a mature example of a species of *Jovites* with a very distinct umbilical opening of the body-chamber whorl. It is partly crushed and no exact idea of its transverse section can be formed from its outlines. It could, however, be ascertained that the greatest inflation corresponds to the commencement of the last volution, and that in the vicinity of the aperture a second, although moderate, inflation and widening of the external part follows the compression of the body-chamber.

The siphonal part has not been preserved well enough to state whether or not a keel-like median projection has been developed. The opening of the umbilicus is very considerable. Indeed, it expands so considerably that the height of the transverse section does not increase throughout one half of the entire length of the last volution.

In its general character and ornamentation this specimen agrees very closely with the type from the Tropites limestone of Byans, to which the name *Jovites spectabilis* has been assigned by myself. The sculpture consists of stronger ribs than in *J. dacus* v. Mojs. or *J. daciformis* Diener. Dichotomising ribs are noticed in the posterior half of the last whorl, although rather exceptionally, but are entirely absent in the anterior portion, where they are arranged at larger distances and become sharpened.

Dimensions.

Diameter of the shell	.	.	.	.	.	.	.	.	.	51	mm.
„ „ „ umbilicus	.	.	.	.	.	.	.	.	.	23	„
Height of the last volution	.	.	.	.	.	.	.	.	.	19	„
Thickness at the commencement of the last whorl	.	.	.	.	.	31	„

Sutures.—Not known.

As a specific distinction between *Jovites spectabilis* and *J. boemensis* can only be based on essential differences in the sutural line, I have not ventured on a direct identification of the present specimen, although there are some strong reasons in favour of its identity with the Himálayan species from the Tropites limestone of Byans and Spiti.

JOVITES DACIFORMIS Diener. Pl. V, fig. 2; VII, fig. 8.

1905. *Jovites daciformis*, Diener, Fauna of the Tropites limestone of Byans. Pal. Ind. ser. XV, Himál. Foss. Vol. V. Pt. 1, p. 119, Pl. XV, figs. 5-10. XVI, fig. 2.

Two inner nuclei in a satisfactory state of preservation, and two internal casts with fragments of the body-chamber adhering, agree so closely with *Jovites daciformis* from the Tropites limestone of Byans that I do not hesitate in identifying them.

The two figured casts are strongly globose and broader than high, with closed umbilici. The numerous bifurcating ribs cross the keel-like median projection of the siphonal part.

The sutures agree with those of *Jovites daciformis*, not with those of *J. dacus* v. Mojs. The lateral leaves accompanying the stem of the high and slender saddle are arranged obliquely to the axis of the latter.

The presence of *Jovites daciformis* in the fauna of the exotic blocks of Malla Johar having been established, there is some probability of the two fragments collected by myself in the Kiogarh range, south of Sangcha Talla, in 1892, and described as *Jovites nov. sp. ex aff. boemensis* by E. v. Mojsisovics (upper triassic cephalopod fauna of the Himálayas, Pal. Ind. ser. XV, Himálayan Foss. Vol. III, Pt. 1, p. 18, Pl. IX, figs. 4, 5), belonging also to this species. Their only difference consists in the smaller number and somewhat greater strength of the ribs, but the range of variation with reference to this feature is so large in *J. daciformis*, as well as in its European representative, *J. dacus*, that it can barely be considered sufficient for justifying a specific separation.

JOVITES nov. sp. ind. Pl. IX, fig. 6.

The figured fragmentary cast belongs to all appearance to a species of *Jovites* which is distinguished from all congeneric forms by the absence of any external ornamentation. It was provided with low and moderately inflated whorls. The egression of the umbilicus, which is not as widely expanded as in the group of *Jovites dacus* Mojs., corresponds to the beginning of the last volution.

In the transverse section the presence of obtusely rounded siphonal shoulders is a remarkable feature. The siphonal area is broad and flattened. The faint, keel-like projection in the middle of the siphonal part extends to the end of the last whorl.

The surface of the cast is entirely smooth and devoid of any kind of sculpture.

Dimensions.

Diameter of the shell												44 mm.
" " umbilicus												6 "
Height } of the last volution											19 "
Thickness											ob.	20 "

Sutures.—Not known.

Remarks.—There is no European species of *Josites* known to me to which the present specimen might be compared. Its incomplete state of preservation prevents me from introducing a new specific designation.

Gen.: JUVAVITES v. Mojsisovics.

JUVAVITES KRAFFTI nov. sp. Pl. VII, fig. 3.

This species belongs to a group of *Juvavites interrupti*, which is distinguished by a gradual obliteration of the sculpture in the body-chamber of adult individuals. Its nearest allies in the Alpine trias are *Juvavites nepotis* v. Mojsisovics (Cephalopoden der Hallstaetter Kalke, l. c. VI-2, p. 92, Taf. LXLi, fig. 12) and *J. Ellæ* v. Mojsisovics (l. c. p. 93, Taf. CXXIX, fig. 20). There is also a great external resemblance between the figured specimen and *Juvavites (Anatomites) Halavatsi* v. Mojsisovics (l. o. p. 90, Taf. CXXIX, fig. 22), but the absence of any paulostome furrows or ribs in the Himálayan type peremptorily forbids its identification with representative of the subgenus *Anatomites.*

At the beginning of the last volution the whorls are semigloboso, being considerably broader than high. In the anterior half the height of the volution increases more rapidly than its width. Thus the transverse section of the aperture appears to be compressed and provided with a very narrow external part, whereas it is broadly rounded in the chambered portions of the shell.

The sculpture on the two halves of the shell corresponds exactly, but is interrupted along the median line of the external part by a narrow, smooth zone, which is reached by the lateral ribs with very steep angles. This interruption of the sculpture does not result from an obliteration of the ribs, but from a gradual elevation of the intercostal furrows. The median zone of the external part is therefore elevated above the general level of the shell. Where the elevation of the intercostal furrows is not sufficiently high to obscure the ribs, the latter are seen to cross the external part, meeting from both sides at very obtuse angles.

The posterior part of the last volution is covered with numerous high and narrow ribs, which are slightly curved and separated by deep and broad intercostal valleys. Their arrangement is rather irregular. Two stem-ribs are simple but accompanied on either side by intercalated ribs. The rest are forked ribs. They are either dichotomous or united into fasciculi with a tripartite arrangement.

The change in the transverse section of the last whorl coincides with a complete change of the sculpture. The numerous and strong ribs become obliterated rather rapidly. The ornamentation of the anterior half of the shell, which probably belongs to the body-chamber, is reduced to a small number of broad and indistinct folds. Otherwise this portion of the shell is entirely smooth. No traces of tubercles have been noticed, either in the umbilical or in the marginal region.

Dimensions

Diameter of the shell	60	mm.
" " umbilicus	4.5	"
Height of the , above the umbilical suture	41	"
last volution { " " preceding whorl	20	"
Thickness of the last volution	56	"
Height } at the beginning of the last volution	21	"
Thickness }	28	"

Sutures.—Not known.

JUVAVITES DOGRANUS nov. sp. Pl. V, fig. 3.

This species, which is represented in the Himálayan collection by the figured, incomplete specimen only, is very nearly allied to the preceding one, from which it differs by its smaller size, more strongly compressed shape, and more closely set ribs.

The transverse section of the last volution changes considerably, being elliptical at the beginning, whereas it is of nearly triangular shape at the aperture, faint rounded angles forming the boundary of the gently curved external part. The umbilicus is deep, but very narrow.

The ribs are more numerous and developed less strongly than in *Juvavites Kraffti*. The smooth band, which interrupts the sculpture along the median zone of the external part, is comparatively broad. Its elevation does not reach the height of the lateral ribs.

The ribs are arranged almost symmetrically on both sides of the shell. Some of them cross the external part in uninterrupted curves, which are slightly turned forward. The ornamentation is rather irregular. At the beginning of the last volution simple ribs alternate with bipartite, tripartite, and quadripartite ones; but in the anterior portion of this whorl the majority of the ribs are dichotomous, and a simple rib, which is restricted to the marginal region, is often intercalated between two stem-ribs.

As in *Juvavites Kraffti*, the sculpture becomes obliterated completely in the vicinity of the aperture. But this obliteration does not coincide exactly with the beginning of the body-chamber, the last two septa being situated within the smooth region of the shell.

Dimensions.

Diameter of the shell	62 mm.
„ „ „ umbilicus	4 „
Height of the { above the umbilical suture	34·5 „
last volution { „ „ preceding whorl	20·5 „
Thickness of the last volution	24·5 „
Height } at the beginning of the last whorl	16 „
Thickness }	17 „

Sutures.—The sutures, as far as known, agree well with those in typical species of *Juvavites.* Saddles elongated and dolichophyllic. Siphonal saddle terminating in rounded phylla, imparting to the apex a rounded aspect, whereas in the principal lateral saddle the median phyllum is considerably longer than the adjoining lateral leaves.

Principal lateral lobe bipartite, its two short diverging points being divided by a mesial indentation. Auxiliary series not accessible to observation.

Remarks.—A species, which is probably very nearly allied to the present one, is *Juvavites tonkinensis* Diener (Note sur deux espèces d'Ammonites triasiques du Tonkin, Bull. Soc. géol. de France, 3e. sér. XXIV, 1896, p. 883), from the upper basin of the Black River in Tonkin. There is a very close agreement in the sculpture, with the only exception that the ribs are curved forward more strongly on the external part of the Indo-Chinese form. Near the aperture of *Juvavites tonkinensis* the ornamentation is also obliterated, but the transverse section of the last volution is not subject to any change throughout the entire length of the last volution, the anterior half of which belongs to the body-chamber.

A species nearly allied to *Juvavites tonkinensis* as well as to the present form has also been noticed from the upper Daonella beds of the Central Himálayas, but its unsatisfactory state of preservation does not allow a closer comparison.

JUVAVITES sp. ind. ex aff. SUBINTERRUPTO v. Moj. Pl. VII, figs. 4, 6.

Two fragmentary casts belong to a globose form from the group of *Juvavites interrupti,* and, chiefly by the character of their sculpture, they recall *J. subinterruptus* v. Mojsisovics (Cephalopoden der Hallstaetter Kalke, l. c. VI-2, p. 90, Taf. LXXXIX, fig. 13; XC, figs. 2, 3; CXXVI, fig. 16) from the carnic stage of the Hallstatt limestone.

No traces of septa having been preserved, I cannot decide whether or not the two specimens were provided with their body-chambers, although from the character of their ornamentation I am inclined to consider them as nuclei rather than as adult individuals. They are robust and strongly globose, much more so indeed than any of the Alpine or Sicilian species of this group. There is no change in the shape of the transverse section from the beginning of the last volution to the aperture.

The direction of the ribs is nearly radial on the lateral parts, but faintly curved or flexuous. The majority of the ribs are dichotomous, bifurcating at a point situated about one-third of the height of the whorl, but in a small number of ribs a second

bifurcation is noticed in the marginal region. In the middle part of the last volution of the specimen illustrated in fig. 6 a bundle of ribs is formed by the coalescence of two forked stem-ribs near the umbilical margin, but this bundle is not
accompanied by any paulostome furrow. Some intercostal valleys appear, indeed,
to be a little larger than the adjoining ones, but they are not true paulostome furrows.
The difference between the broader and narrower intercostal furrows is certainly
smaller than in *Juvavites anatomitoides* Gemmellaro (I cefalopodi del trias superiore
della parte occidentale della Sicilia, p. 185, Tav. XVIII, fig. 50), which is, nevertheless, a typical species of *Juvavites s. s.*

The ribs are not arranged symmetrically on both sides of the shell and are
interrupted along the middle line of the external part by a comparatively broad,
smooth band. They terminate very abruptly, the smooth zone not corresponding to a gradual elevation of the intercostal valleys.

Dimensions.

		Fig. 4.
Diameter of the shell	. . . .	. 45 mm.
„ „ „ umbilicus	. . . .	. 6 „
Height of the ⎰ above the umbilical suture	. . . . .	. 27.5 „
last volution ⎱ „ „ preceding whorl	. . . .	. 16 „
Thickness of the last volution	. . . .	. 10 „

Sutures.—Not known.

JUVAVITES nov. sp. ind. (GROUP OF CONTINUI). PL. IX, fig. 8.

The sphæroidal shell represents the chambered nucleus of a small species of
Juvavites. The last septum corresponds to the aperture; little fragments adhering
to the external part at the beginning of the last volution form a part of the body-
chamber.

The whorls are considerably wider than high and are covered with numerous
forked ribs. The majority of the ribs are dichotomous. One bundle formed by the
coalescence of two forked stem-ribs immediately outside the umbilical margin and
two simple, undivided ribs have also been noticed. The dichotomous ribs bifurcate
at a point about one-third of the height of the whorl.

The ribs are arranged symmetrically on both sides of the shell, and are united
in the middle of the broad siphonal part without any interruption. In crossing the
siphonal part they describe a flat curve, with its convexity turned forward, but
there is no geniculation at the place where they meet from both sides, as in
Juvavites gastrogonius v. Mojsisovics, or in the unnamed species from the carnic
Hallstatt limestone which has been illustrated by E. v. Mojsisovics on Pl.
LXXXIX, fig. 0, of the "Cephalopoden der Hallstaetter Kalke" (Abhandl. K. K.
Geol. Reichsanst. VI-2).

Although the ribs are not interrupted along the median zone of the external
part, they appear to be considerably lower than in the flanks, on account of a
gradual flattening of the intercostal valleys.

Dimensions.

Diameter of the shell	. . . .	23 mm.
„ „ „ umbilical suture	. . . .	3 „
Height of the { above the umbilicus	. . . . .	11 „
last volution { „ „ preceding whorl	. . . .	7 „
Thickness of the last volution	. . . .	20·5 „

Sutures.— Not known.

Remarks.—There is no European species to which the present fragment could be more closely compared. I have mentioned it here especially in order to prove the presence of a form of the group of *Juvavites continui* in the fauna of exotic block No. 2. But no stratigraphical evidence can be gathered from its presence, typical species of this group having been distributed both in carnic and noric beds of the Alpine trias.

Subgen. : *GRIESBACHITES* v. Mojsisovics.

1896. *Griesbachites* v. Mojsisovics, Denkschr. Kais. Akad. d. Wiss. math. nat. Kl. Bd. LXIII, p. 643.
1899. *Griesbachites* v. Mojsisovics, Himālayan Foss. Pal. Ind. ser. XV, Vol. III, Pt. I, p. 35.

The subgeneric denomination of *Griesbachites* has been proposed by E. v. Mojsisovics for a small number of species of *Juvavites*, in which marginal tubercles are developed both on the body-chamber and in the inner whorls of the shell. There were only three typical species of this subgenus known to him, *Griesbachites Medleyanus* Stoliczka (Memoirs Geol. Survey of India, Vol. V, p. 54, Pl. IV, fig. 5) from Spiti, of unknown age, *G. Hanni* v. Mojsisovics, from the upper Daonella bed of the Bambanag and Lauka sections, and *G. Kastneri* v. Mojs., from the carnic Ellipticus beds of the Salzkammergut. The first of these three species is considered as the prototype of the subgenus.

In his preliminary notes on the cephalopod faunæ of the Himālayan trias (Sitzungsber. Kais. Akad. d. Wiss., 1892, Bd. CI, p. 65) E. v. Mojsisovics united *Juvavites Medleyanus* with the genus *Sagenites* v. Mojs., considering the development of stout marginal tubercles as a feature of generic importance. In his memoir on the Cephalopoda of the Hallstatt limestone (Abhandl. K. K. Geol. Reichsanst. VI-2, p. 157) he compares the Indian species to *Sagenites Schaubachi* v. Mojs. of the group of *reticulati* and even notices traces of a longitudinal striation on the siphonal part of the cast.

From his later description in the Palæontologia Indica it is, however, evident that he had been misled by an external similarity between *Griesbachites Medleyanus* and *Sagenites*. That there is, indeed, no real affinity between this species and *Sagenites* is obvious from my recent examination of A. v. Krafft's materials. In his collections from exotic block No. 2 six specimens of *Griesbachites* are represented, which are so nearly allied to the prototype of the subgenus, *G. Medleyanus*, that they had been identified with the latter by A. v. Krafft. In several specimens large fragments of the shell have been beautifully preserved. In none of them does the test show any trace of reticulation or longitudinal striæ. Thus the morphological agreement with Sagenites is not connected with any closer affinity.

All characters of importance lead to the suggestion of a very close affinity of *Griesbachites* with *Jueavites*. Among knob-bearing forms with distinct marginal ears *Juvavites Chamissoi* v. Mojsisovics (l. c. p. 94, Taf. LXXXVII, fig. 2) has been left with this genus by E. v. Mojsisovics, whereas a second one, *Juvavites fulminaris* v. Dittmar (E. v. Mojsisovics, l. c. p. 130, Taf. LXXXVII, fig. 1), has been grouped with the subgenus *Anatomites*, on account of some differences in the strength of the lateral ribs.

To me both species seem to be related most intimately to the prototype of the subgenus *Griesbachites*. The only feature of distinction, which has been pointed out by E. v. Mojsisovics, is the confinement of marginal tubercles to the bodychamber of adult individuals in the two last-mentioned species, whereas in *Griesbachites Medleyanus* such marginal tubercles already make their appearance on the chambered whorls. But this feature of distinction I cannot consider to be a very safe one. In *Juvavites fulminaris* the inner whorls are not known. According to the description, the commencement of the last volution corresponds exactly with the last septum. The entire periphery of this volution being adorned with marginal tubercles, E. v. Mojsisovics himself declares it to be an open question whether or not such tubercles may be developed on the inner nucleus. On the other hand it is barely possible to ascertain the position of the last septum in the type-specimen of *Juvavites Kastneri*. As marginal ears make their appearance in the anterior half of the last volution only, a very small part of the chambered portion of the shell only can be distinguished by their presence. A similar difficulty of ascertaining the boundary between air-chambers and body-chamber is met with in the three Himâlayan specimens, which I have united provisionally with *Juvavites Kastneri*. Nor have any sutures been noticed by E. v. Mojsisovics in his fragments of *Griesbachites Hauni*, and his suggestion that they were completely chambered still lacks proof.

Although I am at variance with the views of E. v. Mojsisovics on the systematic value of the relative position of marginal tubercles on the body-chamber or on the last air-chambers, I fully concur in his opinion that the forms grouped around *Griesbachites Medleyanus* are distinguished by characters sufficiently important to justify a proper subgeneric designation. But I prefer to propose a slight change in the circumscription of this subgenus, and to include in it all forms of *Juvavites* which are distinguished by the development of strong marginal tubercles or ears.

Whether all forms distinguished by the presence of marginal tubercles are also connected genetically is a question which cannot be answered at present. Some knob-bearing species of *Juvavites* from the triassic rocks of Sicily have been grouped with the subgenus *Anatomites* v. Moja. by Gemmellaro. Such species are *Anatomites elegans* Gemmellaro (I cefalopodi del trias superiore della parte occidentale della Sicilia, p. 241, Tav. XX, figs. 14-16; XXIII, fig. 9), *A. Bukowskii* Gemm. (l. c.

p. 34, Tav. XI, figs. 7, 8; XXIII, fig. 8), *A. Timaei* Gemm. (l. c. p. 236, Tav. XVII, figs. 10, 11; XXIII, fig. 7). All of them are provided with deep paulostome furrows and have the marginal tubercles restricted to the body-chamber. In *A. Timaei* they are even confined to the posterior portion of the body-chamber only, whereas in the vicinity of the aperture the ornamentation becomes altogether obsolete.

But paulostome ribs, the characteristic feature of the subgenus *Anatomites*, have also been noticed in one of the most typical species of *Griesbachites*, in *G. Hanni*, by E. von Mojsisovics. They are not known either in *G. Medicyanus* or in *G. Kastneri*, but it must be remarked that in neither species have inner nuclei as yet been available for examination.

If the examination of new materials should prove the inner nuclei of these two species to be devoid of paulostomes, we should be obliged to suggest that marginal tubercles have been acquired by different lines of *Jucavites*. In this case it would become necessary to restrict the name *Griesbachites* to the descendants of *Jucavites s. s*, and to introduce a new subgeneric designation for the knob-bearing types of *Anatomites*. But then *G. Hanni* could no longer be left within the subgenus *Griesbachites*, even if the latter were taken in the narrow original interpretation which has been proposed by E. v. Mojsisovics.

In the fauna of exotic block No. 2 the subgenus *Griesbachites* is represented by two typical species, one of them with Alpine and the second with Indian affinities. Together with *Cladiscites* they are the most remarkable and characteristic elements in the fauna of exotic block No. 2.

GRIESBACHITES cf. KASTNERI v. Mojsisovics. Pl. VI, figs. 1, 2, 3.

1893. *Jucavites Kastneri* E. v. Mojsisovics, Cephalopoden der Hallstätter Kalke. Abhandl. K. K. Geol. Reichsanst. VI-2. p. 45, Taf. CLXXXVI, fig. 5.

Three specimens, in which all characters have been excellently preserved with only the exception of the sutural line, are related so closely to the Alpine *Jucavites Kastneri* v. Mojs., from the carnic Ellipticus beds of the Salzkammergut, that I cannot find any distinctive feature which might justify a specific separation.

It is especially the smallest specimen (fig. 1) which in its shape and proportions exactly agrees with the type specimen of *J. Kastneri*. Its whorls are moderately inflated, their width being slightly inferior to the height, and they overlap one another as far as the umbilical margin, which is steeply rounded and separates a low but vertical umbilical wall from the gently curved flanks. Siphonal part rounded, and marked off from the lateral parts by an indistinct marginal shoulder.

The two other specimens differ from the smallest one by their more compressed shape. The whorls are considerably higher than broad. The lateral parts are flattened and they converge less strongly towards the external part; the marginal shoulder is more distinctly marked. In the specimen illustrated in fig. 3 the cross-section of the aperture is almost rectangular in its outlines, the siphonal part widening into a flattened area.

As the specimen illustrated in fig. 2 is a truly transitional shape between the two other types, I cannot consider these differences as characters of specific importance, but only as marks of individual variability

There are also some individual variations of minor importance shown in the sculpture of the three figured specimens. The specimen illustrated in fig. 2 in its ornamentation agrees most nearly with the Alpine type. It consists of broad and low folds or ribs, which are directed radially in the lower half of the lateral parts, but show a sigmoidal curve in the vicinity of the marginal shoulder. Their number is increased towards the external part, either by bifurcation of the stem-ribs or by intercalation of secondary ribs. Some of the primary ribs are more strongly marked than the rest, but the difference in strength is only insignificant.

Along the median line of the siphonal area the ribs are interrupted by a narrow, smooth band. Their arrangement on both sides of this median zone is rather irregular, alternation and correspondence of opposite ribs occurring indiscriminately. In the anterior half of the last volution the development of marginal tubercles sets in. Six marginal tubercles are counted in the last half volution. In the majority of tubercles a bifurcation of ribs is noticed. As a rule every fourth or fifth rib is adorned by a tubercle. The majority of tubercles are arranged symmetrically with the median line of the shell, but there are some exceptions from this rule. The ribbing weakens gradually towards the aperture, whereas the tubercles become stronger.

The specimen illustrated in fig. 1 possesses broader ribs separated by wide intercostal furrows, but a large number of marginal tubercles, eight of which are counted in the circumference of the last half volution. Thus only one or two simple ribs are intercalated between such ribs as bear marginal tubercles. The smooth zone of the siphonal part is very narrow. Some of the ribs, which are arranged almost symmetrically on both halves of the shell, even close together on the external area in the vicinity of the aperture.

In the specimen fig. 3 there are very numerous, chiefly forked ribs. Tubercles are developed in a rather advanced stage of growth, and in small numbers only. The ribs continue increasing in strength, although becoming less numerous, to the very aperture of the shell. Coming alternately from both halves of the last volution, they are interrupted by a narrow, smooth zone along the external part, with the exception of the last three ribs, which close together.

All the specimens differing somewhat from each other in their external shape and ornamentation, but being linked together by the agreement of all characters of importance, they are to be regarded as varieties of one species.

Dimensions.

	Fig. 1.	Fig. 2.	Fig. 3.
Diameter of the shell	48·5 mm.	61 mm.	62 mm.
" " umbilicus	5 "	6 "	6 "
Height of the { above the umbilical suture	28 "	31·5 "	31 "
last volution { " " preceding whorl	14 "	50 "	35 "
Thickness of the last volution	23 "	20 "	23·5 "

Sutures.—Not known. I have entirely failed in discovering any trace of septa. It is consequently impossible for me to decide whether my specimens are yet entirely chambered or provided with parts of their body-chambers. E. v. Mojsisovics declared his type-specimen of *J. Kastneri* to be a chambered nucleus, but was obliged to state that its sutural line was not known to him in detail.

Remarks.—Although the establishment and limitation of the present species offer considerable difficulties on account of its individual variability, I consider the specific identity of the three figured examples as sufficiently well established by their agreement in all characters of importance. The species is so closely allied to the European *Juvarites* (*Griesbachites*) *Kastneri*, that I have not thought it appropriate to introduce a new specific denomination, although the question of identity cannot be decided without a thorough knowledge of their sutural lines.

Among Himálayan species *Griesbachites Hanni* v. Mojsisovics (Himálayan Foss., Vol. III, Pt. 1, p. 39, Pl. X, figs. 3-5) from the upper Daonella beds of the Bambanag cliffs and of Lauka might be taken into consideration for a closer comparison. Of this interesting species specimens more or less fragmentary only are, unfortunately, known to us. But, as we may judge from the beautiful illustrations given by E. v. Mojsisovics, the large size of the umbilicus seems to be a remarkable feature of distinction. In the ornamentation the two species resemble each other very closely, although marginal tubercles seem to be developed more numerously in *G. Hanni*. On the outer volution an alternation of the marginal tubercles occurs by the swelling of each second rib along the marginal shoulder. In some fragments even marginal tubercles occurring regularly on each rib have been noticed.

The Tibetan species from exotic block No. 2 cannot, therefore, be united with *G. Hanni* from the triassic belt of the main region of the Himálayas.

GRIESBACHITES PSEUDOMEDLEYANUS nov. sp.   Pl. VII. figs. 1, 2; IX, fig 7.

1902. *Juvarites* (*Griesbachites*) *Medleyanus* (Stoliczka) A. v. Krafft, Mem. Geol. Surv. of India. Vol. XXII. Pt. 3. p. 143.

There seems to exist a similar relationship between this species and *Griesbachites Medleyanus* Stol. from the upper trias of Spiti as between *G. cf. Kastneri* and *G. Hanni* v. Mojs. They are, indeed, very closely allied, and their identification was only a slight mistake of A. v. Krafft's, which is easily understood if we consider that the sutural line of his Tibetan specimens was not known to him.

Two well-preserved examples of large size are before me. In the smaller one (fig. 2) one quarter of the last volution belongs to the body-chamber. In the larger specimen (fig. 1) I have not been able to discover any trace of septa, but from its size and sculpture we may assume that at least one half of the last volution forms part of the body-chamber.

In their external shape both specimens agree exactly with the type-specimen of *Griesbachites Medleyanus* Stoliczka (Mem. Geol. Surv. of India, Vol. V, p. 54,

Pl. IV, fig. 5) as figured anew and redescribed by E. v. Mojsisovics (Himálayan Foss. Vol. III, Pt. 1, p. 38, Pl. X, fig. 2). The whorls overlap one another up to the rounded umbilical margin, and are higher than wide. The lateral parts are gently curved and pass into the regularly rounded siphonal part without the intervention of any marginal shoulder. The umbilicus is deep, but very narrow.

In contrast to the faintly marked lateral ornamentation of *G. Medleyanus* the sculpture is strongly developed in the present species, and not restricted to the chambered portions of the shell. In my smaller specimen it is continued as far as the aperture, although with diminishing strength, whereas in the larger example (fig. 1) the lateral parts in the anterior half of the last volution are nearly smooth.

The sculpture consists of numerous broad and low folds, which are of an approximately radial direction, not sigmoidal as in *G. Kaitneri* or in *G. Hanni*. The majority of ribs are dichotomous. In the smaller specimen they are interrupted along the middle of the external part, alternating distinctly on the two halves of the shell. In the larger specimen, where the ornamentation on both sides of the last whorl corresponds pretty well, a large number of the ribs are not interrupted, but closing together on the siphonal area.

Fourteen marginal tubercles are counted within the circumference of the last volution in my smaller specimen. They are already developed at the beginning of the last whorl and, consequently, are not confined to the body-chamber. Whether or not this has also been the case in the larger specimen cannot be decided, but there is no doubt that in the latter the development of marginal tubercles sets in at later stages of growth only, the posterior quarter of the last volution being still free from any marginal sculpture. In this specimen the marginal tubercles attain very considerable dimensions, swelling out into high and elongated " ears " in the vicinity of the aperture. In my smaller specimen the tubercles in the anterior portion of the shell are also distinguished by their remarkable size.

Dimensions.

	Fig. 1.	Fig. 2.
Diameter of the shell	122 mm.	113 mm .
„ „ „ umbilicus	9 „	7 „
Height of the { above the umbilical suture	67 „	61 „
last volution { „ „ preceding whorl	41 „	32 „
Thickness of the last volution	54 „	56 „

Sutures. —The sutural line, as far as known, differs remarkably from that in *G. Medleyanus* and bears a greater similarity to the sutures of *Austomites* than to the group of *Juvavites interrupti.* The main saddles are not as narrow and slender as in *G. Medleyanus*, but rather broad and of nearly equal height. The second lateral saddle is provided with three large lateral lappets on its umbilical slope. The third lappet is separated from the preceding one by a deep incision, which might be considered eventually as a rudimentary auxiliary lobe. The actual auxiliary lobe terminates in two sharp points and is followed by a large auxiliary saddle. Thus the second lateral saddle and the auxiliary series are united into a sort of suspensive lobe.

A similar intimate connection of the second lateral and auxiliary saddles is noticed in several species of *Anatomites*, especially in *A. Bacchus* v. Mojsisovics (Cephalopoden der Hallstaetter Kalke, Abhandl. K. K. Geol. Reichsanst. VI-2, p. 143, Taf. LXXXVII, fig. 20) or in *A. Philippii* v. Mojsisovics (l. c. p. 128, Taf. LXLIII, fig. 3).

This difference in the character of the sutural line justifies a specific separation of the present form from *Griesbachites Melleyarus*.

GRIESBACHITES nov. sp. ind.   Pl. VII, fig. 7.

The generic position of the only specimen available for examination cannot be fixed with full certainty. This specimen, which is incomplete, badly preserved, and somewhat distorted by pressure in the rocks, belongs to an undeterminable species, which seems to be related to the group of *Juvacites continui*. All lateral ribs cross the external part without any interruption. On the other hand they are not forked three or four times, as in the Alpine representatives of the group of *Juvacites continui*, but are only dichotomous. Simple, intercalated, and forked ribs occur in almost equal numbers. This is a kind of sculpture which reminds us more strongly of the group of *Griesbachites Hanni* than of *Juvacites s. s.* In the vicinity of the aperture traces of marginal tubercles are noticed, but so indistinctly that their presence could not be ascertained.

I have provisionally referred this specimen to the subgenus *Griesbachites*. Provided this determination were correct, it should certainly be considered as a new species, although, on account of its incompleteness, it is preferable to refrain from the imposition of a new specific name.

Dimensions.—Not measurable.

Sutures.—Not known.

Subgen. : *ANATOMITES* v. Mojsisovics

ANATOMITES sp. ind. aff. CAMILLI v. Mojs.   Pl. VII, fig. 5.

A single specimen consisting of air-chambers only agrees pretty well in its outlines and sculpture with *Anatomites Camilli* v. Mojsisovics (Cephalopoden der Hallstaetter Kalke, VI-2, p. 103, Taf. XCI, fig. 3) from the carnic Hallstatt limestone of Aussee.

The shell is of very globose shape, more strongly inflated than in *A. Camilli*, and is provided with a robust sculpture. In the vicinity of the umbilicus the number of broad and coarse stem-ribs is comparatively small, but towards the marginal shoulders their number increases considerably by bifurcation, which is repeated in such ribs as precede the deep paulostome furrows. These paulostome-furrows are turned forward more strongly than the normal intercostal valleys, which are directed radially.

Three paulostome-furrows are counted within the circumference of the last volution. This is probably also the normal number of paulostomes in *A Camilli*, the posterior part of the last volution having been so considerably injured in the Alpine type-specimen that the presence of a third paulostome could not be ascertained by E. v. Mojsisovics.

The ribs are not arranged symmetrically to the median plane of the shell, although an exact correspondence in all details does not exist. They cross the siphonal part without any distinct interruption, but some of them are turned very low at the place where they meet from both sides of the shell.

Dimensions.

Diameter of the shell	29 mm
" " umbilicus	3 "
Height of the { above the umbilical suture	14·5 "
last volution { " " preceding whorl	8·5 "
Thickness of the last volution	24 "

Sutures.—Not known.

Remarks.—Although this specimen agrees very remarkably with *Anatomites Camilli* in its sculpture, especially in the division of ribs and in their arrangement in reference to the paulostome-furrows, I dare not venture on an identification on account of its more strongly inflated shape. Among the numerous species of *Anatomites* from the upper-triassic rocks of Sicily, as described and figured by Gemmellaro, there is none which might put in a claim for a closer comparison.

ANATOMITES sp. ind. ex aff. HENRICI v. Mojs. Pl. VII, fig. 9.

This species may be looked upon as a Himálayan representative of the Alpine group of *Anatomites Bacchus* v. Mojsisovics (Cephalopoden der Hallstaetter Kalke l. c. VI-2, p. 143, Taf. LXXXVI, fig. 73 LXXXVII, figs. 14-21). It is closely allied to this form or perhaps still more so to *A. Henrici* v. Mojsisovics (l. c. p. 146, Taf. LXXXVIII, figs. 11. 12), agreeing with the latter not only in the absence of any distinct sculpture, but also in the compressed shape of its transverse section.

From *Anatomites Bacchus*, and from its Indian representatives in the Tropites beds of Spiti which have been illustrated on Pl. XIV, figs. 6, 7, of the third part of Vol. V of this series, it differs by its narrow whorls, which are but slightly thicker than high, and by the absence of any keel-like elevation in the middle of the regularly rounded siphonal part. An identification with either *Jocites* or *Isculites* is at once excluded on account of the absence of an expanding umbilicus. Although the general shape might at first suggest an affinity of our species with *Arcestes* rather than with *Anatomites*, the arrangement of the sutural line proves it to be a representative of the family of *Juraeitinae*.

The cast of the nucleus being strongly weathered, no traces of ornamentation have been noticed.

Dimensions.

Diameter of the shell	.	35 mm.
„ „ „ umbilicus	ab.	„
Height of the valve to the umbilical suture		19 „
last volution { „ „ preceding whorl		8 „
Thickness of the last volution		21·5 „

Sutures.—As far as known, exhibiting the general characters of the sutural line in the genus *Juvavites*. Details not accessible to examination.

ANATOMITES sp. ind. ex aff. CRASSEPLICATO v. Mojs.

A fragmentary cast, unworthy of illustration, belongs to a large species of the subgenus *Anatomites*, which reminds us of *A. crasseplicatus* v. Mojsisovics (Cephalopoden der Hallstaetter Kalke l. c. VI-2. p. 139, Taf. XCIV, figs. 6-10) from the carnic Subbullatus beds of the Salzkammergut, on account of its globose shape and of its broad folds which are restricted to the outer half of the shell and are separated by narrow intercostal furrows. Two low paulostome-furrows are noticed in the vicinity of the aperture, exactly as in the Alpine specimen of *A. crasseplicatus* illustrated by E. v. Mojsisovics in fig. 10 on Pl. XCIV of his memoir.

With this specimen our fragment agrees in its size, but it does not possess an expanding umbilicus.

Gen. : GONIONOTITES. Gemmellaro.

GONIONOTITES cf. ITALICUS Gemmellaro. Pl. IX, fig. 9 ; V, figs. 6, 7.

1884. *Gonionotites italicus* Gemmellaro, I cefalopodi del trias superiore della regione occidentale della Sicilia. p. 158, Tav. V, figs. 6, 7 ; IX. figs. 6, 7 ; XXI, figs. 4-6 ; XXX, fig. 8.

This species is represented by two inner nuclei (Pl. V, figs. 6, 7) and a larger fragment in which the last septum is situated close to the aperture. That the two nuclei and the larger specimen do really belong to the same species is obvious from a comparison of the inner whorls of the latter, which agree in every respect with the figured nuclei.

The inner volutions are strongly compressed, discoidal, with a high and narrow cordiform transverse section. The small umbilicus is surrounded by a steep wall and separated from the lateral parts by a rounded-off margin. The greatest transverse diameter coincides with the umbilical margin. From this place the flattened sides converge very gradually towards the siphonal part, which is regularly rounded.

The sculpture consists of numerous sigmoidal folds, which are developed more strongly in the vicinity of the marginal shoulders than in the umbilical region. They are either simple or dichotomous. Exceptionally a first bifurcation is noticed at a point about one-third of the height of the whorl, and a second one near the marginal shoulders. The ribs are interrupted along the middle of the external part.

In the two nuclei the ornamentation becomes gradually obsolete in the vicinity of the aperture. Constrictions or paulostome ribs have not been noticed.

In the larger fragment, which consists of an inner nucleus and the six last air-chambers preceding the body-chamber, the ornamentation of the test has been completely obliterated. At the same time the transverse section becomes high and very strongly compressed. The external part is sharpened into a steeply rounded and narrow ridge. The inflation of the shell in the vicinity of the peristome, as it has been described in the type-specimen of *Gonionotites italicus* by Gemmellaro, could not be noticed in this fragment, whose aperture is situated immediately in front of the last septum.

Dimensions.

	Larger fragment. PL. IX, fig. 9.	Inner nucleus. PL. V, fig. 6.
Diameter of the shell	?	30 mm.
" " " umbilicus	?	7·5 "
Height of the { above the umbilical suture	45 mm.	17 "
last volution { " " preceding whorl	23 "	10 "
Thickness of the last volution	23·5 "	12 "

Sutures.—There are five saddles outside the umbilical suture, but the two auxiliary saddles could not be examined in detail.

Siphonal lobe broad and nearly as deep as the principal lateral lobe. It is divided by a broad and pyramid-shaped median prominence, whose apex is flatly rounded and accompanied by a small indentation on each side. The median prominence does not reach half as high as the siphonal saddle.

Principal lateral lobe deeply serrated, terminating in two sharp points which are separated by a median indentation. Second lateral lobe considerably shorter, ending with an elongated terminal point, with lateral digitations arranged symmetrically on each side. Saddles provided with distinctly individualized, foliaceous branches, especially the siphonal saddle, which is bipartite at its top, the two apical branches being subdivided by secondary incisions. The two lateral saddles, which are also richly ornamented, have their apices divided asymmetrically, the external phyllum exceeding the internal one in size and height.

The sutures do not agree in all their details throughout the chambered outer volution of the fragment illustrated, but show some variations, which may, however, be attributed to an accidental difference in the weathering of the surface of the cast.

Remarks.—This Himálayan species agrees so closely with *Gonionotites italicus* Gemm., that in a better and more complete state of preservation of the larger specimen at hand I should not have hesitated in venturing on a direct identification. The difference in the details of the complicated sutures is so slight that they can scarcely be considered as distinctive features of specific importance.

CERATITOIDEA.

Fam.: *DINARITIDÆ* v. Mojs.

Gen.: TIBETITES v. Mojsisovics.

TIBETITES BHOTENSIS nov. sp. Pl. VI, fig. 5.

The figured specimen, tolerably completely preserved, is an adult individual, more than one-third of its last volution belonging to the body-chamber. The shape and sculpture of the body-chamber agree exactly with those of the chambered parts of the shell. Thus the reference to *Tibetites s. s.* is justified. The complete absence of any notchings in the external ears distinguishes our species from the genus *Cyrtopleurites*, to an Alpine representative of which it is otherwise nearly allied.

Among the hitherto described Himálayan species of *Tibetites* the nearest allies to the present one are *T. Ryalli* v. Mojsisovics (Upper-triassic fauna of the Himálayas, Palæont. Ind. ser. XV, Himál. Foss., Vol. III, Pt. 1, p. 77, Pl. XV, figs. 3, 4) from the Halorites limestone of the Bambanag cliff, and a second unnamed species from the upper Daonella beds of the same locality, which differs from the preceding one by being more delicately ribbed and therefore ornamented more richly.

The chief difference from *T. Ryalli* consists in the entire absence of any lateral tubercles. The stem-ribs are broad and bifurcate at a point situated below the middle of the height of the volution, but the point of bifurcation is not marked by any swelling of the ribs. There are only two rows of tubercles, one of them corresponding to the marginal shoulders, and the second consisting of large external ears. The marginal tubercles are small, of circular outlines, not elongated spirally. The number of intercalated ribs in the upper half of the sides is very small. Two marginal tubercles correspond, as a rule, to one primary stem-rib.

In its general shape this species agrees very nearly with *Tibetites Ryalli* v. Mojs. and with *Cyrtopleurites Herodoti* v. Mojsisovics (Cephalopoden der Hallstaetter Kalke, l. c. VI-2, p. 518, Taf. CLVIII, fig. 10). The whorls overlap one another almost completely and are strongly compressed, the transverse section being considerably higher than broad. The umbilicus is comparatively wide. The narrow and slightly depressed mesial band, which is enclosed within the external ears, undergoes no change from the beginning up to the aperture of the last volution.

Dimensions.

Diameter of the shell	.	.	.	39 mm.
„ „ „ umbilicus	.	.	.	4.4 „
Height of the { above the umbilical suture	.	.	.	22.5 „
last volution { „ „ preceding whorl	.	.	.	17 „
Thickness of the last volution	.	.	.	14 „

Sutures.— Not known in detail. A more exact obaracterisation is not possible on account of the defective preservation of the sutural line.

NAUTILOIDEA.

Fam. : *ORTHOCERATIDÆ.*

Gen. : ORTHOCERAS Breyn.

ORTHOCERAS div. sp. ind.

A few fragments of *Orthoceras*, unworthy of illustration, are only sufficient to determine the genus. There are at least two species present. The fragment of a large body-chamber, with parts of the striated shell adhering to the cast, belongs to the group of *Orthocerata striata*. The majority of isolated air-chambers belongs to a species of the group of *O. lævis.* Their dimensions and the small angle of divergency remind one of *Orthoceras triadicum* v. Mojsisovics (Cephalopoden der Hallstaettor Kalke, l. c. VI-1, p. 5, Taf. I, figs. 2, 3), but their incomplete state of preservation renders them unfit for specific determination.

Fam. : *CLYDONAUTILIDÆ* v. Mojs.

Gen. : PROCLYDONAUTILUS v. Mojs.

PROCLYDONAUTILUS TRIADICUS. v. Mojsisovics. Pl. I, fig. 1.

1873. *Nautilus triadicus* E. v. Mojsisovics, Cephalopoden des Hallstaetter Kalke, l.c. VI-1, p. 27, Taf. XIV, figs. 1-4.

1902. *Proclydonautilus triadicus* v. Mojsisovics, *ibidem*, Supplemental. p. 209.

1904. *Clydonautilus triadicus* Gemmellaro, I cefalopodi del trias sup. della regione occid. della Sicilia, p. 7, Pl. I, figs. 14, 15.

1904. *Proclydonautilus triadicus* P. Smith, Comparative stratigraphy of the marine trias of Western America, Proceed California Acad. of Sciences, 3d ser. Vol. 1, p. 401. Pl. XLVII, fig. 2.

1905. *Proclydonautilus triadicus* Hyatt et Smith, Triassic cephalopod genera of America, U. S. Geol. Surv. Profess. Papers No. 40, p. 204. Pl. XLIX, figs. 1-3 ; L, figs. 1-17.

The only specimen available for examination agrees so closely with the type-specimen of *Proclydonautilus triadicus* from the carnic Hallstatt limestone of the Salzkammergut, that a direct identification cannot be avoided.

My specimen is of moderate size, and provided with the beginning of the body-chamber. It has slowly increasing whorls, which overlap one another completely, and a closed umbilicus. The transverse section is ovoid, the flatly curved lateral parts passing into the steeply rounded external part without intervention of any marginal shoulder. The aperture is of equal height and width.

The shell, which has been preserved entirely on one side of the last volution, is smooth, without any trace of a longitudinal or transverse sculpture. Not even striæ of growth have been noticed.

Dimensions.

Diameter of the shell	.	.	.	.	.	.	.	.	.	. 42	mm.
„ „ „ umbilicus										. 0	„
Height of the { above the umbilical suture										. 27	„
last volution { „ „ preceding whorl										. 10	„
Thickness of the last volution	.									. 27	„

Sutures.—The septa are situated very close to each other, especially those preceding the last septum, the lateral walls of the saddles even touching one another occasionally.

Siphonal lobe small and narrow, not divided by a median prominence. Lateral lobe broad, deep, and tongue-shaped. A small umbilical lobe follows outside the large and regularly rounded lateral saddle

Siphuncle.—Not known.

Remarks.—The specimen of *Proclydonautilus* from the carnic stage of Castronuovo in Sicily, which has been identified with *P. triadicus* by Gemmellaro, differs slightly from the Alpine and Himâlayan examples of this species by its whorls increasing still more slowly. But as it agrees with them in the remaining characters, especially in the arrangement of the sutural line, there is no sufficient reason for a specific separation.

The specimens from the Subbullatus beds of California, as described and illustrated by J. Perrin Smith, seem to agree with the Alpine type as closely as my Himâlayan example.

Proclydonautilus triadicus must be counted among the most wide-spread cephalopoda of the carnic stage, being almost universally distributed throughout the triassic seas.

PROCLYDONAUTILUS BUDDHAICUS nov. sp.   Pl. I, figs. 2, 3, 7.

This species is very nearly allied to *Proclydonautilus Griesbachi* v. Mojsisovics (Upper-triassic faunæ of the Himâlayas, Pal. Ind. ser. XV, Himâl. Foss. Vol. III, Pt. 1, p. 123, Pl. XXII, fig. 1) and to *P. Griesbachiformis* Diener (Fauna of the Tropites limestone *ibid.* Vol. V, Pt. 1, p. 15, Pl. XVII, fig. 2).

My type-specimen, which is provided with its body-chamber, attains large dimensions. It is strongly involute, provided with a very small umbilicus and with high, laterally compressed whorls. The inflated siphonal part is separated from the converging lateral parts by distinct marginal shoulders. In the inner volutions these shoulders form sharp angles, which are slightly elevated above the general convexity of the shell. These acute rims, which recall the blunt keels in *Norites*, make their appearance at an early stage of development. At the beginning of the last volution the elevated sharp rim is reduced to an acute edge, which passes gradually into a rounded angle. Near the aperture this broadly rounded angle forms the boundary of the external part, which is inflated still more strongly than in the adolescent stage.

This mode of development is just the contrary of what is seen in *P. Griesbachi* v. Mojs. In *P. Griesbachi* the inner volutions possess a regularly cordiform transverse section, their inflated external part merging into the sides with a continuous swelling, without even the indication of an angle. It is only in later stages of growth that angles are developed, which are most conspicuous in the vicinity of the aperture. In its developmental features our species agrees better with *P. Griesbachiformis* from the Tropites limestone of Byans, which is also distinguished by the presence of sharp marginal angles at very early stages of growth. But in this species the elevated keel-shaped rims persist also in the gerontic stage.

Another feature of distinction between *Proclydonautilus buddhaicus* and *P. Griesbachiformis* is the shape of the transverse section in the siphonal region. In *P. buddhaicus* the siphonal area is strongly inflated, whereas it is depressed and even deeply excavated between the marginal keels in *P. Griesbachiformis*. Otherwise the transverse section is very similar in both species. The lateral parts regularly converge from the place where the shell reaches its greatest transverse diameter, *i.e.*, from the vicinity of the umbilical margin. There is no umbilical edge present, but the flanks descend in a strongly bent curve from the region of the greatest inflation to the umbilical suture.

The sculpture of the test consists of delicate transverse striæ which are approximately parallel to the septa.

Dimensions.

Diameter of the shell	. . . .	110 mm.
„ „ „ umbilicus	. . . .	8 „
Height of the } above the umbilical suture	. . . .	75 „
last volution } „ „ preceding whorl	. . . .	52 „
Thickness of the last volution	. . . .	67 „

Siphuncle.—In the last air-chamber the orifice of the siphuncle is situated above the middle of the distance between the external parts.

Sutures.—Agreeing with those of *P. Griesbachi.* Siphonal lobe not divided by a median prominence.

Remarks.—There seems to exist a very near affinity between *Proclydonautilus buddhaicus* and the American genus *Cosmonautilus* Hyatt et Smith (Triassic cephalopod genera of America, U. S. Geol. Surv. Prof. Pap. No. 40, p. 207). The typical species, *Cosmonautilus Dilleri* (l. c. p. 207, Pl. LI, fig. 1 ; LII, fig. 1 ; LIII figs. 1, 2 ; LIV, figs. 1-4 ; LV, figs. 1-11) develops marginal edges at a very early stage of growth, but these edges become adorned with tubercles, until the shape and sculpture are very like those in *Metacoceras* Hyatt. At the diameter of 35 mm. the tubercles become obsolete and the marginal shoulders lose their angularity.

Having succeeded in chiselling out the inner nucleus of a second specimen of *P. buddhaicus* with a diameter of 25 mm., I could ascertain the entire absence of any marginal tubercles. This species does not, therefore, go through a *Metacoceras* stage, as does *Cosmonautilus*. The absence of *Metacoceras*-characters in the young stage, and the persistence of marginal edges into more advanced stages of growth—they do not become obsolete, before a diameter of 55 mm. has been reached—distinguishes our Indian shell from *Cosmonautilus*, which, according to the present stage of our knowledge, is confined to the American trias.

P. Smith has suggested that *Clydonautilus biangularis* v. Mojsisovics (Himálayan Foss., l. c. Vol. III, Pt. 1, p. 134, Pl. XXII, figs. 2, 3) from the Halorites limestone of the Bambanag section might belong to his genus *Cosmonautilus*, " as may also some of the European species assigned by E. v. Mojsisovics to *Clydonautilus*." But with this view I cannot agree, because the two forms differ widely by their mode of development. Young specimens of *Clydonautilus biangularis* have a rounded siphonal part. The two marginal angles, which are not combined with any tubercles, make their appearance simultaneously with the flattening and individualisation of the external area only at a diameter of about 16 mm. Even in old age the external part remains flattened and is not inflated.

Gen.: Styrionautilus v. Mojsisovics.

STYRIONAUTILUS nov. sp. ind.   Pl. II, fig. 1.

Of this interesting species only a single, fragmentarily preserved cast of the body-chamber has been found, with the last air-chamber and a small part of the inner nucleus adhering. The reference to the genus *Styrionautilus* v. Mojsisovics has been established with full certainty, the siphonal saddle having been observed crossing the external area without any indication of a siphonal lobe.

There is no species to which the present one appears to be nearly allied. From the Alpine representatives of the genus *Styrionautilus* it is distinguished by the biangular shape of its cross-section and by its very high and strongly compressed whorls. In the inner volutions the sides converge from the place of the greatest inflation outside the deep umbilicus towards the narrow external part in a very flat curve. The siphonal area is truncated and bordered by sharp marginal angles. In the body-chamber these marginal angles become obsolete, and the flanks pass gradually into the external part which is no longer truncated but steeply rounded. The deep umbilicus is surrounded by a perpendicular wall which unites with the lateral parts in a sharply rounded edge.

Dimensions.—Not measurable, on account of the defective state of the figured specimen. In the body-chamber a height of the last volution of 40 mm. corresponds to a transverse diameter of 29 mm.

Siphuncle.—Elongated elliptically in the direction of the radius. Its position is approximately central in the last air-chamber.

Sutures.—The siphonal saddle crosses the external area in a straight line which is not interrupted by any mesial depression. The present species is therefore a typical representative of the genus *Styrionautilus*, in the sutures of which no transitional stage is marked to *Proclydonautilus* or *Clydonautilus*. The deep lateral lobe is tongue-shaped, as in *Styrionautilus styriacus* v. Mojsisovics (Cephalopoden der Hallstaetter Kalke, Abhandl. K. K. Geol. Reichsanst. VI-1, p. 27, Taf. XIV, fig. 7) or in *St. Sauperi* v. Hauer. It is followed by a large lateral saddle and by a rounded umbilical lobe. An internal or annular lobe has not been observed.

Fam.: GRYPONAUTILIDÆ.

Gen. : GRYPOCERAS HYATT.

GRYPOCERAS SUESSIIFORME nov. sp. Pl. X, fig. 1.

This is a very interesting species, recalling *Gryponautilus Suessii* v. Mojsisovics (Cephalopoden der Hallstaetter Kalke l. c. VI-1, p. 26, Taf. VI, fig. 11, Taf. XIII, fig. 2), but distinguished by its wider siphonal area and by the persistence of an open umbilicus in advanced stages of growth.

There is only one specimen available for examination. It is a cast, with some fragments of the shell adhering to the siphonal and lateral parts. More than three-quarters of the last volution consist of air-chambers. As the beginning of the body-chamber has been preserved, the complete example must have been considerably inferior in size to full-grown specimens of *G. Suessii*.

In its shape, sculpture and sutures the figured specimen is very similar to *G. Suessii*. The whorls are very thick, their transverse section being twice as broad as high. The greatest transverse diameter is situated in the lower third of the lateral parts. From this place the strongly inflated flanks converge both towards the abdominal and umbilical margins. The external area is broad, gently arched and separated from the lateral parts by sharp edges, which are accompanied by a row of small, spirally elongated, very low and blunt tubercles along their external slopes. About 25 tubercles are counted within the circumference of the last volution.

The width of the external area is considerably larger than in *G. Suessii*. In my type-specimen an external area of 18 mm. in width corresponds to a transverse diameter of 37 mm., whereas in the Alpine species the respective proportions are as 12·5 to 34 or as 21 to 67 mm. Another feature of difference is the flattened shape of the external area in *Gryponautilus Suessii*, whereas it is slightly convex in the Himálayan species.

The umbilical margin is steeply rounded and is bordered by a perpendicular umbilical wall. The umbilicus is comparatively wide, wider than in the smaller of the two Alpine type-specimens of *G. Suessii*, illustrated by E. v. Mojsisovics, which consists of air-chambers only. As the beginning of the body-chamber has been preserved in my Himálayan example, the umbilicus cannot have been closed, as in full-grown specimens of *O. Suessii*. The present species must consequently be grouped with the genus *Grypoceras s. s.*, not with the subgenus *Gryponautilus* v. Mojs.

The lateral parts of this species exhibit the delicate transverse sculpture which has been described in *Gryponautilus Suessii* by E. v. Mojsisovics. It consists of sharp, crescentic ledges with their convexities turned forward, which originate in the blunt tubercles of the marginal edges.

Dimensions.

Diameter of the shell	.	.	.	.	. 66 mm.
„ „ umbilicus		.	.	.	. 8·5 „
Height of the { above the umbilical suture	.	.	.	. 24 „
last volution { „ „ preceding whorl	.	.	.	. 17·5 „
Thickness of the last volution	.	.	.	. 37 „

Sutures.—Agreeing with those of *O. Suessii*. External saddle divided by a very flat lobe. Lateral lobe gently curved and followed by an equally flat lateral saddle.

Siphuncle—Not known.

Fam. : *TEMNOCHEILIDÆ.*

Gen. : Mojsvarockras Hyatt.

Mojsvarockras sp. ind. ex aff. Turneri Hyatt et Smith. Pl. I, fig. 4.

This is a single fragment of the outer volution of a *Nautilus* consisting of four air-chambers and of the beginning of the body-chamber. It was strongly evolute and provided with little embracing whorls and with a wide umbilicus. The transverse section is subquadrangular and considerably wider than high. The lateral parts are flattened and separated from the steep umbilical wall by a sharply rounded margin and from the gently curved siphonal area by acute abdominal edges.

Surface nearly smooth, but ornamented with two rows of faint tubercles outside the umbilical margin and on the abdominal edge.

Dimensions.—Not measurable.

Siphuncle.—Below the centre of the whorl.

Sutures.—Septa very simple, slightly sinuous, with very shallow siphonal and lateral lobes and with a broadly curved external saddle. The presence of an internal annular lobe could not be ascertained.

Remarks.—This species is more nearly allied to *Mojsvaroceras Turneri* Hyatt et Smith (Triassic cephalopod genera of America, U. S. Geol. Surv. Prof. Pap. No. 40, p. 200, Pl. XLVIII, figs. 6-11) from the *Tropites* beds of California, than to any of the Alpine forms. A more detailed comparison is, however, difficult, on account of the incomplete state of preservation of both the American and Himalayan types.

DIBRANCHIATA.

Fam.: *BELEMNITIDÆ.*

Gen.: ATRACTITES Guembel.

ATRACTITES sp. ind.

A slab of rock contains numerous fragments of rostra, which are perfectly smooth, without any ribs or furrows. They are elongated, with a circular transverse section. The lumen of the funnel does not change throughout the entire length of the rostrum, as far as known.

Among my materials there is no example fit for illustration nor for a specific determination.

Gen.: DICTYOCONITES v. Mojsisovics.

DICTYOCONITES sp. ind. ex aff. HAUERI v. Mojs.   Pl. II, fig. 2.

The genus *Dictyoconites,* as introduced by E. v. Mojsisovics, comprises such forms of *Aulacoceratinæ,* in which the conotheca of the phragmacone is ornamented externally with numerous raised longitudinal lines and with delicate asymptotic ribs, which in the guard or rostrum correspond with deep dorso-lateral grooves.

Among the materials available from the triassic limestone of exotic block No. 2 there is a fragment of the phragmacone of *Dictyoconites,* with a short part of the surrounding guard. It cannot serve for the establishment of the species, although it seems to indicate a form of the group of *D. striati,* nearly allied to *D. Haueri* v. Mojsisovics (Cephalopoden der Hallstätter Kalke, l. c., VI-1, Supplementbd. p. 187, Taf. XIV, figs. 15, 16) from the carnic Hallstatt limestone of Aussee.

The divergent angle of the phragmacone is very small. The conotheca, which has been partly preserved, is ornamented with numerous and delicate longitudinal striæ, among which the asymptotic striæ are not remarkable in any way, either by their strength or by their position. They can only be recognised near the beginning of the calcareous sheet of the rostrum, where deep dorso-lateral furrows are noticed.

The transverse section of the phragmacone is elliptical. A section through the apical region of the phragmacone together with the surrounding guard shows the deep dorso-lateral furrows shifted towards the flattened antisiphonal side. In the conotheca of the phragmacone no transverse ornamentation has been noticed. Both in this absence of a reticulate sculpture and in the slender shape of the phragmacone the illustrated fragment agrees better with *Dictyoconites Haueri* v. Moja. than with *D. reticulatus* v. Hauer.

Notwithstanding its unsatisfactory state of preservation this fragment deserves to be mentioned, as it is the first representative of the genus *Dictyoconites* hitherto known in the Indian triassic province.

CONCLUSIONS.

By far the richest among the mesozoic faunæ collected by A. v. Krafft in the district of the exotic blocks in Malla Johar is that from exotic block No. 2. It consists of the following species :—

(a) *Lamellibranchiata.*

1. *Cassianella sp. ind.*

(b) *Gasteropoda.*

2. *Lozonema (Polygirina) cf. elegans* Hoernes.
3. *Sagana cf. geometrica* Koken.
4. *Capulus (Phrys) johorensis* nov. sp.
5. *Naticopsis sp. ind. ex aff. obvallata* Koken.

(c) *Ammonoidea.*

6. *Cladiscites crassestriatus* v. Moja.
7.   „   *cf. Gorgia* Gemm.
8.   „   *cf. pusillus* v. Moja.
9.   „   *sp. ind. cf. coracis* Gemm.
10. *Hypocladiscites subcarinatus* Gemm.
11.   „   *subaratus* v. Moja.
12. *Arcestes cf. periolena* v. Moja.
13.   „   *cf. Richthofeni* v. Moja.
14.   „   *sp. ind. aff. decipiens* v. Moja.
15.   „   *cf. placenta* v. Moja.
16. *Proarcestes Gaytani* v. Klipst.
17.   „   *sp. cf. Ausseanus* v. Hauer.
18.   „   (?) *sp. ind. ex aff. Burrandei* Lbe.
19.   „   *sp. ind.* (group of *extralabiati*).
20. *Discophyllites Floweri* nov. sp.
21. *Pinacoceras sp. ind. aff. ers* v. Moja.
22. *Placites cf. perauctus* v. Moja.
23. *Discotropites cf. sandlingensis* v. Hauer.
24. *Tropites cf. subbullatus* v. Hauer.
25.   „   *sp. ind. aff. acutangulo* v. Moja.
26.   „   *sp. ind. aff. Wodani* v. Moja.
27. *Anatropites cf. spinosus* v. Moja.
28.   „   *Pulgrimii* nov. sp.
29. *Margarites irregularicostatus* nov. sp.
30. *Jovites cf. spectabilis* Dien.
31.   „   *daciformis* Dien.
32.   „   *nov. sp. ind.*

33. *Jeuavites Krafti* nov. sp.
34.   „   *Dograuma* nov. sp.
35.   „   nov. sp. ind. ex aff. subinterrupto v. Mojs.
36.   „   nov. sp. ind. (group of continui).
37. *Griesbachites* cf. *Kastneri* v. Mojs.
38.   „   *Pseudomedleyanus* nov. sp.
39.   „   nov. sp. ind.
40. *Anatomites* sp. ind aff. *Camilli* v. Mojs
41.   „   „   .. ex aff. *Heurici* v. Mojs.
42.   „   „   .. ex aff. *crassiplicato* v. Mojs.
43. *Cosionotites* cf. *itabeus* Gerum.
44. *Tibetites bhotensis* nov. sp.

(d) *Nautiloidea.*

45. *Orthoceras* sp. ind. (group of *O. laeia*).
46.   „   sp. ind. (group of *O. striatu*).
47. *Proclydonautilus triadicus* v. Mojs.
48.   „   *buddhaicus* nov. sp.
49. *Syrionautilus* nov. sp. ind.
50. *Gryporcras turriniforme* nov. sp.
51. *Mojsvaroceras* nov. sp. ind ex aff. *Tacneri* Hyatt et Smith.

(e) *Dibranchiata.*

52. *Atractites* sp. ind.
53. *Dictyoconites* nov. sp. ind. aff. *Haueri* v. Mojs

Altogether 53 species—among which the Cephalopoda, numbering 48 species, by far predominate, both in species and in individuals. The red limestone of exotic block No. 2 may consequently be termed a cephalopod-bearing facies with equal reason with the Halorites limestone of the Bambanag range or the Tropites limestone of Byans. In the richness of species it is but little inferior to the Halorites limestone, from which 69 species (62 ammonites) have been described hitherto by E. v. Mojsisovics and by myself.

Leaving out of discussion those forms which do not admit of specific determination or of a closer comparison with any species hitherto described, there remain 45 species.

The most important fact appearing on a first glance at the preceding list is the very large percentage of species nearly allied to European forms. The number of faunistic elements peculiar to the Indian triassic province is comparatively small. There is only one single genus of exclusively Indian habit (*Tibetites*) represented in this fauna. A special stress must be laid on the very close affinity with the carnic faunæ of the Alpine Hallstatt limestone, because in none of the faunæ from the triassic belt of the main region of the Himálayas are such affinities indicated as clearly and strongly. In all triassic horizons of the main region of the Himálayas which are rich in Cephalopoda species representing types which differ widely from Alpine forms occur in considerable numbers, whereas in this fauna they are of very rare occurrence.

This close affinity of the fauna of exotic block No. 2 with the faunæ of homotaxial beds in Europe is so much the more important, as the determination of its geological age must be based on palæontological evidence only, the exotic block itself exhibiting no stratigraphical connection with the surrounding beds.

With regard to its general character, the fauna of exotic block No. 2 bears the stamp of the *carnic* age so indubitably that it seems to me superfluous to discuss its correlation with the carnic Hallstatt limestone of the Alps. There is not a single form among the Cephalopoda which might point to either ladinic or noric (juvavic) affinities, not even *Tibetites bhotensis*, which belongs to a genus appearing already in carnic beds although it reaches its chief development in the lower noric stage only.

It is more difficult to establish a more exact determination of the geological age. An analysis of the fauna shows that it has relations both with the julic and tuvalic faunas. It remains therefore for me to decide whether it should be correlated with the zone of *Trachyceras Aonoides* (julic substage) or with that of *Tropites subbullatus* (tuvalic substage). A. v. Krafft, from a cursory examination of his fossil materials, has decided in favour of a correlation with the zone of *Tropites subbullatus*, but from a detailed analysis we shall learn an almost equal distribution of the elements of the two Alpine zones in our Himálayan fauna.

It will be found convenient to treat the affinities of our fauna with each carnic fauna of extra-Himálayan districts separately, although the palæontological evidence must rely chiefly on a comparison with the julic and tuvalic faunæ of the Alpine Hallstatt limestone.

The assemblage of genera being almost the same in both zones, we are obliged to investigate the specific affinities of the Cephalopoda, which by far predominate over all the rest of the organic remains. All forms, which have not been determined specifically but designated only according to their relationship with Alpine types, are of very little service for an identification of the exact geological horizon. For this purpose such species only can be taken into consideration as are either directly identical with, or so closely allied to, Alpine types that they could be referred to such as *cf.*

The following species are identical or probably identical with those from the Hallstatt limestone which in Europe connect the faunæ of the julic and tuvalic substages :—

> *Lecanites cf. elegans* Howen.
> *Sageceras cf. geometricn* Koken
> *Cladiscites crassestriatus* v. Mojs.
> *Placites cf. perauctus* v. Mojs.
> *Discotropites cf. sandlingensis* Hau.
> *Proclydonautilus triadicus* v. Mojs.

The following species are identical or probably identical with species which in the Alpine Trias are restricted to the julic substage :—

> *Cladiscites cf. pusillus* v. Mojs.
> *Arcestes cf. peridens* v. Mojs.

> *Arcestes cf. Richthofeni* v. Moja.
> *Progrestes Gaytani* v. Klipst.
> ,, *cf. Auseanus* v. Hau.
> *Anatropites cf. spinosus* v. Moja.
> *Griesbachites cf. Kastneri* v. Moja.

As elements pointing to a close affinity with the fauna of the tuvalic substage of the Salzkammergut, the following two species only can be quoted :—

> *Arcestes cf. placenta* v. Moja.
> *Tropites cf. subbullatus* v. Hau.

Among fifteen species identical or probably identical with Alpine forms, seven belong to the julic, two to the tuvalic substage, six are common to both substages. It is evident from this proportion that the greater number of relationships are in favour of a correlation with the julic substage. The circumstance that a species referable to *Tropites subbullatus*, the leading fossil of the tuvalic substage of the Hallstatt limestone, is represented in this fauna, loses much of its importance, as this species is exceedingly rare in the red limestone of exotic block No. 2, and as types allied very closely to *T. subbullatus* make their first appearance in the julic substage (*T. Quenstedti* v. Moja.). On the other hand those carnic elements which are most conspicuous for their fecundity in species and individuals, especially *Cladiscites*, *Placites*, and *Discophyllites*, are of a rather indifferent habit and do not indicate exclusively either julic or tuvalic affinities.

With the upper-triassic rocks of Sicily the fauna of exotic block No. 2 has six species in common. These are the following :—

> *Cladiscites cf. Gorgia* Gemm.
> ,, *cf. coraris* Gemm.
> *Hypocladiscites antecurinatus* Gemm.
> *Gonionotites cf. italicus* Gemm.
> *Discotropites cf. sandlingensis* v. Hau.
> *Proclydonautilus triadicus* v. Moja.

All these species have been found in the carnic limestone of Modanesi (Castronuovo) or Votano (San Stefano Quisquina). *Cladiscites cf. Gorgia* occurs also at Madonna del Balso, where carnic and noric elements have been mixed together.

The carnic age of the fauna of Votano and Modanesi is obvious from Gemmellaro's lists, but an exact correlation with carnic fauna of the Eastern Alps has not yet been attempted. G. Di Stefano, it is true, considers the limestone of Votano and Modanesi as a homotaxial equivalent of the Alpine Subbullatus beds, but G. v. Arthaber (Die Alpine Trias des Mediterrangebietes, Lethæa geognostica, 2 Theil, Bd. I, p. 461) believes with equal reason that both the *Aonoides* and *Subbullatus* zones are represented in the fauna of those two Sicilian localities.

The region which is geographically least distant from the exotic blocks of Malla Johar is the main region of the mesozoic belt of the Central Himálayas. There are three districts in this region with fossiliferous triassic beds, the fauna of which show relations to that of exotic block No. 2. But those relations are less close than the affinities between the present fauna and the carnic faunæ of the Alpine Hallstatt limestone.

With the Daonella beds of Lauka in Kumaon the present fauna has only one species in common, *Hypocladiscites subaratus* v. Moja, an Indian representative of the group of *Cladiscites sublornati*, which differs from the European *Hypocladiscites sublornatus* by some very subordinate details.

The number of species probably identical in the faunæ of exotic block No. 2 and of the Tropites limestone of Byaos is five. These species are the following :—

> *Proarcestes cf. Gaytani* v. Klipst.
> *Jovites cf. spectabilis* Dien.
> ,, *aociformis* Dien.
> *Tropites cf. subbullatus* v. Hauer.
> *Discotropites cf. sandlingensis* v. Hauer.

With the Tropites shales of Spiti the present fauna has the following species in common :—

> *Jovites cf. spectabilis* Dien.
> *Tropites cf. subbullatus* v. Hau.
> *Proarcestes cf. Gaytani* v. Klipst.

As species indicating very close affinities the four following might be mentioned :— *Anatomites sp. ind. ex aff. Henrici* v. Moja., which is certainly very nearly allied to *Anatomites sp. ind. cf. Bacchus* v. Moja. from the Tropites shales of Lilang, *Tropites sp. ind. aff. acutangulo* v. Moja., *Juvavites (Griesbachites) Pseudomedleyanus*, which belongs to the same group of *Juvavitinæ*, as *Griesbachites Medleyanus* of unknown geological age, and *Discophyllites Floweri* Dien., which agrees with *D. Ebneri* from the Daonella beds of Lauka in all characters, except some differences in the arrangement of the sutural line.

The presence of *Tropites aff. acutangulo, Jovites spectabilis* and *Jovites aociformis* increases the number of species with decidedly tuvalic affinities in the present fauna, and consequently reduces the preponderance of julic elements. To the species connecting the faunæ of the julic and tuvalic substages in the Indian triassic province *Proarcestes Gaytani* has to be added. Thus the majority of species (7), which are referable to forms previously described, are distributed through both the julic and tuvalic substages. Six species point to a closer relation with julic and four with tuvalic faunæ.

This analysis seems to show that in the fauna of exotic block No. 2 there is an assemblage of species indicating nearly equal affinities with the zones of *Trachyceras Aonoides* and of *Tropites subbullatus*. It cannot therefore be correlated with either of them directly or exclusively, but must be considered as a homotaxial equivalent of *both* substages.

III — FOSSILS FROM EXOTIC BLOCK NO. 5 (MALLA KIOGARH E.G.)

Four exotic blocks of very small size were discovered by A. v. Krafft in the igneous rocks and black shales of the upper Flysch near Malla Kiogarh encamping ground and marked on the map accompanying A. v. Krafft's memoir (Mem. Geol. Surv. of India, Vol. XXXII, Pt. 3) as E. B. 4, 5, 6, 7. Blocks 6 and 7 yielded some fossils of liassic age. In the concretionary limestones of block 4 no fossils were found. Block 5 is described by A. v. Krafft as a massive, much altered red limestone. Among the small number of fossils collected, one Ammonite has been noticed, strongly resembling *Sageceras* (l. c. p. 162). A. v. Krafft consequently considered this block to be of middle or upper triassic age.

The number of fossils suitable for a determination is exceedingly small. The following two species are represented among my scanty materials :—

CARNITES sp. ind. Pl. XVI, fig. 1.

The fragments, which have been compared to *Sageceras* by A. v. Krafft, belong to a chambered whorl with a narrow and deeply excavated siphonal furrow, which is bordered by sharp, marginal keels. In its external characters this fragment agrees equally well with representatives of the genera *Sageceras* v. Moja. and *Carnites* v. Moja., but the character of its sutural line does not admit of any doubt that we have to deal with a species belonging to the latter genus.

Of the sutural line two lobes and saddles only are accessible to examination. The saddles have reached a stage of development transitional between the brachyphyllic and dolichophyllic stages. The outer saddle corresponds with the adventitious saddle, the inner one with the principal lateral saddle in *Carnites floridus* Wulf. The resemblance of the sutures to those of *Carnites floridus* is very striking, but a specific identification of my fragment is, nevertheless, impossible, on account of its too incomplete state of preservation.

PROARCESTES sp. ind. ex aff. AUSSEANO v. Hauer. Pl. XVI, fig. 2.

A large specimen of *Arcestes* is lying before me, showing a diameter of nearly 100 mm. Notwithstanding its remarkable dimensions it is entirely chambered. The breadth of the whorls surpasses the height considerably. The well-rounded external part passes gradually into the similarly rounded lateral parts. The umbilicus is comparatively broad and surrounded by a high and steep wall. The umbilical margin is obtusely rounded.

In the circumference of the last volution three varices are faintly developed which are directed radially, and cross the external part without being turned forward. This character distinguishes our species from the group of *Arcestes intuslabiati*, which it recalls otherwise by the shape of its umbilicus. The varices are flat and low and disposed at regular distances.

Dimensions.

Diameter of the shell	. . . .	97 mm.
„ „ „ umbilicus	. . . .	12·5 „
Height of the { above the umbilical suture	. . . . .	48 „
last volution { „ „ preceding whorl	. . . . .	31 „
Thickness of the last volution	. . . . . .	60 „

Sutures.—The external part of my specimen having been injured by weathering, it is not possible to examine the details of the siphonal prominence, the most characteristic element in the sutures of *Arcestes.* Otherwise the sutures do not differ from those in the group of *Arcestes bicarinati.* The principal lateral saddle stands on the convexity, by which the external part merges into the sides. There are altogether five saddles outside the umbilical margin.

Remarks.—The determination of chambered nuclei of *Arcestidæ* is, as a rule, uncertain, the chief features of distinction having been made on differences of nuclei and body-chambers by E. v. Mojsisovics. Although the present specimen shows in its general characters a great resemblance to *Proarcestes Aussecanus* v. Hauer (Cephalopoden von Aussee, Haidinger's Naturwiss. Abhandl. I, 1877, p. 268, Taf. VIII, figs. 6-8), it can be included in the group of *Arcestes bicarinati* with some reserve only.

CONCLUSIONS.

The carnic type of the fauna of exotic block No. 5 is obvious from the few remains quoted above, in spite of the great deficiency of the materials. Both *Carnites* and the group of *Arcestes bicarinati* (*Proarcestes*) are restricted to the carnic stage in the Mediterranean region, of which they are characteristic.

IV.—FOSSILS FROM EXOTIC BLOCKS NOS. 16 AND 17 (KIOGARH HIGH PLATEAU).

South of the Kiogarh high plateau two exotic blocks containing fossils of lower liassic age have been discovered by A. v. Krafft and marked as Nos. 16 and 17 on the map accompanying his memoir. In one place only the limestone was found in *situ.* A. v. Krafft (l. c. p. 106) describes the rock as bedded, concretionary, chiefly of red colour, but with a few grey layers, thicker than the red beds. "It is impossible to say how many liassic blocks were originally present, as they have all been more or less decomposed into large patches of débris. We can distinguish two main occurrences, one (E. B. 17) situated near E. B. 18, a large number of permo-carboniferous crags, and the other (E. B. 16) somewhat higher up near the crest of a ridge running from south to north."

The fossils, which were collected by A. v. Krafft in the accumulation of débris at both localities, have not been kept separate, the labels attached to the slabs of rock being marked "B. B. 16 and 17." I have consequently treated the fauna of the two blocks as a single one. Although not rich in well-preserved specimens, it is very interesting on account of its affinity with European faunæ of lower liassic age.

DIBRANCHIATA.

ATRACTITES sp. ind. Pl. XIV, fig. 1.

In two Himálayan materials a species of *Atractites* is represented by two fragments of phragmacones. They are casts without any trace of the test. The transverse section is circular. Angle of emergency very small. Distance of septa considerable, equal to three-quarters of the diameter of the anterior septum.

The scarcity of my materials excludes any attempt at a specific determination. Phragmacones similar to the present ones have been described from the Margaritatus-beds (middle Lias) of the Schafberg by Geyer (Die mittelliasische Cephalopodenfauna des Hinter Schafberges, Abhandl. K. K. Geol. Reichsanst. XV, Bd. p. 65, Taf. IX. fig. 3). In the lower lias this group of *Atractites* is represented by *A. liasicus* Guembel, and by an unnamed species of Valence (Bukowina), which has been mentioned by Uhlig (Über eine unterliasische Fauna aus der Bukowina. Abhandl. des deutsch. naturwiss. med. Ver. Lotos, Prag, 1900, Bd. II, p. 31). *Atractites italicus* Mich. (=*orthoceropsis* Savi et Menegh.) is distinguished from our species by its elliptical cross-section.

AMMONOIDEA.

Fam.: *PHYLLOCERATIDÆ*, v. Zittel.

Gen.: PHYLLOCERAS, Suess.

PHYLLOCERAS MONTGOMERYI nov. sp. Pl. XIII, fig. 1; XI, figs. 3, 4.

This beautiful species is represented in A. v. Krafft's collection by a large, fairly well-preserved cast, consisting entirely of air-chambers, and by several examples of smaller size. In its shape and sutures it shows a very great resemblance to *Phylloceras persanense* Herb., from which it differs only by some characters of minor importance.

The shell consists of very stout, rather rapidly increasing whorls, which overlap one another to more than one half of their height, and leave a comparatively wide umbilicus open. The cross-section is rectangular with rounded-off margins. The broad siphonal area is flatly arched. The greatest transverse diameter is situated in the middle of the height. The lateral parts are almost flat and pass into the vertical umbilical wall by an obtusely rounded edge. The surface of the cast has been slightly injured by weathering. It cannot be decided therefore whether or not a delicate ornamentation was present. But the absence of constrictions has been ascertained indubitably.

That this species is a Himálayan representative of the European group of *Phylloceras persanense*, is obvious from a comparison of the illustrations given by Uhlig and Fucini. Among the genus *Phylloceras* this group forms a very characteristic section, distinguished by its stout shape and rectangular cross-section and by the remarkable development of the principal lateral lobe among its sutural elements. Among the species from the lower lias belonging to this group there is unfortunately only a single one, *Phylloceras persanense* Herbich (Szeklerland, Mitteil. aus dem Jahrb. der Kgl. Ungarischen Geologischen Anstalt, V, p. 111, Taf. XX, E. fig. 3; XX, F. fig. 1), of which satisfactory information is given by the memoirs of Uhlig (Ueber eine unterliassische Fauna aus der Bukowina, Abhandl. Deutsch. naturwiss. med. Ver. Lotos, Prag. 1900, Bd. II, p. 15, Taf. I, fig. 1) and Fucini (Cefalopodi liasici del Monte di Cetona I, Palæontographia Ital. VII. 1901, p. 22, Tav. IV, figs. 1, 2). To this species our Himálayan type is certainly very nearly allied, although it is specifically different. Both species agree in the shape of the cross-section, which is almost identical, but *Phylloceras Montgomeryi* has more rapidly increasing whorls, which overlap one another less strongly, a wider umbilicus, and no constrictions. The absence of constrictions distinguishes our species likewise from *Ph. Calais* Meneghini (Fossiles du Medolo, Paléont. Lombarde 4e sér. p. 24, Tav. III, figs. 1, 2), *Ph. dubium* Fucini (l. c. Paléont. Ital. VII, p. 27, Tav. V, figs. 5, 6) and *Ph. microgonium* Gemmellaro (Sui fossili degii strati à *Ter. Aspasia* etc., p. 10, Tav. I, figs. 4-6). In the relative size of the umbilicus our Himálayan form agrees better with the latter species than with *Ph. persanense*.

European species of this group, which are conspicuous by the absence of paulostomatic constrictions, are *Phylloceras leptophyllum* v. Hauer, *Ph. subcylindricum* Neumayr, and perhaps *Ph. Hebertinum* Reynès.[*]

Phylloceras leptophyllum v. Hauer is only known to us by the unsatisfactory description and illustration given by Herbich (l. c. p. 112, Taf. XX, II. fig. 1). It differs from *Ph. persanense* by its whorls increasing more rapidly. In this character it approaches our Himálayan specimens very closely. No front view of P. v. Hauer's type having been figured by Herbich, a closer comparison is, unfortunately, not possible. Provided a complete agreement in their external characters should be proved, no specific identity of *P. leptophyllum* and *P. Montgomeryi* could be established, regarding the difference in the arrangement of their sutures.

Phylloceras subcylindricum Neumayr (Zur Kenntnis der Fauna des untersten Lias in den Nordalpen, Abhandl. K. K. Geol. Reichsanst. VII, p. 22, Taf. I, fig. 15) is distinguished by higher, more strongly compressed and more slowly increasing volutions but agrees with *Ph. Montgomeryi* in the width of the umbilicus.

---

[*] *Phylloceras elatinum* Gemmellaro (Sui fossili degli strati à *Terebratula Aspasia*, etc., Palermo, 1887, p 9, Tav. I, fig. 7, 11, figs. 13-15) cannot be counted among these species, as might be suggested from a cursory examination of Gemmellaro's illustration on 11, 1, fig. 7. The smaller example is provided with four or five constrictions, as is obvious from Gemmellaro's description. Pompeckj has discovered a species either identical with or very closely allied to *Ph. elatinum* in the liassic strata of Kosak tash, Asia Minor. The species shows deep constrictions with their directions turned very strongly forward (Palæontologische und stratigraphische Notizen aus Anatolien, Zeitsch. Deutsch. Geol. Ges. 49, Bd. 1897, p. 735, Taf. XXIX, figs. 6-9).

Phylloceras Hebertinum Reynès (Essai de géol. et paléont. Aveyronnaises, p. 94, Pl. II, fig. 3) is a dwarf species, with its lateral parts more strongly arched than in *Ph. ptersonense*. In general its cross-section seems to be less distinctly rectangular, especially in the types from the Medolo, which have been illustrated by Meneghini (Fossiles du Medolo, l. c. p. 30, Pl. III, fig. 6), which is provided with whorls of nearly elliptical outlines. The species from the Kemik tash (Asia Minor), which has been referred to *Ph. Hebertinum* by Pompeckj (Zeitschr. Deutsch. Geol. Ges. XLIX, p. 730, Taf. XXIX, fig. 10), shows a closer affinity to *Ph. persanense* than the types from the Medolo. The greatest transverse diameter is situated in the upper portion of the height, the lateral parts and siphonal area are flattened less distinctly and the umbilicus is narrower than in *Ph. Montgomeryi*

Dimensions.

Diameter of the shell	.	.	.	.	.	.	.	.	42 mm.
„ „ „ umbilicus		.	.	.	.	.	.	.	11 „
Height of the { above the umbilical suture	.			.	.	.	.	42.5 „
last volution { „ „ preceding whorl	.			.	.	.	cca 26 „
Thickness of the last volution	.	.	.	.	.	.	.	26 „

Sutures.—The sutural line is not entirely known to me. As far as accessible to examination it agrees pretty well with the sutures of *Phylloceras persanense*. The less rich ramification of the branches and the more massive shape of the saddles may be partly due to a stronger weathering of the casts in my Himálayan specimens. In the chief characters of the sutural line, namely, in the high position of the siphonal lobe and in the width of the lateral lobes, especially of the principal lateral lobe, there is a complete agreement between the two species.

The principal lateral lobe is nearly twice as deep as the siphonal lobe and is tripartite at its base. The following lobes diminish gradually in depth. The siphonal lobe is narrow, bifid and divided by a median prominence with entire borders. The siphonal saddle is diphyllic. Of its two lateral branches the inner one projects strongly beyond the inner terminal leaf. The lateral saddles are also diphyllic. The principal lateral and siphonal saddles are of nearly equal height.

The sutural line can only be traced as far as the outer wall of the first auxiliary saddle, but from its position we are allowed to suggest that the number of auxiliary elements must have been comparatively small, smaller probably than in the types of *Ph. persanense* illustrated by Uhlig and Fucini.

The sutural line of *Phylloceras leptophyllum* v. Hauer, although imperfectly known to us by Herbich's illustration, differs certainly from the sutures of *Ph. persanense* and *P. Montgomeryi* by the shape of the siphonal saddle. Whether or not Fucini is right in uniting *P. leptophyllum* with *P. convexum* de Stefani (Lias inferiore ad Arieti dell' Appennino settentrionale, Atti Soc. Toscana di scienze nat. in Pisa, Memorie, VIII, 1887, p. 40, Taf. 1, fig. 14; II, fig. 16), cannot be decided until larger materials of both species are available for examination.

Remarks.—The propriety of uniting the group of *Phylloceras perannense* with the subgenus *Geyeroceras* Hyatt, which has been proposed for the accommodation of *Ph. cylindricum* Sow. and its allies, is questionable. The two groups agree in the rectangular shape of their transverse sections and in the generality of their external characters, but the sutural line of *Ph. cylindricum* is distinguished from the sutures of *Ph. perannense* and its allies by the triphyllic arrangement of the siphonal saddle and by the deep position of the siphonal lobe. To me both characters appear to be of sufficient importance for separating the present species from Hyatt's subgenus *Geyeroceras*.

PHYLLOCERAS SCLATERI nov. sp.   Pl. XII, fig. 2; XIII, fig. 3.

This species, which is represented by two nearly complete and well-preserved casts and by some fragmentary examples, is closely allied to *Phylloceras Lipoldi* v. Hauer, from the lower lias of the Mediterranean province. The Mediterranean species, which is well known to us from the memoirs of F. v. Hauer (Beiträge zur Kenntnis der Heterophyllen in den oesterr. Alpen, Sitzgsber. Kais. Akad. d. Wiss. XII, 1854, p. 884, Taf. III, figs. 8-10), Reynès (Monographie des Ammonites du Lias inférieur, Atlas Pl. XLIV, figs. 27-31), Geyer (Ueber die liassischen Cephalopoden des Hierlatz bei Hallstatt, Abhandl. K. K. Geol. Reichsanst. XII, p. 220, Taf. I, figs. 13, 14) and Fucini (Cefalopodi liasici del Monte di Cetona, Paleont. Ital. VII, 1901, p. 24, Tav. IV, fig. 9) is distinguished by its comparatively wide umbilicus and ovoid cross-section. In both characters my Himálayan species agrees with the European type, from which it differs, however, by its higher, more strongly compressed whorls, which increase more slowly. But in general the two species approach each other so closely in their external features, that with transitional shapes at hand, the Himálayan form might be termed a large and compressed variety of *Ph. Lipoldi*.

In my larger type-specimen exactly one-half of the last volution belongs to the body-chamber. The umbilical wall is very steep and separated from the flanks by an obtusely rounded edge.

Fucini considers *Phylloceras Hebertianum* Reynès to be the nearest ally to *Ph. Lipoldi*. This near affinity is, however, restricted to the type from Medolo, as illustrated by Meneghini (Fossiles du Medolo, Paléontologie Lomb. IV, Appendice, Pl. III, fig. 6) and not extended to the types from Aveyron as described by Reynès, or from Asia Minor, which have been discovered by Pompeckj. From the present species Meneghini's examples of *Ph. Hebertianum* differ remarkably by their whorls increasing more rapidly, by the absence of a distinctly defined umbilical margin, by their inflated and regularly elliptical—not ovoid—cross-sections, and by their smaller umbilicus.

In *Phylloceras Wachneri* Gemmellaro (Sui fossili degli strati à *Ter. Aspasia*, p. 11, Tav. I, figs. 1-3) the whorls increase still more rapidly, the umbilicus is considerably smaller, and the greatest transverse diameter is situated above the middle of the height.

Phylloceras ancylonotos de Stefani (Lias inferiore ad Arieti del Appennino settentrionale, Atti Soc. Toscana scienze nat. Pisa, Mem., Vol. VIII, 1887, p. 50, Tav. II, fig. 15), which is considered as identical with *Ph. Lipoldi* by Fucini, has been based on a fragment too incomplete to warrant a certain identification. It shows no closer affinity with our Himálayan species, having more elliptical outlines and a narrow umbilicus. What is seen of the sutural line of *Ph. ancylonotos* in de Stefani's illustration does not agree with the sutures of *Ph. Lipoldi*, the siphonal saddle appearing to be considerably larger than the principal lateral one.

Dimensions.

Diameter of the shell	.	.	.	.	.	.	.	.	.	74 mm.
„ „ „ umbilicus	.	.	.	.	.	.	.	.	.	13 „
Height of the last volution { above the umbilical suture	.	.	.	.	.	.	.	30·5 „
	preceding whorl	.	.	.	.	.	.	20 „
Thickness of the last volution	.	.	.	.	.	.	.	.	26·6 „

Sutures.—The illustrations of the sutural lines of *Phyll. Lipoldi* in the memoirs of F. v. Hauer, Geyer, and Fucini do not agree in a satisfactory manner. The differences are too remarkable to be explained by the different state of preservation of the examples examined. The sutures illustrated by F. v. Hauer and Geyer show massive saddles with broad stems, whereas the lobe line figured by Fucini is conspicuous by very slender saddles with elongated branches and small terminal phylls. I have had the opportunity of examining Geyer's type-specimen from the Hierlatz and of convincing myself of the absolute correctness of his drawings.

The sutures of *Phyll. Sclateri* take an intermediate position between those in Geyer's type from the Hierlatz and in Fucini's examples from the lower lias of Monte di Cetona, exhibiting saddles which are deeply incised, but provided with large terminal leaves. All saddles are diphyllic, as in *Ph. Lipoldi*. The siphonal lobe is as deep as the second lateral one, not remarkably inferior in length to the principal lateral lobe, and very narrow. All the lateral lobes are tripartite at their base, the median point being the longest.

There are four auxiliary lobes and three corresponding saddles developed in the last septum preceding the body-chamber.

The siphonal saddle has four branches, including the two terminal phylls. It is considerably shorter than the principal lateral saddle.

PHYLLOCERAS sp. ind. aff. SCLATERI, Dien.   Pl. XIII, fig. 2.

This species, which is represented by a single, imperfectly preserved cast, recalls still more strongly *Phylloceras Lipoldi* than the typical *Ph. Sclateri*. It has the lateral parts not flattened, but regularly although moderately arched, and a comparatively high and steep umbilical wall, which is separated from the flanks by an obtusely rounded edge. The umbilicus is wider than in *Phylloceras Lipoldi* and in *Ph. Sclateri*. The whorls envelop one another for two-third parts of their entire height.

Dimensions.

Diameter of the shell	. . . .	56 mm.
„ „ „ umbilicus	. . . .	13 „
Height of the (above the umbilical suture	. . . .	34 „
last volution („ „ preceding whorl	. . . .	30 „
Thickness of the last volution	. . . . .	21 „

Sutures.—The sutures are distinguished from those of *Phylloceras Sclateri* by a smaller number of auxiliary lobes, corresponding to the larger diameter of the umbilicus. There are only three auxiliary lobes and two saddles developed. The umbilical margin divides the second auxiliary saddle, whereas in *Ph. Sclateri* three auxiliary saddles are exposed within the distance between the second lateral saddle and the umbilical edge.

PHYLLOCERAS HORSEFIELDII nov. sp. Pl. XII, fig. 3.

The specimen illustrated, a cast consisting of air-chambers only, agrees in its external features with *Phylloceras oenotrium* Fucini (Cefalopodi liasici del Monte di Cetona, Palæont. Ital. VII, 1901, p. 34, Tav. V, figs. 8, 9 ; VI, fig. 1) in such a remarkable way, that I should not have hesitated to identify it with this characteristic species from the lower lias of Italy but for the fact that the less complicated structure of its sutural line required a specific separation of the Himálayan form.

It is provided with rapidly increasing, very high and strongly compressed whorls, which include a comparatively wide umbilicus. The lateral parts are very flatly curved, reaching their greatest transverse diameter below the middle of their height. The umbilical margin is narrowly rounded.

From the typical shape of *Phylloceras Zetes* d' Orb. (= *Ammonites heterophyllus awaitkei* Quenstedt, Cephalopoden, p. 100, Taf. VI, fig. 1) our specimen differs chiefly by its wider umbilicus. It is distinguished both from the species from Enzesfeld, which has been united with *Ph. Zetes* by F. v. Hauer (Cephalopoden aus dem Lias der nordöstlichen Alpen, Denkschr. Kais. Akad. d. Wissensch. XI, 1855, p. 56, Taf. XVIII) and from *Ph. psilomorphum* Neumayr (Zur Kenntnis der Fauna des untersten Lias in den Nordalpen, Abhandl. K. K. Geol. Reichsanst. VII, p. 21, Taf. II, fig. 4) by a more regularly oval shape of its cross-section. In those two species the lateral parts converge as flat planes from the place corresponding to the greatest transverse diameter, towards the rounded external part, whereas they are curved very distinctly in the direction of the umbilicus. This imparts to their cross-sections a sagittal or subsagittal shape, whereas in my Himálayan species the lateral parts are distinctly and regularly, although discretely, arched.

Phylloceras globerrimum Neumayr (l. c. p. 20, Taf. II, figs. 2, 3) is too imperfectly known to permit of a closer comparison with the present species.

Dimensions.

Diameter of the shell	.	.	.	.	.	.	.	.	. 67	mm.
„ „ „ umbilicus	.	.	.	.	.	.	.	. 0·1	„
Height of the { above the umbilical suture	.	.	.	.	. 39	„
last volution { „ „ preceding whorl	.	.	.	.	. 29	„
Thickness of the last volution	.	.	.	.	.	.	. 20	„

Sutures.—The sutures of this species are very much like those in *Phylloceras dubium* Fucini (l. c. p. 27, Tav. V, figs. 5, 6) and in the Italian types of *Phylloceras Lipoldi* described by that author.

The siphonal lobe is shorter than in *Ph. dubium*, as it does not reach the length of the second lateral lobe. The principal lateral lobe is very broad, nearly as broad as in the group of *Ph. persuanense* Herb. All the lobes as far as known are tripartite at their base, but with the internal branches arranged asymmetrically to the median indentation. The saddles are diphyllic, with large terminal leaves. The lateral branches are especially well developed in the siphonal saddle, which is provided with a narrower stem than the rest of the saddles.

There are two auxiliary saddles present within the distance from the second lateral saddle to the umbilical margin.

From the sutures of *Phylloceras oenotrium* the sutural line of this species differs so remarkably by its more simple structure, that a close affinity of the two forms is rather doubtful, notwithstanding their great external similarity. For its nearest relationship we will perhaps have to look among the group of *Ph. Lipoldi*. From *Ph. Sclateri* the present species is distinguished not only by its external features, but also by considerable differences in its sutural line, especially by the large size of its siphonal saddle, which is ramified more richly.

PHYLLOCERAS CALDWELLII nov. sp.   Pl. XIII, fig. 4.

This species, which is represented by a single, almost entirely chambered cast in A. v. Krafft's collection, belongs to the relationship of *Phylloceras Horsefieldii*, as is obvious from the similar structure of its sutural line.

In its external shape and involution it is distinguished from *Ph. Horsefieldii* by its whorls being more strongly convex and arched less regularly. The greatest transverse diameter is situated in the umbilical region. In the posterior half of the last volution it coincides with the umbilical margin, which is rounded off sharply. Near the aperture it is shifted somewhat towards the lower part of the flanks, which pass into the high and steep umbilical wall in a more regularly rounded curve. Thus the transverse section is of a decidedly subsagittal shape, much more so than in *Phylloceras oenotrium* Fucini.

This character of the cross-section distinguishes our species from the group of *Ph. frondosum* Reynès (Essai de géol. et paléontol. Aveyronnaises, Paris, 1868, p. 08, Pl. V, fig. 1). In the typical shapes of *Ph. frondosum* the cross-section is, according to Pompeckj (Zeitschr. Deutsch. Geol. Ges. 49, Bd. 1897, p. 729) of regularly elliptical outlines, whereas it is slightly oval in some forms described by Pucini (Cefalopodi liassici del Monte di Cetona Pte. 1, Pal. Ital. VII, 1901, p. 43) as transitional shapes connecting the typical *Ph. frondosum* and *Ph. Wæhneri* Gemm. But in some of them the greatest transverse diameter is situated as close to the umbilical region as in our Himalayan species.

The siphonal part is narrowly rounded at the beginning of the last volution, but becomes curved more flatly in the vicinity of the aperture.

The umbilicus is as wide and deep as in specimens of *Ph. ocnotrinm* of equal size.

Dimensions.

Diameter of the shell	.	.	.	.	.	.	.	.	.	.	. 92 m.m.
„ „ „ umbilicus	.	.	.	.	.	.	.	.	.	.	92 „
Height of the } above the umbilical suture	.	.	.	.	.	.	.	.	48 „
last volution } „ „ preceding whorl	.	.	.	.	.	.	.	.	35 „
Thickness of the last volution	.	.	.	.	.	.	.	.	.	. 32 „

Sutures.—Agreeing in general with those of *Phylloceras Horsefieldti.* All main saddles diphyllic. Three auxiliary lobes and saddles outside the umbilical suture. The second auxiliary lobe coincides with the umbilical margin. The two inner auxiliary saddles are monophyllic.

The most characteristic feature in the sutural line is the shape of the principal lateral lobe, which is very large at its base and very narrow in its upper portion, where the branches of the bordering saddles approach one another considerably. A strong divergence of the basal branches of this lobe has also been noticed in *Phylloceras Horsefieldti* and in the group of *Ph. frondosum*, but in none of the species belonging to the latter group is it developed as strongly as in *Ph. Caldwellii*, in which the distance of the extreme basal points of this lobe is equal to two-fifths of the entire radius.

This remarkable enlargement of the basal region is restricted to the principal lateral lobe. The second lateral lobe is of normal shape. Its median terminal point reaches considerably deeper than the two lateral points, whereas in the principal lateral lobe the difference in depth between the three basal points is almost insignificant.

The terminal branches of the two lateral lobes approaching one another very closely, the stem of the principal lateral saddle is laced at its base more strongly than in *Phylloceras Horsefieldti.*

Remarks.—Among the undescribed species of *Phylloceras* from the lower Lias of Adneth there is one, which, from its external similarity, might be supposed to be nearly allied to the present form.

PHYLLOCERAS sp. ind. ex aff. DIENERI Rosenbg. Pl. XII, fig. 5.

This is a very remarkable species of *Phylloceras*, which deserves mentioning notwithstanding the very incomplete and unsatisfactory state of preservation of the only specimen available for description.

It is a widely umbilicated *Phylloceras* with flatly arched lateral parts and with a nearly elliptical cross-section. The greatest transverse diameter corresponds to the middle of the height. The siphonal part has not been preserved nor am I able to give any exact measurements of the cross-section.

The most remarkable feature of this species is the development of distinct narrow paulostomatic folds, which are directed radially and show an inverse imbrication. The two folds, which are clearly exposed in the figured fragment, are disposed at right angles. Traces of a third radial fold have been noticed exactly opposite the anterior one. The folds were probably connected with faintly marked paulostomatic constrictions, but the poor state of preservation of my only type-specimen is not sufficient for a positive conclusion.

In the lower lias of the Kratzalpe (Salzburg) there is an undescribed species, for which the name *Phylloceras Dieneri* will be proposed by Rosenberg, who is studying the fauna of this locality, which recalls the present cast in its external features. It is also provided with radial folds, which are disposed at right angles, each quadrant of the last volution being separated from the neighbouring one by a radial fold, which is accompanied by a low constriction.

Dimensions. —Not measurable.

Sutures. —Not known.

Subgen.: SCHISTOPHYLLOCERAS Hyatt (Group of *Phylloceras Cermœsense* Herb.)

PHYLLOCERAS (SCHISTOPHYLLOCERAS) MONGOLICUM nov. sp.
Pl. XI, fig. 2; XII, fig. 1.

This species is a representative of a very remarkable group of *Phylloceratidæ* which is distinguished by wide umbilici and by a triphyllic termination of the principal lateral saddle. Two Alpine species of this group have been studied most carefully, namely, *Phylloceras Cermœsense* Herb, and *Ph. planispira* Reynès. To both of them our Himálayan form is very closely allied, without, however, being actually identical with either of them.

The new name *Phylloceras mongolicum* is proposed for two specimens. One of them is well preserved and nearly complete, one half of its last volution belonging to the body-chamber. The second one is a fragment of the last volution, comprising the last air-chambers and a portion of the body-chamber. In this specimen the sutural line has been excellently preserved.

The nearly complete specimen illustrated on Pl. XI, fig. 2, agrees in its shape and dimensions very closely with the type-specimen of *Phylloceras Cermœsense*, Herbich (Szeklerland, Mitteilungen aus dem Jahrb. d. ungar. Geol. Anst. V, p. 113, Taf. XX, K. fig 1). The slowly increasing whorls leave a wide umbilicus open. The shape of the shell is disciform, with strongly compressed volutions.

The transverse section is irregularly ovoid, with the largest transverse diameter situated in the lower part of the height.

In their involution the two specimens do not agree exactly. In my Himálayan type the whorls overlap one another to more than one half their height. In Herbich's type-specimen the rate of involution is less considerable, but among the Alpine representatives of *Phylloceras Uermœsense*, illustrated and described by Waehner (Beitraege zur Kenntnis der tieferen Zonen des Lias in den nordœstlichen Alpen, Beitraege zur Palæont. und Geol. Œsterr.-Ungarns, etc., Bd. XI, p. 173, Taf. XXIII, figs. 3-5, Taf. XXIV, figs. 1-8), there are some examples in which the overlap of the two last volutions is scarcely inferior to that in *Ph. mongolicum*.

The siphonal part is sharply rounded, more sharply even than in any of the Alpine examples of *Phylloceras Uermœsense* illustrated by Waehner, but never acute. The lateral parts are marked off from a steeply inclined umbilical wall by an umbilical edge, which is obtusely rounded. But the umbilical wall is neither as steep nor separated from the lateral parts as sharply as in the majority of the Alpine types of *Ph. Uermœsense*. It is especially in the inner volutions that the low umbilical wall passes into the lateral parts more gradually. In this character my specimen might be compared with the European type illustrated by Waehner on Pl. XXIII, fig. 3.

The peristome has not been preserved, but a deep contraction preceding the aperture of my type-specimen might perhaps indicate the vicinity of the actual peristome, provided it were not accidental.

As has been demonstrated by Waehner, in *Phylloceras Uermœsense* an internal shelly ridge is occasionally developed along the median line of the siphonal part. In casts the presence of this internal ridge is marked by a deep, angular depression or furrow. Specimens showing this external furrow have been described as *Ph. autonotum* by Herbich (Szeklerland, l. c., p. 115, Taf. XX, G. fig. 2). It is worth mentioning that among the materials collected by A. v. Krafft there is also a fragment with the trace of a siphonal furrow, recalling somewhat casts of *Ph. autonotum*. It is too fragmentary to permit of a specific identification.

Waehner's amalgamation of *Phylloceras Uermœsense* and *Ph. autonotum* has been doubted by Prinz (Centralblatt f. Mineral. etc., 1906, p. 238), who considers the latter species as the prototype of a new subgenus *Kochites* (= *Schistophylloceras* Hyatt), but I am not inclined to follow his view, agreeing entirely with Uhlig (Centralblatt, l. c., 1906, p. 421), that no valid arguments have as yet been raised against the correctness of the results of Waehner's careful studies.

It is not impossible that a broad and low keel occurs in the vicinity of the aperture of my Himálayan type-specimen, as has been described in large examples of *Phylloceras Uermœsense* by Waehner, but its presence cannot be ascertained, this region of the external part having suffered from weathering.

Paulostomatic constrictions or folds are entirely absent in all my specimens.

Fragments of the shelly substance, as far as preserved, are nearly smooth. Of

M

the radial stripes and striæ, which have been described and illustrated in *Phyllo-ceras Uermæsense* by Wachner, no traces have been discovered, but this fact may be partly due to the imperfect condition of the test.

Phylloceras Uermæsense is certainly more nearly allied to *Ph. mongolicum* than any congeneric species of this group. There exists also a close similarity with *Ph. planispira* Reynès. The chief character of distinction between the two species is the difference in size, *Ph. planispira* being a small form which never exceeds 50 mm. in diameter. As has been remarked by Uhlig (Ueber eine unter-liasische Fauna aus der Bukowina, Lotos, Prag., l. c., p. 17), the umbilical wall of *Phylloceras planispira* always slopes at flat angles, and has a broadly rounded margin. In these two features our species agrees more closely with *Ph. Uermæ-sense*.

Dimensions.

Diameter of the shell	67 mm.
„ „ „ umbilicus	30 „
Height of the { above the umbilical suture	35 „
last volution { „ „ preceding whorl	23 „
Thickness of the last volution	29 „

Sutures.—The sutural line shows a remarkable similarity with the sutures of *Phylloceras Uermæsense*, from which it differs, however, by the more robust shape of the saddles, the phylla being considerably smaller in comparison to the stems, and by the smaller number of auxiliary elements. In both characters it seems to agree, perhaps, somewhat more nearly with the sutures of *Ph. planispira*, although a closer comparison is rendered difficult by the incorrect illustration in Reynès' memoir and by the very small size of the sutural lines which have been illustrated by Uhlig and Geyer.

The siphonal lobe is very short, reaching less deeply than the external branch of the principal lateral lobe. The second lateral lobe is shorter than the principal one, but longer than the first auxiliary lobe—which should, perhaps, be designed more exactly as third lateral lobe. From the first auxiliary lobe the line con-necting the basal points of the following two lobes runs in a radial direction toward the umbilical suture. A third lobe is situated on the umbilical wall.

An equal number of auxiliary elements has been counted in small examples of *Phylloceras Uermæsense* by Wachner, whereas full-grown types reaching the dimensions of our specimen (Pl. XII, fig. 1) are provided with six auxiliary lobes.

The siphonal saddle is diphyllic, the inner terminal leaf being the higher one. The second inner leaf following underneath the terminal phyllum projects rather strongly, making the principal phylla of this saddle approach a tripartite arrangement. The principal lateral saddle is distinctly triphyllic, the middle leaf being the highest, the external phyllum standing a little deeper than the internal one, and at a nearly equal level with the inner terminal phyllum of the siphonal saddle. It is considerably larger than its neighbours. The two following saddles exhibit a diphyllic arrangement. The second auxiliary saddle has one single terminal leaf.

It is obvious from a comparison of the sutural lines in *Phylloceras Verusacsense* and *Ph. mongolicum* that there is an almost complete agreement in the arrangement of the phylla of the saddles, even in the minor details. Nevertheless there exists a decided difference in the general shape of the saddles, the lobes being comparatively broad, the stems massive, the phylla small and slender in my Himalayan species.

Should the differentiation of the present species not be considered justified by its external features of distinction—which in this as in so many other cases is only a matter of individual conception or personal judgment—the structure of its sutural line would, according to my opinion, require its separation from *Phylloceras Verusacsense*.

Remarks.—There is some difference of opinion among palæontologists regarding the systematic position of *Phylloceras Verusacsense*.

If all *Phylloceratidæ* with a wide umbilicus are included in the genus (or subgenus) *Rhacophyllites*, as interpreted originally by K. v. Zittel (Handbuch der Palæontologie, II, p. 430), *Ph. Verusacsense* must certainly be grouped with this genus. The interpretation of *Rhacophyllites* proposed by E. v. Mojsisovics in 1902 would not give us any clue for a decision, because, of the two characters of subgeneric importance, namely, presence of a short siphonal lobe and diphyllic or triphyllic arrangement of the main saddles but monophyllic auxiliary saddles, the first character is developed in *Ph. Verusacsense*, whereas the second is not.

If the subgeneric designation of *Rhacophyllites* is restricted to those forms of *Phylloceras* in which the body-chamber differs from the chambered parts of the shell in shape and sculpture, or in which the auxiliary series is united into a suspensive lobe, as has been proposed by E. v. Mojsisovics in 1882 and by Geyer in 1886, it is equally difficult to decide whether or not *Ph. Verusacsense* should be grouped with *Rhacophyllites*. Shape and sculpture of the body-chamber do not agree exactly with those of the chambered parts of the shell, because occasionally a shelly internal ridge and a low keel are developed in the body-chamber. The auxiliary elements are not united into a sloping suspensive lobe, but show a serial arrangement, decreasing in size quite regularly from the second lateral lobe to the umbilical suture.

Hyatt (Zittel's Text-book of Palæontology, English edition, Cephalopoda, p 568) has elevated *Phylloceras autonotum* Herb., which, according to Wæhner, is identical with *Ph. Verusacsense*, to the rank of a proper subgenus, *Schistophylloceras*. To the same group of *Phyllocerota* Prinz has applied the subgeneric designation of *Kochites*, which, regarding the law of priority in palæontological nomenclature, cannot be accepted.

The subgeneric value of a character, which is rather faintly marked and not even developed in all specimens of *Ph. Verusacsense*, might be questioned. It might also be taken into consideration, as it has been remarked by Uhlig, that the development of a keel in *Phylloceras* is a very subordinate feature in the history of its evolution, the carinate forms disappearing without having given rise to any progeny. If a subgeneric rank should, notwithstanding these objections, be attributed to the group of *Phylloceras Uermannense-anlonotum*, the present species from the exotic blocks of Malla Johar ought, probably, to be grouped with *Schistophylloceras*, although the presence of a keel has not been ascertained.

Subgen.: *Rhacophyllites* v. Zittel.

Rhacophyllites cf. gigas Fucini. Pl. XI, fig. 1.

1901. *Rhacophyllites gigas* Fucini, Cefalopodi liassici del Monte di Cetona, Pte. 1, Palaeont. Ital. VII, p. 56, Tav. IX, figs. 2-5.

This species is represented in A. v. Krafft's collection by a well-preserved cast without any trace of its shelly substance. It consists both of air-chambers and the body-chamber, to which exactly one half of the last volution belongs. It is an Indian representative of a group of *Rhacophyllites* which is widely distributed in the lower and middle lias of Europe, and is very nearly allied to *Rhacophyllites gigas* Fucini, *Rh. transsylvanicus* Hauer and *Rh. diopsis* Gemm. I have referred it to the first of these three species as *cf.*, although it is perhaps not exactly identical with it.

With the type of *Rh. gigas* illustrated by Fucini on Pl. IX, fig. 4, of his above-quoted memoir, my Himálayan specimen agrees in all its characters of specific importance. The whorls are strongly compressed and include a wide umbilicus. The lateral parts are very gently arched and separated from the high, steeply inclined umbilical wall by a distinct, obtuse edge, which becomes gradually rounded in the body-chamber only. The largest transverse diameter is perhaps situated a little higher than in *Rh. gigas*, but the difference can be insignificant only. The external part is regularly rounded, neither inflated nor truncated.

The surface of the chambered parts of the shell is smooth. Near the beginning of the body-chamber an indistinct sulcus or constriction has been noticed crossing the sculpture in front at oblique angles. The sculpture, which is restricted to the body-chamber, consists of strong and moderately sharp ribs, which are separated by broad intercostal valleys. Fifteen ribs are counted altogether within the circumference of the body-chamber. The ribs are slightly falciform and describe a crescent-shaped curve, with its convexity turned forward in crossing the siphonal area. This is the place where they reach their maximum strength, whereas they are obliterated gradually in the lower part of the flanks. None of them reaches the umbilical region.

From the Italian type of *Rhacophyllites gigas* our specimen differs in some subordinate details of its shape and sculpture. Its whorls increase more rapidly, as is obvious from a comparison of the dimensions in the two following examples :—

						Himálayan type-specimen.			Fucini's type-specimen (Pl. IX, fig. 4).
Diameter of the shell		.	.	.	.	99 mm.	.	.	72 mm.
„ „ „ umbilicus.		.	.	.		39 „	.	.	17·5 „
Height	} of the last volution		.	.	.	87·5 „	.	.	36 „
Thickness			.	.	.	29·5 „	.	.	17·5 „

The difference in the width of the umbilicus is larger than in the height of the last volution. Nor do the proportions of height and thickness in the transverse section agree exactly, our Himálayan form being comparatively thicker than the Italian type-specimen illustrated by Fucini on Pl. IX, fig. 4. But there are some other European examples of *Rh. gigas* which seem to agree better with our specimen in this respect. Fucini did not succeed in ascertaining the presence or absence of a constriction in the body-chamber of his specimens, although their absence in the chambered parts of the shell was made certain.

If Fucini's suggestion, that the ribs originate in the vicinity of the umbilical margin, could be proved to be correct, this character might be counted among the subordinate features of distinction, as in our specimen none of the lateral ribs comes near the umbilical margin.

Among Fucini's illustrations of *Rhacophyllites gigas* no view, unfortunately, has been given of the external part. Thus it is impossible to say whether or not there exists a complete agreement in the direction of ribs in that region between the Himálayan and Italian species.

A second species, to which the present one appears to be very closely allied, is the group of forms from the lower lias of Monte di Cetona, which have been united with *Rhacophyllites transsylvanicus* v. Hauer by Fucini (l. c. p. 52, Tav. VIII, figs. 1-7).

As has been demonstrated by Uhlig (Ueber eine unterliassische Fauna aus der Bukowina, Abhandl. d. deutsch. naturwiss. Med. Ver. Lotos, Prag, 1900, II, Bd. p. 90), the name *Rhacophyllites transsylvanicus* must be assigned to the specimen illustrated by Herbich on Pl. XXI of his memoir "Das Szeklerland" (Mitt. aus d. Jahrb. d. Ungar. Geol. Anst. V, 1878). From this type of the species our Himálayan specimen differs considerably by the smaller number of its ribs, which are curved more strongly along the siphonal area. But among Fucini's materials numerous forms have been assigned to the Hungarian species, which might be considered as shapes intermediate between *Rh. transsylvanicus* and *Rh. diopsis* Getum., some of them bearing a greater affinity to our species than the typical *Rh. transsylvanicus*. It is especially the var. *dorsocavata* Fucini (Pl. VIII, fig. 7) which approaches our specimen in the shape of its cross-section and in the direction of the lateral ribs crossing the siphonal area in crescent-shaped curves. But the number of ribs is considerably larger in all the Italian types illustrated by Fucini. To this distinctive feature the presence of faintly marked paulostomatic constrictions in the chambered parts of the Italian shells must be added, whereas such are certainly absent in my Himálayan specimen.

A third species which might put in a claim for closer comparison with the present one is *Rhacophyllites diopsis* Gemmellaro (Sugli fossili degli strati à *Terebratula Aspasia* della contrada Rocce rosse presso Galati, Giorn. di sci. nat. ed econ. Palermo, 1881, p. 6, Tav. II, figs. 6-8; VI, figs. 1, 2).

The specific independence of *Rhacophyllites diopsis* has been questioned by C. de Stefani, Greco, Fucini and Uhlig, who advocated its amalgamation with *Rh. Nardii* Menegh. But in his above-quoted memoir (p. 50) Fucini, disagreeing with his former view, insists again on a separation of the two species, restricting the name of *Rh. Nardii* to Meneghini's type-specimen from Campiglia (l. c. Pl. VII, fig. 1).

A comparison between Gemmellaro's type-specimen of *Rh. diopsis* (figs. 6, 7) and my Himálayan example is rendered difficult by the fact that the first is provided with the test, whereas the latter is a cast. The number and strength of ribs are somewhat larger in *Rh. diopsis*, and the ribs are turned forward more strongly. The entire absence of constrictions has been remarked expressly by Gemmellaro. From our species *Rh. diopsis* is, moreover, distinguished by the opening of the umbilicus near the aperture of its body-chamber whorl, where its umbilical suture leaves the normal spiral.

Two body-chamber fragments from the Hierlatz, which have been described as *Rhacophyllites cf. diopsis* by Geyer (Ueber die liasischen Cephalopoden des Hierlatz bei Hallstatt, Abhandl. K. K. Geol. Reichsanst. XII, p. 225, Taf. I, fig. 20), approach our species in the strength and direction of the ribs more closely than Gemmellaro's type, but are distinguished by the smaller height and greater width of their transverse sections.

From other congeneric species with ribbed body-chambers our specimen is easily distinguished. *Rhacophyllites Nardii* Meneghini —in the narrow circumscription proposed by Fucini (l. c. p. 48, Tav. VII, figs. 1-7)—and *Rh. libertus* Gemmellaro (l. c. p. 4, Tav. II, figs. 1-5) have wider umbilici and deep constrictions. *Rh. luneusis* de Stefani (Lias infer. ad Arieti, l. c. p. 57, Tav. III, figs. 1, 2) and *Rh. Quadrii* Menegh. are provided with a more delicate ornamentation and with paulostomatic constrictions affecting both the chambered parts of the shell and the body-chamber.

Sutures.—The sutural line of the present specimen differs from the sutures of *Rhacophyllites gigas* or *Rh. transsylvanicus* by some insignificant details only.

The siphonal lobe is bifid, very narrow and short, not reaching deeper than the middle of the length of the principal lateral lobe. The lateral lobes are tripartite. As in *Rh. gigas*, the arrangement of basal branches is different in the two lobes, the denticulations adjoining the stem of the principal lateral saddle being less deep than the opposite ones. From the first auxiliary lobe the sutural line descends towards the umbilicus, thus exhibiting the arrangement of sutures characteristic in typical representatives of *Rhacophyllites*. The number of auxiliary lobes could not be ascertained, but I do not think that more than three could be developed in the last septum preceding the body-chamber.

The siphonal saddle is lower than the principal lateral one. It is diphyllic. The inner terminal branch is higher and larger than the outer one. Fucini describes this saddle as triphyllic, counting the large lateral inner branch among the terminal leaves. With this view I am, however, obliged to disagree, the two real terminal phylla being distinctly laced at their base and thus separated from the next lower phylla, which must consequently be considered as lateral, not as terminal ones.

The lateral saddles are diphyllic. The second lateral saddle differs from the corresponding element in *Rh. gigas* by the larger size of the outer lateral branch. Another subordinate difference consists in the shape of the main saddles, which are laced at their bases more strongly in the present example.

Rhacophyllites schofariformis nov. sp.   Pl. XII, fig. 4.

This species of *Rhacophyllites* represents a very peculiar type, which is distinguished by the trumpet-shaped enlargement of its body-chamber in the apertural region.

My type-specimen is a somewhat fragmentary cast, but sufficiently well preserved for allowing a satisfactory reconstruction, which is enough to render conspicuous all its external features. In its involution it agrees almost equally well with some widely umbilicated species of *Phylloceras* (*Ph. Sclateri* Dien., *Ph. peregrinum* Herb.), as with some types of *Rhacophyllites stella* Sow., possessing comparatively narrow umbilici. It has been grouped with *Rhacophyllites* in this memoir on account of its abnormal body-chamber. The remarkable change in the cross-section of the last whorl is obvious from the following measurements :—

Diameter of the shell	. . . . .	. .	96 mm.
„ „ „ umbilicus	. . . . .	. .	15 „
Height } at the beginning of the last volution Thickness }	. .	. .	20 „ 11·5 „
Height } corresponding to a diameter of 65 mm. Thickness }	. .	. .	33 „ 21·5 „
Height } near the aperture Thickness }	. . . .	. ab	29 „ 47 „

The last volution begins with a high and strongly compressed cross-section of a subsagittal shape, the lateral parts converging from the umbilical region, which corresponds to the greatest transverse diameter, as very flatly arched planes towards the narrowly rounded siphonal part. The whorl increases rather rapidly as far as the middle of the last volution, changing its transverse section very slowly, which turns gradually from a subsagittal into a more regularly oval shape. In the vicinity of the aperture the width of the cross-section increases far more considerably than its height, and the greatest transverse diameter is shifted towards the middle of the volution. Thus a trumpet-shaped inflation of the

apertural region is produced, recalling a similar trumpet-shaped opening of
the aperture in some species of *Lytoceras* (Neumayr).

The actual peristome has not been preserved, but cannot have been situated
considerably in front of the aperture in my cast.

The surface is entirely smooth, without any traces of folds or paulostomation
constrictions.

Sutures.—Not known.

Fam.: *PLEURACANTHIDÆ* Hyatt.

The careful researches of Waehner have acquainted us with a very interesting stock of Ammonites, which seem to mark transitional stages connecting
Phylloceras, *Lytoceras* and *Psiloceras*, thus proving the common origin of all the
widely different families of liasic Ammonoidea, the roots of which must probably
be looked for in the triassic genus *Monophyllites* (*Mojsárites*).

To the groups of transitional forms, combining characters of *Phylloceras*,
Lytoceras and *Psiloceras*, a special systematic position should be attributed. By
uniting one of them (*Euphyllites* Waehner) with the *Phylloceratidæ*, a second
one (*Ectocentrites* Waehner) with the *Lytoceratidæ* and a third one, consisting
of two more genera (*Pleuracanthites* Can. and *Analytoceras* Hyatt) in the new
family of *Pleuracanthidæ* Hyatt has not taken sufficient care of their natural
connection. I should prefer to accept the new family of *Pleuracanthidæ*, which
has been proposed by Hyatt, as a descriptive term for all groups of forms by which
the gaps between *Phylloceras*, *Lytoceras* and *Psiloceras* are bridged over in the
lower liassic age.

It is very interesting to find representatives of this remarkable family in the
lower lias of the Himálayas. The presence of two genera at least, *Pleuracanthites*
and *Analytoceras*, has been ascertained. The presence of *Euphyllites* and
Ectocentrites is as yet doubtful, although very probable.

Gen.: ANALYTOCERAS Hyatt.

1900, *Analytoceras* Hyatt in K. v. Zittel, Text-book of Palæontology, Vol. I. Cephalopoda, p. 568

ANALYTOCERAS sp. ind. aff. ARTICULATO Sow. Pl. X, fig. 4.

Hyatt has been fully justified in elevating *Lytoceras articulatum* Sow. (A.
d'Orbigny, Paléont. franç. Terrains jurass. I, p. 312, Pl. 97. figs. 10-13) to the
rank of a proper genus. Although this species approaches a typical *Lytoceras* in
its external characters, its sutural line differs so remarkably by the tripartite
arrangement of its lateral lobes that it cannot be left in that genus.

In the Himálayan collection a single specimen has been found, which exhibits
so close a relationship to *Analytoceras articulatum* that in a better state of
preservation it might, perhaps, have led to an identification with the European
fossil. Only the last volution has been preserved, and even this has been seriously

injured by weathering. It has the external shape of a typical *Lytoceras*, with whorls increasing as rapidly as in the specimen illustrated by Waehner on Pl. VIII, fig. 2, of his monograph of lower liassic Ammonites of the eastern Alps (VII. Teil. Beiträuge zur Palæont. u. Geol. Oesterr. Ungarns, etc., Bd. IX, 1895). Although the surface of the whorl is badly corroded, traces of the original ornamentation are visible in the umbilical region of the apertural portion of the body-chamber. The sculpture consists of numerous, thin, radiating lines which are occasionally interrupted by stronger folds. These folds are smooth, as in *Analytoceras articulatum*, never fimbriate, as in the group of *Lytoceras fimbriatum*.

Dimensions.— Not measurable.

Sutures.—The sutural line agrees very closely with that in full-grown specimens of *Analytoceras articulatum*. The siphonal lobe is very short, bifid and divided by a low median prominence. The most characteristic element is the principal lateral lobe, which shows a tripartite arrangement of its branches, two lateral branches being disposed symmetrically to a median one. This arrangement of the lateral lobes differs widely from what is seen in a typical *Lytoceras*, where the lobes are divided into two branches by a median secondary saddle rising from the base.

In this character *Analytoceras* very closely approaches *Pleuracanthites* and *Ectocentrites*, and thus enters into the series of forms connecting the two families of *Lytoceratidæ* and *Psiloceratidæ*.

The siphonal and principal lateral saddles are diphyllic. Their details have been partly destroyed by weathering, but what remains is yet sufficient to show that they were arranged on the same plan as the corresponding sutural elements in *A. articulatum*.

I have not been able to trace the sutural line beyond the first auxiliary lobe, which is united with the second lateral saddle into a suspensive lobe.

One half of the last volution belongs to the body-chamber. The sutural line illustrated in the figure corresponds to the last septum.

Gen.: PLEURACANTHITES Canavari.

PLEURACANTHITES sp. ind. aff. BIFORMIS Sow. Pl. XV, fig. 3.

Lytoceras biforme Sow. has been elevated to the rank of the prototype of a proper genus by Canavari (Atti. Soc. Toscana sci. nat. Proc. verb. III, 1883, p. 270). The distinctive characters of this new genus, for which the name *Pleuracanthites* was proposed by its author, have been studied in detail by Waehner (Beiträege zur Kenntnis der tieferen Zonen des unteren Lias, etc., VII. Teil, Beitr. Palæont. u. Geol. Oesterr. Ung., etc., IX, 1895, p. 27). According to his diagnosis this genus is distinguished from *Lytoceras*, with which it agrees in its general shape and involution, by the presence of a long body-chamber comprising more than one entire volution, deep lateral sinuses or crescentic tubercles, and indistinctly developed siphonal crests.

A species agreeing in all these characters of generic importance with the European prototype of *Pleuracanthites* is represented in A. v. Krafft's collection by a single specimen, which, although incomplete, is sufficiently well preserved to permit a safe determination.

In its involution the present example agrees best with the Alpine type, distinguished by whorls increasing rather rapidly in height and thickness. The cross-section of the inner whorls recalls very strongly that in Waehner's specimen from Schreinbach illustrated in Pl. IV, fig. 1, of his memoir. It is nearly elliptical, with an obtusely rounded siphonal edge, but is considerably thicker than high. In the last volution the height increases more rapidly than the width. At the same time the siphonal crest becomes obsolete and the greatest transverse diameter is shifted gradually from the middle of the flanks towards the umbilical region. This change in the shape of the body-chamber is restricted to the vicinity of the aperture, and is a feature of specific distinction between our form and *Pleuracanthites biformis* (Sow.) Canavari (Beitr. z. Fauna d. unteren Lias von Spezia, Palaeontographica, XXIX, p. 156, Taf. XVII, figs. 8-11), in which the shape of the cross-section is not subject to any variation throughout its entire length.

The actual peristome cannot have been situated far from the aperture of my specimen, the last traces of septa being visible in the third quarter of the penultimate whorl. My specimen was certainly provided with a long body-chamber, exceeding the last volution in length considerably

The ornamentation is not very strongly marked in my specimen, but this character may be due to its state of preservation, the surface of the cast having been injured by weathering. Sinuses corresponding to paulostomes are, however, noticed at several places in the penultimate and last volutions. They agree exactly with the corresponding sculptural elements in *Pleuracanthites biformis*. In the penultimate whorl they are even combined with broadly elevated, crescent-shaped knobs or folds, exactly like those in the inner volutions of Waehner's type-specimen from Schreinbach illustrated on Pl. III, fig. 2.

In the body-chamber a gradual obliteration of the sculpture is marked by the absence of concentric folds. In the vicinity of the aperture, where the surface of the cast has been preserved satisfactorily, two delicate paulostomatic stripes are exhibited, forming deep sinuses with their convexities turned backwards.

Dimensions.

Diameter of the shell	70 mm.	
„ „ „ umbilicus	5½ „	
Height } of the last volution	5½ „	
Thickness	34 „	
Height } at the beginning of the last volution .	16 „	
Thickness	31 „	

Sutures.—Of the sutural line the siphonal and principal lateral lobes and the siphonal saddle only are known to me. They seem to agree very closely with the corresponding elements in *Pl. biformis*.

The siphonal lobe is short, very narrow and provided with a high median prominence. The siphonal saddle is richly ramified and its terminal phylla are comparatively small.

Gen.: EUPHYLLITES Waehner.

EUPHYLLITES sp. ind. (?) Pl. XV, fig. 5.

The materials of the genus *Euphyllites* in A. v. Krafft's collections are very scanty indeed, and I have as much hesitation in placing them in this genus as in venturing on their generic determination at all. Nevertheless I think it advisable to mention them here as by them the presence of an Ammonite in the liassic crags of Malla Johar is indicated, which in its external characters seems to show a close affinity to the smooth variety of *Euphyllites Struckmanni* Neumayr (Zur Kenntnis der Fauna des untersten Lias in den Nordalpen, Abhandl. K. K. Geol. Reichsanst. VII, p. 36, Taf. VI, fig. 5).

My specimen is a cast of the body-chamber comprising exactly one half volution, with fragments of the penultimate whorl adhering to it. In its involution it takes a position intermediate between some triassic species of *Discophyllites* (*D. neojurensis* Quenst.) and *Euphyllites Struckmanni*. A reconstruction of its outlines makes me suppose that the width of its umbilicus was 32 mm., corresponding to a diameter of 85 mm. But in the shape of the cross-section there is a complete agreement with the type-specimens of *E. Struckmanni*, as illustrated by Waehner (Beiträge zur Kenntnis der tieferen Zonen des unteren Lias in den nordöstl. Alpen, Beitr. Palæont. u. Geol. Oesterr. Ungarns, etc., XI, p. 170, Taf. XXII, figs. 1-8 ; XXIII, fig. 1). The lateral parts are perfectly flat, running parallel and passing gradually into the rounded siphonal part, whereas they are separated from the steep umbilical wall by an obtusely rounded edge. The proportion of height and thickness is 34 : 24 mm.

The surface of the cast, which has been partly injured by weathering, is entirely smooth.

The present cast has been broken off, unfortunately, in front of the last septum. Thus the terminal leaves of the two lateral saddles only have been preserved. They are large and rounded elliptically as in *Psiloceras* or in *Schistophylloceras Germanense* Herb.

Gen.: ECTOCENTRITES Waehner.

ECTOCENTRITES sp. ind. (aff. ALTIFORMIS Bon.?).

A body-chamber fragment of large dimensions reminds me of the equally-sized example of *Ectocentrites altiformis* Bonarelli (Cefalopodi sinemuriani del

Appennino centrale, Palæont. Ital., 1899, V, p. 73, Tav. IX, figs. 4-6) from the lower lias of Monte di Cetona, which has been illustrated by Fucini (Palæont. Ital., VII, 1901, Tav. XIV, fig. 1). The cast is strongly weathered and not worthy of illustration. The cross-section is suboval and compressed, a height of 55 mm. corresponding to a width of 39 mm. The external part is regularly rounded and passes into the flattened lateral parts without any intervention of a distinct umbilical shoulder. The umbilical wall is high and steep and unites with the flanks in a sharply rounded margin.

The sculpture consists of numerous ribs, which are disposed rather irregularly. The presence of marginal tubercles could not be ascertained. The siphonal part was undoubtedly smooth.

A more minute description of the present cast is precluded on account of its insufficient state of preservation. But it is nevertheless interesting to establish the fact, that the genus *Ectocentrites* is probably also represented in the Himálayan lias.

Fam.: *AMALTHEIDÆ* Fisch.

Gen.: OXYNOTICERAS Hyatt.

OXYNOTICERAS sp. ind. ex aff. GREENOUGHI Sow. Pl. X, fig. 5.

The fragment of a body-chamber is comparable by its shape and sculpture with *Oxynoticeras Guibalianum* D'Orbigny (Paléont. française, Terrains jurass, I, p. 259, Pl. 73) or with *O. Greenoughi* Sow. (Wright, Lias Ammonites, Palæont. Soc., p. 397, Pl. XLIV, XLV). It belongs to a moderately compressed and sharply carinated ammonite, whose surface was covered with numerous undivided ribs. The transverse section agrees best with that of the species from Adneth, identified with *O. Greenoughi* by F. v. Hauer (Cephalopoden aus dem Lias der nordwestlichen Alpen, Denkschr. Kais. Akad. d. Wiss. XI, 1856, Taf. XII, fig. 1). It is lanceolate and compressed less strongly than in *O. oxynotum* Quenst., from which it differs also in its ornamentation. The flanks, which are moderately convex, meet at an angle of 60 degrees. The keel is acute.

The umbilical region having been completely destroyed in my fragment, nothing can be said about its involution. Notwithstanding its defective condition it is sufficient for certifying the presence of the genus *Oxynoticeras* in the liassic fauna of the exotic blocks of Malla Johar. A specific determination is impossible, because the fragment represents a stage of growth in which nearly all species of *Oxynoticeras*, as distinguished by Hyatt (Genesis of Arietidæ, Smithson. Instit., 1885, p. 214), are still provided with acute keels.

Fam.: *ÆGOCERATIDÆ* Neum.

Gen.: SCHLOTHEIMIA Bayle.

SCHLOTHEIMIA sp. ind. ex aff. TRAPEZOIDALIS (Sow.) Can. Pl. XV, fig. 2.

This is a typical species of the genus *Schlotheimia*, with slowly increasing whorls, a wide umbilicus and with radial ribs, which on the siphonal part are interrupted along a median furrow where they terminate from either side in stout, knob-shaped elevations. The ribs are not turned forward in the vicinity of the siphonal margin, as in the majority of species belonging to the group of *Schl. angulata*, but meet one another from either side of the external furrow in a straight line. This character is known in *Schl. Charmassei* d'Orb. of extra-Alpine, in *Schl. trapezoidalis*, *Schl. ventricosa* Sow. and *Schl. posttaurina* Wachn. of Alpine species.

Among these species *Schlotheimia Charmassei* shows a less near affinity to the present form than the Alpine types, especially *Schl. trapezoidalis* (Sow.) Canavari (Unt. Lias von Spezia, Palæontographica, XXIX, p. 165, Taf. XVII, figs. 8, 9). Although the only Himálayan specimen available for examination is, unfortunately, fragmentary, it permits a satisfactory reconstruction of its shape and sculpture, sufficient for a closer comparison with *Schl. trapezoidalis*, which is well known to us from the careful studies of Wachner (Beitraege zur Kenntnis der tieferen Zonen des unteren Lias, etc., Beitr. z. Palæont. Oesterr. Ungarns, etc., IV, p. 185, Taf. XXIII, figs. 1-4; XXI, fig. 6).

In its general shape and involution my Himálayan specimen differs from *Schl. trapezoidalis* by its more slowly increasing whorls and by its wider umbilicus. In these external features it reminds us rather of *Schl. Donar* Wachner (l. c. p. 172, Taf. XIX, figs. 4; XXI, figs. 1, 2) or of *Schl. extranodosa* Wachner (l. c. p. 168, Taf. XX, figs. 7-11), than of *Schl. trapezoidalis*, but from both species it differs essentially in its sculpture. The lateral and siphonal parts are nearly flat, imparting to the cross-section a rectangular outline, with rounded-off angles.

The sculpture consists of single ribs. As far as the defective state of preservation renders the ornamentation visible, no dichotomous ribs have been noticed. The ribs are of moderate strength, almost perfectly radial, with a very small forward-turned geniculation near the siphonal margin. They originate at the umbilical margin as delicate folds, but increase in strength considerably, while crossing the lateral parts, till on the siphonal area they swell into stout, knob-shaped elevations, which stand opposite each other on both sides of the external furrow. It is this character of ornamentation which imparts to the Himálayan species a similarity with *Schl. trapezoidalis*.

Dimensions.

Diameter of the shell	.	.	.	.	.	.	.	.	.	61 mm.
„ „ umbilicus			.	.	.	.	.	.	.	24 „
Height		of the last volution						.	.	20 5 „
Thickness						.	.	.	.	15 „

Sutures.—The last septum separating the body-chamber from the preceding air-chamber is exposed in rough outlines. Only the siphonal lobe and saddle are known in detail. Siphonal lobe narrow and divided by a high median prominence, each wing terminating in a single sharp point, very similar to the corresponding sutural element in *Schlotheimia Donar*, as illustrated on Taf. XIX, fig. 4*d.* of Wæhner's memoir. Principal lateral saddle united with the following elements into a suspensive lobe.

One-third of the last volution belongs to the body-chamber.

SCHLOTHEIMIA nov. sp ind.   Pl. XI, figs. 5, 6; XV, fig. 4.

This species is nearly allied to the preceding one, from which it differs chiefly by its strongly inflated whorls and by its coarser sculpture. The ribs are high, acute and rarely dichotomising. As in the preceding species, they are turned forward very little in the vicinity of the marginal shoulder.

The cross-section is oval, the lateral parts being gently curved, not flattened. The whorls increase slowly and surround a wide, open umbilicus. A more minute description of this species is precluded, on account of the fragmentary state of the casts available for examination.

Dimensions.—Not measurable

Sutures.—Not known.

SCHLOTHEIMIA sp. ind. aff. MARMOREA Opp. Pl. XIV, fig. 5.

A fragment of a chambered cast of a large *Schlotheimia* shows its affinity to the Alpine group of *Schl. marmorea* Oppel (Palæontologische Mitteil. aus dem Museum des bayr. Staates, 1862, p. 139) by its sutural line, which is extremely complicated, much more so than in any of the extra-Alpine representatives of this genus. It only needs a comparison of the sutural line of my specimen with the illustration given by Wæhner on Pl. XXII, fig. 1*c,* of his "Beitraege zur Kenntnis der tieferen Zonen des unteren Lias in den nordwestlichen Alpen" (Beitr. Palæont. Oesterr. Ungarns, etc., IV, 1886), to see that we have to deal here with a species which is very nearly allied to *Schl. marmorea.* That the lobes are less broad in proportion to the saddles, cannot be considered as a distinctive feature of importance, as this character is subject to a remarkable variability in *Schl. marmorea.* As has been stated by Wæhner, individuals with low whorls are, as a rule, provided with more slender and deeper lobes than types with strongly compressed and high volutions.

The median prominence is high, broad and strongly serrated. The siphonal lobe is deeper than in *Schl. marmorea*, taking an intermediate position between the median and lateral indentations of the principal lateral lobe. The two wings of the siphonal lobe diverge very strongly. From the base of the siphonal saddle an accessorial branch is cut off by a very deep indentation. This branch is conspicuous by being less deeply serrated than the rest of the branches and by the rounded off lines of its secondary indentations. It corresponds exactly to the massive branch at the base of the outer margin in the siphonal saddle of *Schl. marmorea*, but it is provided with larger denticulations.

The siphonal saddle is very richly ramified and reclines towards the principal lateral saddle, which is the higher one. In both saddles the branches of the outer margin are considerably larger and developed more richly than those of the inner one.

The second lateral lobe is very short and united with the following sutural elements into a suspensive lobe. The sutural line, as illustrated in fig. 5, is not accessible beyond the second lateral lobe, but on the other side of my cast two auxiliary lobes are exposed, although the sutural line has there been strongly injured by weathering.

In its external features this fragment does not differ considerably from equal-sized specimens of *Schlotheimia marmorea*. But it must be borne in mind that a closer comparison is, unfortunately, excluded by its fragmentary condition. Neither the rate of involution nor the cross-section of the whorls are known to me. The lateral parts are flatly curved and covered with numerous ribs of moderate strength, which are provided with acute edges. In the marginal region they are directed forward more strongly than in *Schl. Charmassei* d'Orb., but a little less strongly than in the typical *Schl. marmorea*. There is no distinct siphonal furrow developed in the middle of the external part, but only a smooth zone, along which the ribs are interrupted. They do not correspond exactly along both sides of this zone.

Among the specimens of *Schlotheimia marmorea* illustrated by Waehner it is the middle-sized type from the Kammerkaralpe (l. e. Taf. XXII, fig. 2), which agrees best with our example in the shape of the siphonal part.

The fragmentary condition of this specimen does not allow of any measurements of its parts.

Gen.: *ÆGOCERAS* Waagen.

.EGOCERAS sp. ind. (ex aff. BIFER Quenst. ?) Pl. XIV, fig. 6.

I have mentioned this poorly preserved fragment of the side of an outer whorl as the only representative of the genus *Ægoceras* Waag. in the fauna of exotic blocks Nos. 16 and 17. It is, however, not sufficient to enable me to come to a tolerably certain conclusion about its specific position, although it might perhaps be placed near *Æg. bifer* Quenstedt (Cephalopoden, p. 83, Taf. IV, fig. 14, Der Jura, p. 103, Taf. XIII, figs. 11-13, Ammoniten des schwabischen Jura, I, Lias, p. 160, Taf. XXII, figs. 7-27).

The whorl is subquadrangular, with the external part well rounded. Number of ribs small, about fifteen in the circumference of an entire volution. Ribs short, straight and stout, elevated into obtuse knobs near the siphonal margin. Knobs connected by low bridges, which cross the siphonal area, but not by distinctly developed external rhombi. The fragment cannot be attributed therefore to the group of *Æg. planicosta* Sow.

The regular character of ribs in our fragment seems to distinguish it from full-sized specimens of *Æg. bifer* as illustrated by Wright (Monograph on the Lias Ammonites, Palæontograph. Soc. London XXXIV, Pl. XXVI, figs. 1-4) in a very remarkable way. But these characteristic features of the European species are only developed in later stages of growth. Young examples possess straight, radially directed ribs, which are connected by low, forward-curved bridges across the siphonal part.

A specimen reminding us of the present fragment in its external features and sculpture has been described from the lower lias of the Hierlatz as *Æg. bifer* by Geyer (Ueber die liasischen Cephalopoden des Hierlatz bei Hallstatt, Abhandl. K. K. Geol. Reichsanst. XII, p. 260, Taf. III, figs. 18, 19). From this and from the equal-sized specimen of *Ægoceras bifer annulosus* Quenstedt (Ammoniten, etc., Taf. XXII, fig. 20) our specimen seems to differ only by its external ridges being lower and by its ribs being stouter and broader.

If we take into consideration that this habit represents a mode of variation, which is likely to be developed still more strongly in later stages of growth, the independent position of the present form cannot well be doubted.

Fam.: *ARIETIDÆ* v. Zittel.

Gen.: ARIETITES Waagen.

I am in accordance with the views of Wachner, K. v. Zittel, Boese and Uhlig in rejecting the subgeneric divisions of the genus *Arietites*, which have been proposed by Hyatt (Genesis of Arietidæ, Smithson. Instit., 1889) and in retaining the name *Arietites* for all groups of liassic forms, which are provided with strong radial ribs and keel furrows accompanying a distinctly developed median keel.

This genus is rather richly represented in the Himálayan lias. In the following descriptions four species have been enumerated, but their actual number is probably larger, the defective state of preservation not permitting of specific determination of several examples.

All species are closely allied to types from the lower lias of the Eastern Alps, which are well known to us from the careful researches of Wachner.

ARIETITES cf. COREGONENSIS (Sow.) Can. Pl. XIV, fig. 2 ; XV, fig. 1.

1882. *Ægoceras Coregonense* Canavari, Beiträge zur Fauna des unt. Lias von Spezia, Palæontographica XXIX, p. 173, Taf. XIX, figs. 12-15.

1886. *Arietites Coregonensis* Wæhner, Beiträge zur Kenntniss der tiefsten Zonen des unt. Lias in den nordöstl. Alpen, V. Theil, Beiträge zur Palæont. Oest err. Ungarns, etc., VI, p. 311, Taf. XXI, figs. 1-3 ; XXII, figs. 1-3, XXIII, figs. 1-4 ; XXIV, figs. 1-6.

This remarkable species is represented by two fairly complete specimens. The smaller example is able to make up for the deficiency of the inner volutions in the larger one. For their comparison with the Alpine form numerous examples from the lower lias of Adneth, Breitenberg and Enzesfeld have been available to me, among them some of Wæhner's type-specimens.

In the larger specimen the whorls overlap one another but very little. The last and penultimate whorls differ considerably in the shape of their cross-sections. At the beginning of the last volution the transverse section is broader than high and provided with rounded sides. In the last volution the lateral parts become flattened gradually, and the height increases more rapidly than the width. In the penultimate whorl the greatest transverse diameter is situated in the middle of the flanks, whereas in the last volution it is shifted gradually towards the umbilical margin. At the same time the siphonal area is reduced in width and its marginal shoulders are formed by the lateral keels, accompanying the two deep external channels, whereas at the beginning of the last whorl the median keel and the accompanying channels occupy only the middle zone of the siphonal area.

In all these characters my specimen agrees exactly with equal-sized examples of *Arietites Coregonensis*.

Throughout the entire last volution the median keel and the external furrows or channels are well defined. The latter are bordered on the marginal side by distinctly developed ridges, which become acute in the vicinity of the aperture.

The lateral parts of the last and penultimate whorls—of the inner volutions nothing has been preserved in this specimen (Pl. XIV, fig. 2)—are covered with very numerous, radiating ribs, which are narrower than the intercostal furrows, strongly elevated and sharp. Only in the vicinity of the aperture they are somewhat flattened. Their direction is radial or slightly turned backward. Before reaching the marginal shoulder they describe a flat curve, with its concavity turned forward. There are about seventy ribs counted within the circumference of the last volution, corresponding to a diameter of 125 mm. The same number is given by Wæhner for the last volution of an Alpine specimen of equal size (Pl. XXII, fig. 1, p. 317).

A satisfactory idea of the shape and sculpture of the inner volutions of the Himálayan form may be gathered from my smaller specimen (Pl. XV, fig. 1), which is probably entirely chambered.

In its transverse section this specimen recalls very strongly the variety illustrated by Wæhner on Pl. XXIV, fig. 6. It is comparatively high-mouthed, and provided with a strongly compressed, nearly rectangular cross-section. Both the lateral parts and external area are remarkably flattened. Considering its small dimensions, the keel and the lateral channels are developed very strongly. Secondary keels bordering the marginal sides of the channels are distinctly defined.

The number of ribs is very large. Not less than 64 ribs are counted within the circumference of the last volution, corresponding to a diameter of 67 mm. In the penultimate whorl of the specimen illustrated by Waehner on Pl. XXII, fig. 1, 59 ribs have been counted within a circumference corresponding to the same diameter. The ribs are straight, exactly radial, very slender, and slightly elevated near the marginal shoulder where they become obliterated very rapidly. The external area having been partly injured by weathering in my specimen, I cannot say whether or not they do meet the marginal ridges in a forward-bent curve.

Dimensions.

							Pl. XIV, fig. 2.			Pl. XV, fig. 1.
Diameter of the shell	.	.	.	.	.	.	115 mm.	.	.	64 mm.
		.. „ umbilicus	.	.	.	.	77 „	.	.	36·5 „
Height Thickness	} at the aperture		.	.	.	.	21 „ 21 „	.	.	15·8 „ 16 „
Height Thickness	} at the beginning of the last volution		.	.	.	17·5 „ 21·5 „	.	.	? ?

Sutures.—Not known in detail.

Remarks.—My specimen agree so closely with *Arietites Coregonensis* from the zone of *Schlotheimia marmorea* in the Alpine lower lias, as illustrated by Waehner, that I do not hesitate to unite them with that characteristic species as *cf.*

As this identification is of special importance for a correlation of the Himálayan lias with the deeper stage of the lower lias in Europe, I shall pass in review such species of *Arietites* characteristic of the higher stage of the Alpine lower lias as might put in a claim for a closer comparison. In this respect our examination can be restricted to the group of forms which have been included in the genus *Vermiceras* by Hyatt, and in the subdivision C, by A. v. Sutner in his systematic table of *Arietidæ* as published by Boese (Zeitschr. Deutsch. Geol. Ges. XLVI, 1894, p. 721). In this subdivision the groups of *A. tardecrescens* v. Hauer and of *A. Conybeari* d'Orb. must be taken into consideration, but all species belonging to these groups and characteristic of the higher stage of the lower lias are provided with ribs which are turned forward, even in the inner volutions. This character distinguishes our specimen at once from *A. bavaricus* Boese (l. c. p. 725, Taf. LVI, figs. 1, 2) or from *A. Rothpletzi* Boese (l. c. p. 730, Taf. LVI, figs. 5, 6), which otherwise show a considerable resemblance in the external shape and in the number of ribs.

ARIETITES nov. sp. ind. ex aff. COREGONENSIS Sow.   Pl. X, fig. 8.

This fragment of an outer whorl must be separated from *Arietites Coregonensis* on account of its very low and broad cross-section. A height of 20 mm. corresponds to a transverse diameter of 28 mm. In its outlines the transverse section resembles that of the last volution in large examples of *Arietites Coregonensis*. The greatest transverse diameter coincides with the umbilical margin, which is distinctly defined. From this place the lateral parts converge in flat curves towards the external part, which is remarkably narrow and occupied entirely by the very strongly developed keels and channels. The median keel and the two secondary marginal ridges are of equal height and rounded above, not acute.

The sculpture agrees with the ornamentation in *A. Coregonensis*. The ribs become obliterated very rapidly in the vicinity of the marginal ridges, where they are curved forward very slightly.

The deficient state of this fragment renders an exact determination impossible, but the character of ribs and the shape of the cross-section point to a species closely related to *A. Coregonensis*. The great thickness of the whorls and the strong development of keels and channels might perhaps suggest a comparison with the group of *A. Bucklandi* Sow., but an affinity with that group is excluded by the different shape of the cross-section and by the absence of marginal swelling of the ribs.

ARIETITES HIMALAYANUS nov. sp.   Pl. XIV, fig. 4.

This fragment of the outer whorl and of a small part of the preceding volution belongs to a species which, in the general shape of its disc, is related to *Arietites Boesci* Uhlig, whereas in its sculpture it exhibits a close affinity to *A. Conybeari* Sow. and to *A. Cordieri* Can.

As has been stated by Waehner in his excellent description of *Arietites Cordieri* (Beitraege zur Kenntnis der tieferen Zonen des unteren Lias in den nordwestl. Alpen VI, Theil, Beitr. z. Palæont. Oesterr. Ung. etc., VIII, p. 250, Taf. XVII, figs. 1-4 ; XVIII, figs. 1-6), our knowledge of the British species *A. Conybeari* is restricted to Wright's description and illustration of a single specimen of very large size (Monograph of the Lias Ammonites 1 Pt. Palæontogr. Soc., London, XXXII, 1878, Pl. II, figs. 1-3), rendering a comparison with moderately sized types very difficult. Nevertheless it is obvious from Wright's illustration that our Himalayan specimen agrees with *A. Conybeari*, exactly in such features as are characters of distinction between the British species and its Alpine representative *A. Cordieri*. The ribs, which originate in the umbilical suture, run across the umbilical wall and the lower portion of the flanks obliquely backward, and turn gradually into a radial direction in the marginal region of the lateral parts. Here they swell into knob-shaped elevations, which are not, however, actual tubercles, and terminate abruptly in front of the marginal ridges. The extremities only of these stout ribs are turned forward slightly in uniting with the marginal ridge.

A specimen of *Arietites longidomus* Quenstedt (Ammoniten des schwæbischen

N 2

Jura I, Lias, Taf. VI, fig. 1) in the collections of our University's Palæontological Museum shows exactly the same type of sculpture as my Himálayan example. In both of them the ribs terminate stiffly opposite the marginal ridges, and their extremities are not protruded along the latter for any considerable distance.

The external sculpture is very strongly marked. The marginal ridges are well defined, the channels deeply excavated. The median keel is high and elevated considerably above the marginal ridges.

My specimen differs both from *Arietites Conybeari* and *A. Cordieri* in the shape of its transverse section, which is slender and strongly compressed. In the last volution a width of 19 mm. corresponds to a height of 22 mm. But a character still more remarkable is the compressed shape of the penultimate volution, in which the corresponding measurements are 9 and 12 mm. In order to find similar proportions, we are obliged to look to the group of *Arietites romanicus* Uhlig. In this group it is especially *A. Bocsei* Uhlig (Ueber eine unterliassische Fauna aus der Bukowina, Abhandl. d. med. naturw. Vereins Lotos, II, p. 29, Taf. I, fig. 8) which resembles our species in the shape of the high and oval cross-section. But any closer affinity with the group of *A. romanicus* is excluded by the difference in their sculpture. In *A. Bocsei* the ribs are directed forward from their very origin at the umbilical suture, whereas they are turned backward strongly in the umbilical region of my Himálayan form.

Taking into consideration the remarkable cross-section of the present species, which is combined with a sculpture characteristic of the group of *A. Conybeari*, its independent position cannot well be doubted. On the strength of this fact I have ventured to introduce a new specific name for a fragment, which otherwise should not have been chosen for a type-specimen.

Several fragments of inner volutions are referable with some probability to this new species.

Dimensions.—Not measurable.

Sutures.—Not known.

ARIETITES nov. sp. ind. ex aff. ROTIFORMIS Sow. Pl. XIV, fig. 3.

A small fragment of an outer whorl, consisting of four air-chambers, shows a great resemblance in its shape and sculpture to the group of *Arietites rotiformis* Sow., which, according to Waehner, is represented very richly in the lower lias of the north-eastern Alps.

As there is only a single, small fragment available for examination, nothing definite can be said about the involution of the complete shell. But, to judge by the proportions of the cross-section and by the curvature of the outlines, it seems to have been a type with very low, slowly increasing volutions and with a wide umbilicus, comparable, perhaps, to Waehner's type-specimen of *A. rotiformis var.*

rotator Reynés, illustrated in Pl. II, fig. 1, of his beautiful monograph of the Alpine Ammonites of the deeper zones of the lower lias (Part VII, Beitr. zur Palæont. u. Geol. Oesterr. Ungarns, etc., IX, 1895).

The transverse section is of trapezoidal shape, with flattened lateral and external parts. The greatest transverse diameter corresponds to the siphonal margin. The whorl is thicker than high, if the measurement is taken along the stout marginal tubercles. In my specimen a height of 19 mm. corresponds to a width of 21 mm. taking the distance between two opposite marginal tubercles as transverse diameter.

Keels and channels are well defined. The median keel surpasses the marginal ridges in height.

Six stout transverse ribs are counted in my fragment. Their direction is radial, with the exception of the umbilical wall where they are slightly turned backward. They are elevated into strong tubercles along the external margin. A second elevation on a considerably smaller scale is noticed in the lower third of the lateral parts. The marginal elevations protrude more strongly than in any of the equal-sized specimens of *A. rotiformis*, as illustrated by Wæhner. They are even more stout and massive than in the example illustrated by F. v. Hauer (Cephalopoden aus dem Lias der nordöstlichen Alpen, Denkschr. Kais. Akad. d. Wiss. XI, 1856, Taf. I, fig. 1). The stout tubercles terminate abruptly in front of the marginal ridges.

Sutures —The sutural line has been excellently preserved. It differs from the sutural line of the typical *Arietites rotiformis* by its short siphonal lobe, by the dimoroidic character of the siphonal saddle and by the general height of the saddles.

In all the typical shapes of *A. rotiformis* the siphonal lobe extends beyond the principal lateral by about one-third, whereas in my specimen the two lobes are of equal depth. In this character as well as in the dimeroidic division of the siphonal saddle my Himálayan form appears to be nearly allied to *A. Deffneri* Oppel (Palæont. Mittheil. aus dem Museum d. bayr. Staates, I, p. 131, Taf. X, fig. 1). Their sutural lines agree, indeed, very closely and differ only in some subordinate details, especially in the more equal dimensions of the two secondary saddles, into which the siphonal saddle of the Himálayan species is divided by the secondary median lobe.

Taking into consideration the close agreement of the sutural lines, I should not hesitate in referring our Himálayan form to *A. Deffneri* as *cf.*, had not the differences in the sculpture and shape of the cross-section peremptorily required its isolation. *A. Deffneri* is a high-mouthed species with a rectangular cross-section and more numerous ribs, and is therefore in this respect allied to our species less closely than *A. rotiformis*.

The Himálayan species described here seems to represent a very interesting type of *Arietites*, combining the external characters of *A. rotiformis* with the sutural line of *A. Deffneri*.

CONCLUSIONS.

The fauna of exotic blocks 16 and 17 consists of the following forms—all of them Cephalopoda :—

Dibranchiata.

1. *Atractites sp. ind.*

Ammonoidea.

2. *Phylloceras Montgomeryi* nov. sp.
3. „   *Sclateri* nov. sp.
4. „   nov. sp. aff. *Sclateri.*
5. „   *Horsefieldii* nov. sp.
6. „   *Caldwellii* nov. sp.
7. „   sp. ind. aff. *Dieneri* Rosenbg.
8. „   (*Schistophylloceras*) *mongolicum* nov. sp.
9. *Rhacophyllites cf. gigas* Pocini.
10.   „   *schofariformis* nov. sp.
11. *Euphyllites sp. ind.* (?)
12. *Analytoceras sp. ind. aff. articulato* Sow.
13. *Hetorentrites sp. ind.* (aff. *altiformis* Ben. ?)
14. *Pleuracanthites sp. ind. aff. biformis* Sow.
15. *Oxynoticeras sp. ind. aff. Greenoughi* Sow.
16. *Schlotheimia sp. ind. ex aff. trapezoidalis* Can.
17.   „   sp. ind.
18.   „   sp. ind. aff. marmorea Opp.
19. *Aegoceras sp. ind.* (ex aff. *bifer* Quenst. ?)
20. *Arietites cf. Coregonensis* (Sow.) Can.
21.   „   nov. sp. ex aff. *Coregonensis* Sow.
22.   „   *himalayanus* nov. sp.
23.   „   nov. sp. ind. ex aff. *rotiformis* Sow.

Among the ammonites the genus *Phylloceras* predominates both in the number of species and individuals. It is interesting to compare the distribution of genera among the number of fossils which have been actually collected by A. v. Krafft. His collections from exotic blocks 16 and 17 contain altogether 87 specimens of ammonites which permit of a generic determination. Among them *Phylloceras* is represented by 51, *Rhacophyllites* by 2, *Arietites* by 20, *Schlotheimia* by 8, and the rest of the genera each by one single individual only. This enormous predominance of *Phylloceras* imparts to our Himálayan fauna a decidedly Alpine aspect, and distinguishes it at a single glance from all liassic faunæ of middle-European habit.

The Alpine type of this liassic fauna shows itself immediately and indubitably, not only in the preponderance of *Phylloceras*, but also in the close affinities of the majority of Himálayan species with Alpine forms. There is not one single species in this fauna which bears a distinct local character. Of the striking peculiarities which distinguish the triassic cephalopod fauna of the main region of the Himálayas from the homotaxial faunæ of the Mediterranean region, nothing is seen in the present fauna. No palæontologist would be astonished if it had been found in Sicily or in Greece instead of on the Tibetan frontier.

Among species indicating very close specific relationships with European forms the following may be mentioned :—

> *Phylloceras Montgomeryi*—*Ph. persanense* Herb.
> *Phylloceras Sclateri*—*Ph. Lipoldi* Han.
> *Phylloceras Mongolicum*—*Ph. Cermaxense* Herb.
> *Rhacophyllites cf. gigas* Fucini.
> *Analytoceras sp. ind. aff. articulato* Sow.
> *Pleuracanthites sp. ind. aff. biformis* Sow.
> *Schlotheimia sp. ind. aff. marmorea* Opp.
> *Arietites cf. Coregonensis* (Sow.) Can.
> *Arietites himalayanus*— $\begin{cases} A. Bonari \text{ Uhl.} \\ A. Conybeari \text{ Sow.} \end{cases}$
> *Arietites nov. sp. aff. rotiformis*— $\begin{cases} A. rotiformis \text{ Sow.} \\ A. Deffneri \text{ Opp.} \end{cases}$

This list would probably have been still larger, if the *Phylloceratidæ* of the deeper zones of the Alpine lower lias had been submitted to an examination as detailed and careful as the representatives of other families of Ammonoides. But, unfortunately, a large number of Alpine types of *Phylloceras* of lower liassic age are as yet but very imperfectly known.

Particularly striking is the fact that the genera *Analytoceras*, *Pleuracanthites* and probably even *Euphyllites* and *Ectocentrites*, which are remarkable as connecting links between the families of *Lytoceratidæ*, *Phylloceratidæ* and *Psiloceratidæ*, are also represented in this fauna, although they are of very rare occurrence.

That the fauna of exotic blocks 16 and 17 bears the stamp of lower liassic age, as has been noticed by A. v. Krafft, is shown by the association of genera as well as by the affinities of its species to European ones. But the faunal guides are yet sufficient for a more exact determination of the age.

The specific similarities of the majority of species point to a correlation with the lower stage of the lower lias. I need only emphasise the presence of *Arietites cf. Corrgonensis* Sow., of a second species of *Arietites*, combining characters of *A. rotiformis* and *A. Deffneri*, of a species of *Phylloceras* belonging to the group of *Ph.* (*Schistophylloceras*) *Cermaxense* Herb., of the genera *Analytoceras* and *Pleuracanthites*, and the rich occurrence of *Schlotheimia*, which is represented by at least three species, one of them nearly allied to *Schl. marmorea* Opp.

The relationship to the faunæ of the deeper zones of the lower lias is, as can be easily seen, most clearly marked. On the other hand we cannot exclude peremptorily the possibility of the presence of faunistic elements, which might belong to the higher stage of the lower lias. As such elements, indicating perhaps a relationship with the fauna of the upper division of the lower lias, *Oxynoticeras sp. ind. ex. aff Greenoughi* Sow. and *Arietites himalayanus* might be taken into consideration. The latter, although recalling *A. Conybeari* in its sculpture, approaches in its general shape very closely to *Ar. Boesei* Uhl, which belongs to a group of this genus characteristic of the higher stage of the lower lias.

No great importance should be attributed to the representatives of the genus *Phylloceras*, which do not keep narrow stratigraphical horizons as strictly as might be wished for a more exact correlation. It is, however, noteworthy, that the group of *Phylloceras Partschi* Stur, which in the higher stage of the Alpine lias is invariably the most conspicuous for its frequency, is not represented in the Himalayan lias.

The limited occurrence and small thickness of the liassic strata in the region of exotic blocks in Malla Johar is no argument against the possibility of finding both the deeper and higher stages of the lower lias represented within them. I need only refer the reader to the liassic deposits of Transsylvania, as described by Herbich, which, although being only 3 m. in thickness and very little extended, have yielded numerous ammonites of the deeper and higher stages of the lower lias.

If we summarize the above observations, we come to the conclusion that the fauna of exotic blocks 16 and 17 is to be regarded as homotaxial with the lower stage of the lower lias, whereas the presence of equivalents of the higher stage has as yet not been ascertained.

V. FOSSILS FROM EXOTIC BLOCKS NOS. 6 AND 7 (MALLA KIOGARH).

Among four small exotic blocks occurring within the upper flysch of Malla Johar near Malla Kiogarh encamping-ground,—marked E.B. 4, 5, 6, 7, on the map accompanying A. v. Krafft's memoir—blocks 6 and 7 yielded a small number of ill-preserved specimens of *Phylloceras*, *Eyoceras*, *Aristiles* and *Belemnites*. None of them is sufficiently complete to permit a specific determination.

Phylloceras sp. ind. of. Sclateri Dien.   Pl. XVI, fig. 4.

This chambered fragment belongs to all appearance to a species very closely allied to or identical with *Phylloceras Sclateri*, with which it agrees in its shape and involution. In consequence of the injuries which the surface of the cast has suffered from weathering, the sutures appear somewhat deformed.

Several fragments are also referable with some probability to this species, but their bad state of preservation does not permit of a safe identification.

Phylloceras sp. ind. aff. Montgomeryi Dien.

An imperfectly preserved cast, measuring 58 mm. in diameter, possessing already the commencement of the body-chamber, is comparable by its shape to

Phylloceras Montgomeryi Dien. The sutures are but imperfectly known to me. The specimen is not suitable for illustration.

ÆGOCERAS sp. ind. Pl. XVI, fig. 5.

This is a form (unfortunately only a small fragment of a chambered whorl) which is comparable to *Ægoceras bifer* Quenst, in reference to its general shape. But the marginal swellings of the stout, transverse ribs are but faintly developed, and no external bridges connecting them have been noticed. The sutural line, which has been partly preserved, agrees in general with that of *A. bifer*, as illustrated by Quenstedt in Pl. 22, fig. 11 of his "Ammoniten des schwäbischen Jura" I, Th. (Lias). The arrangement of the sutural line is asymmetrical, the siphonal lobe with its short median prominence being shifted to the right of the median plane of the shell.

ARIETITES sp. ind. ex aff. A. GRUNOWI v. Hauer. Pl. XVI, fig. 3.

The present specimen, which seems to possess the body-chamber, consists of a fairly well-preserved outer volution. The inner whorls have been completely destroyed. It shows a close resemblance to *Arietites Grunowi* v. Hauer (Cephalopoden aus dem Lias der nordöstlichen Alpen, Denkschr. Kais. Akad. d. Wissensch, XI, p. 27, Taf. VIII, figs. 4–6), being distinguished by a very broad transversely elliptical cross-section, and by a low keel, accompanied by shallow and ill-defined channels. The lateral parts are strongly curved and covered with numerous and stout transverse ribs, which are bent forward distinctly in the marginal region and in the external part. Their direction is not exactly radial but slightly turned backward, especially so in the umbilical region. Secondary ridges accompanying the siphonal channels are entirely absent.

The specimen is provided with a very broad umbilicus.

Dimensions.

	mm.
Diameter of the shell	76
„ „ „ umbilicus	44
Height } of the last volution	18
Thickness }	23

Sutures.—Not known.

Remarks.—The new description and illustrations of F. v. Hauer's species, which have been given by Wähner (Beiträge zur Kenntnis den tieferen Zonen des unteren Lias, etc., Beitr. zur Palæont. Oesterr. Ungarns, etc., VI, p. 320, Taf. XXV, figs. 2, 3), enable me to draw a closer comparison between *Arietites Grunowi* and the present specimen. The latter is distinguished from *A. Grunowi*, to which it is certainly closely related, by comparatively lower whorls, by a wider umbilicus and by the more distinct concavity in the direction of the transverse ribs along the siphonal margin. The strength and direction of the ribs on the lateral parts is the

o

same in both species, but on the external part the ribs run much further towards the anterior margin than is the case in *A. Grunowi*.

BELEMNITES sp. ind.

Several fragments of *Belemnites* are only sufficient to determine the genus.

The rostra are very slender, of an elongated conical shape, recalling the liassic *Belemnites acuarius* Schloth. In one of the rostra a deep apical furrow has been noticed.

Ventral or dorso-lateral furrows are not known in any of my specimens.

CONCLUSIONS.

There can be no doubt that the small fauna of exotic blocks Nos. 6 and 7 is of lower liassic age. It is probably homotaxial with the fauna of blocks 16 and 17, although this homotaxis cannot be proved by the presence of identical species, on account of the fragmentary condition of my materials.

SUMMARY.

The discovery of the exotic blocks in Malla Johar by the Himálayan expedition in 1892, in which C. L. Griesbach, C. S. Middlemiss and myself took part, has acquainted us with a region, in which permian and triassic strata show a development differing from that of the corresponding beds observed in the normal sections of the Himálayas. A. v. Krafft's exploration of the Kiogarh range has not only corroborated the facts stated by the Diener expedition, but has filled up two important gaps in the series of the exotic region by the discovery of blocks of lower triassic and liassic age.

His careful examination of the entire territory of Malla Johar enabled A. v. Krafft to draw a closer comparison between the Tibetan series of beds as exposed in the exotic blocks and the normal Himálayan series. This comparison leads to the conclusion that each single subdivision of the Tibetan series known so far from the permian up to the lias differs from the corresponding Himálayan division.

In the mesozoic rocks the lithological difference is most strongly marked in the triassic horizons from the beds with *Flemingites Rohilla* to the top of the carnic stage and in the lias.

In the main region of the Central Himálayas the triassic beds underlying the great mass of the Dachsteinkalk are represented by grey or black limestones and shales, whereas in the Tibetan series they consist of red limestones exhibiting a striking resemblance to the Hallstatt limestones of the Eastern Alps. A. v. Krafft is fully justified in correlating the triassic facies of the Tibetan series, only excepting the noric (juvavic) stage, with the Hallstatt facies. The lithological resemblance of block No. 2 with the carnic Halstatt limestone of the Roethelstein near Aussee is, indeed, so great, that it is no easy matter to distinguish fossils from these two localities without a closer inspection.

In the noric (juvavic) and rhætic stages the difference between the Tibetan and Himálayan series becomes less remarkable. During these stages great masses of grey, dolomitic lime-stones, very poor in fossils, have been deposited in both areas. "Nevertheless there is no complete lithological identity between the two, the Tibetan grey limestone being massive throughout, while the Himálayan Dachsteinkalk is well bedded."

The sharpest contrast between the two series shows itself in the liassic periods. In the main region of the Himálayas (Spiti) the rhætic, liassic and oolitic beds are of uniform lithological character, consisting of dolomitic, thin-bedded, black or grey limestones. The red liassic limestones of blocks 6, 7, 16, 17 are developed in the facies of Adneth, and agree lithologically most closely with rocks composing the liassic crags of the Eastern Carpathians in Transylvania and at Valeasoca (Bukowina).

The faunæ of the exotic blocks of lower triassic and of Muschelkalk age have very small affinities to Alpine faunæ, but are connected very intimately with those of the corresponding divisions in the main region of the Himálayas, although the lithological differences are important. But the carnic fauna of the Tibetan series exhibits remarkable peculiarities pointing in another direction.

As has been remarked by E. v. Mojsisovics, the preponderance of *Ammonea trachyostraca* (*Ceratitoidea* and *Tropitoidea*) with the simultaneous diminution of *leiostraca* (especially *Arcestoidea*) is one of the most striking peculiarities of the upper-triassic cephalopod faunæ of the Indian province. This proportion of the two divisions of triassic *Ammonoidea* is reversed in the carnic fauna of exotic block No. 2. In the Tibetan series the genera *Arcestes* and *Cladiscites*, which are of comparatively rare occurrence in the Himálayan series, appear suddenly in large numbers, imparting to this exotic fauna its special type. With this preponderance of *Arcestoidea*, especially of *Cladiscitidæ*, is united another feature of distinction from the Himálayan series, and this is the very large percentage of species allied to European forms. Referring the reader to the results of my examination of the faunæ of exotic blocks 2 and 5, I need not explain further that those faunæ bear a peculiar character which distinguishes them from the homotaxial fauna of the main region of the Himálayas, but points to a very close affinity with the fauna of the zones of *Trachyceras Aonoides* and *Tropites subbullatus* in the Eastern Alps.

The agreement of the fauna of blocks 16 and 17 with that of the deeper stage of the Alpine lower lias is perhaps still more striking. If no other mesozoic faunæ in the Himálayas were known, our knowledge of this liassic fauna would not justify the establishment of an Indian zoogeographical province. The contrast between the Mediterranean and Indian triassic faunæ in the Himálayan region is nearly obliterated in the Tibetan region of exotic blocks during the liassic period. The difference between the liassic faunæ of Wurtemberg or England and the Alps is, indeed, more conspicuous than that between the Mediterranean and Tibetan faunæ of the lower lias. The affinities of the latter are extraordinarily close, not a single species in the fauna of blocks 16 and 17 bearing a distinct local character.

There can hardly be a doubt that to the north of the main region of the Himálayas there extended from the Eastern basin of the Tethys to the Mediterranean area a zone, in which sediments of a nearly uniform lithological character were deposited during carnic and liassic times, and where the area was inhabited by a fauna with insignificant local peculiarities and barely influenced by the independent development which is noticed in the mesozoic faunæ of the Indian triassic province.

A. v. Krafft and Suess agree in assuming that the Tibetan facies could not be *in situ* below the region of exotic blocks in Malla Johar, but must have been carried there from a territory lying much further to the north. This hypothesis is able to explain the rapid change of facies between the Tibetan and Himálayan series, but it does not explain the striking lithological and faunistic agreement which exists between the carnic and liassic sediments of the Tibetan series, on the one hand, and of the homotaxial beds in the Mediterranean region, on the other. It is the establishment of this agreement which I consider to be the most important result of my examination of A. v. Krafft's Himálayan collections.

PLATE I.

Fig. 1a, b, c, d   PROCLYDONAUTILUS TRIADICUS v. Mojs.

  ,,   2a, b   }   ,,   BUDDHAICUS Dien.
  ,,   3a, b

  ,,   4a, b   MOJSVAROCERAS sp. ind. ex. aff. TURNERI Hyatt and Smith.
  ,,   5a, b, c   NATICOPSIS sp. ind. ex. aff. OBVALLATA Kokon.
  ,,   6a, b   LOXONEMA (POLYGIRINA) cf. ELEGANS Hoevnen.
  ,,   7   PROCLYDONAUTILUS BUDDHAICUS Dien.

All specimens from the exotic block No. 2.

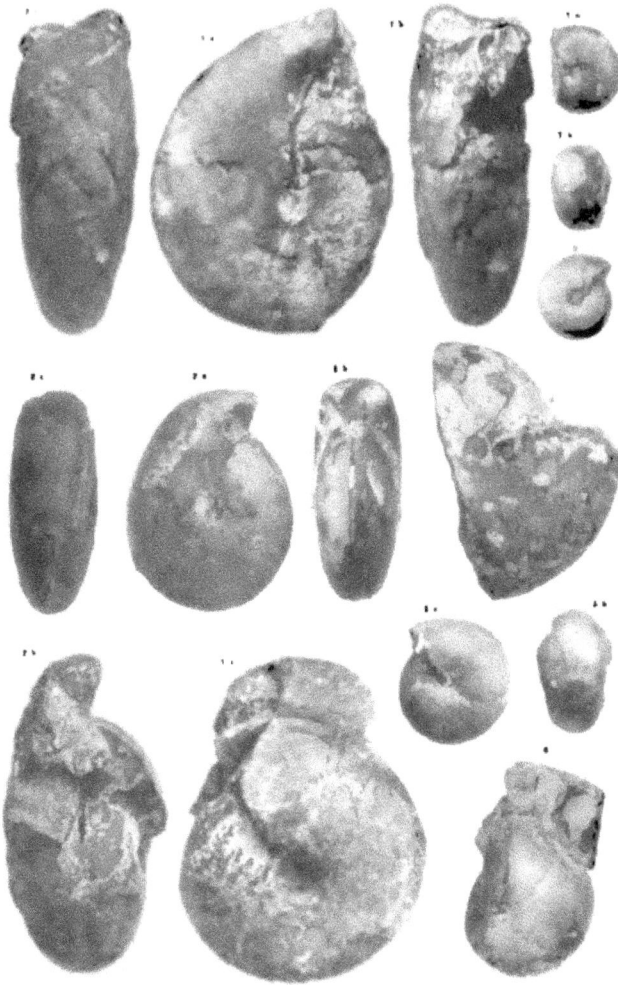

PLATE IV.

Fig. 1a, b, c } CLADISCITES CRASSESTRIATUS v. Moja.
"   2a, b   }
"   3a, b   HALOCLADISCITES SUBRATUS v. Moja.
"   4a, b, c   "   SUBCARINATUS Gemmellaro.
"   5a, b   CLADISCITES cf. SUBQUE Gemm.
"   6a, b   }   "   cf. PUSILLUS v. Moja.
"   7a, b   }
"   8   "   CRASSESTRIATUS v. Moja. Sutures.
                All specimens from the carnic block No. 2.
"   9a, b   CLADISCITES CRASSESTRIATUS v. Moja. Type-specimen from the carnic Hall-
                statt limestone (Ellipticus beds) of the Roethelst in (Salzkammergut).

Pl. V.

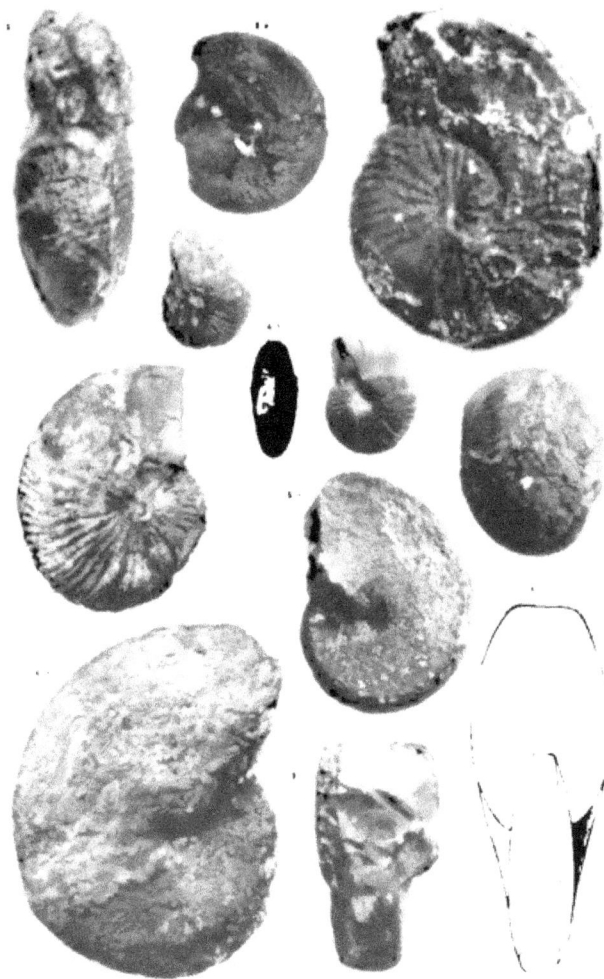

PLATE VI.

Fig. 1a, b, c }
„ 2a, b, c } Juvavites (Griesbachites) cf. Kastneri v. Moja.
„ 3a, b }

„ 4 Discotropites cf. sandlingensis v. Hauer.

„ 5a, b, c Tibetites enotensis Dien.

„ 6a, b } Anatropites Pilokmii Dien.
„ 7a, b }

„ 8a, b Margarites irregularicostatus Dien.

„ 9 Tropites cf. subbullatus v. Hauer.

„ 10a, b, c Anatropites cf. spinosus v. Moja.

„ 11a, b Tropites sp. ind. aff. wodani v. Moja.

All specimens from the exotic block No. 2.

Pl. VI

PLATE VII.

All specimens from the exotic block No. 2.

Ge

PLATE VIII.

Fig. 1   Daonella indica Bittn.
                Halobia, sp. ind.

                This specimen from the exotic block No. 1.

"   2a, b, c   Dicerocardium Flower Dien
"   3a, b   Arcestes sp. ind. aff. decipiens v. Mojs.
"   4a, b   Prodacrytes sp. ind. (Group of Extralabiati).
"   5a, b   (?) sp. ind. ex aff. Basmanops Laube.
"   6a, b   Cassianella sp. ind.

                These five specimens from the exotic block No. 2.

PLATE XI.

Fig. 1 a, b, c   Rhacophyllites cf. Gigas Fucini.

  ,,   2 a, b, c   Phylloceras (Schistophylloceras) mongolicum Dien.

  ,,   3   } Phylloceras Montgomeryi Dien.
  ,,   4

           Siphonal lobe and saddle of the large specimen illustrated on Pl. XIII, fig. 1.

  ,,   5   } Schlotheimia sp. ind
  ,,   6 a, b

All specimens from the exotic blocks 16 and 17.

Pl. IX.

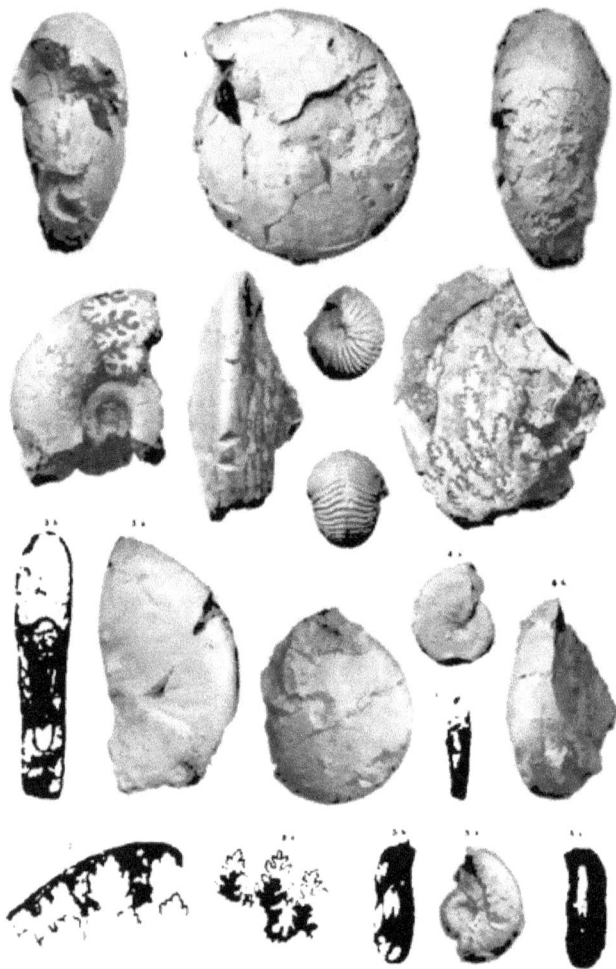

PLATE XII.

All specimens from the exotic blocks 16 and 17.

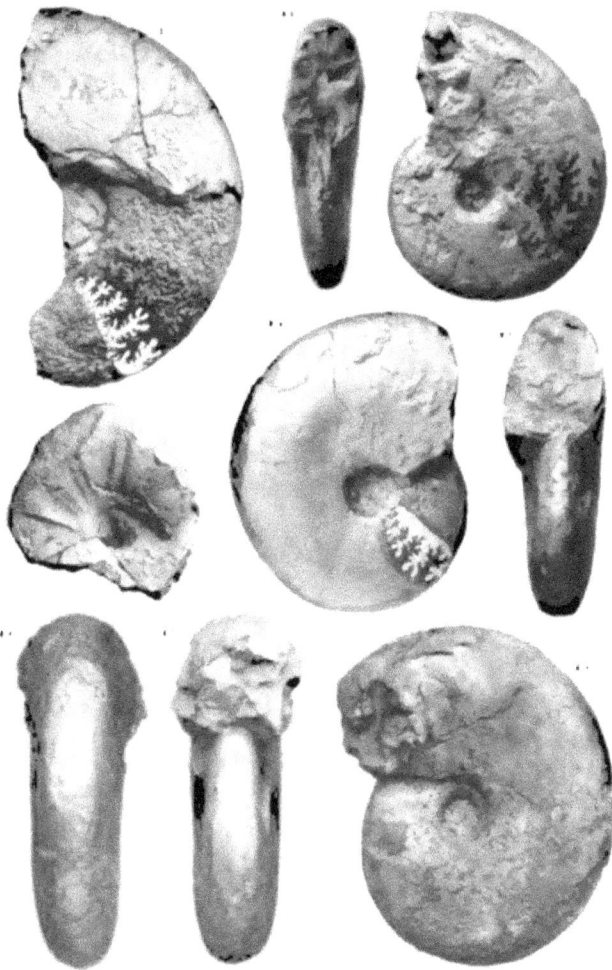

PLATE XIII.

All specimens from the exotic blocks 16 and 17.

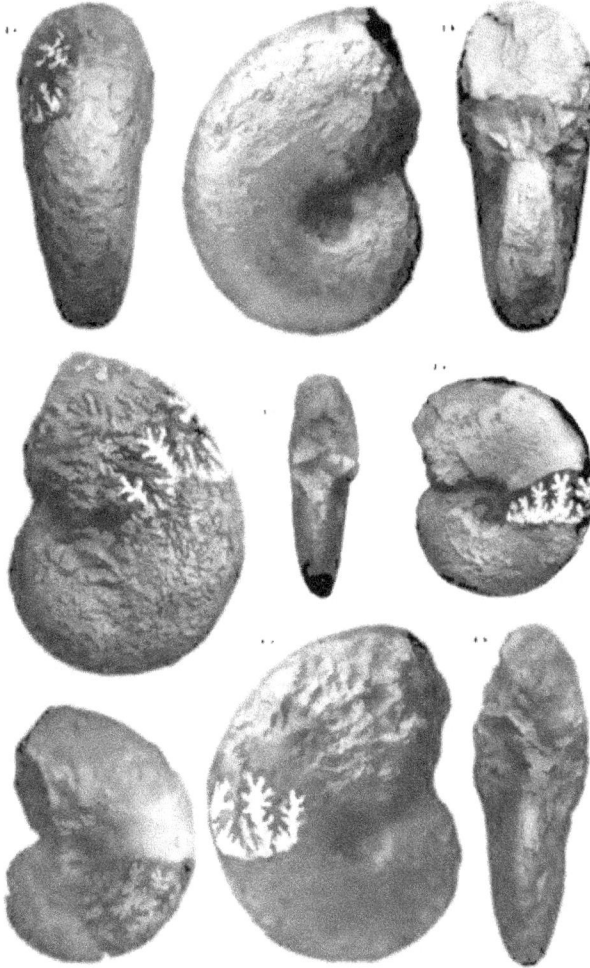

All specimens from the exotic blocks 16 and 17.

Pl. XV.

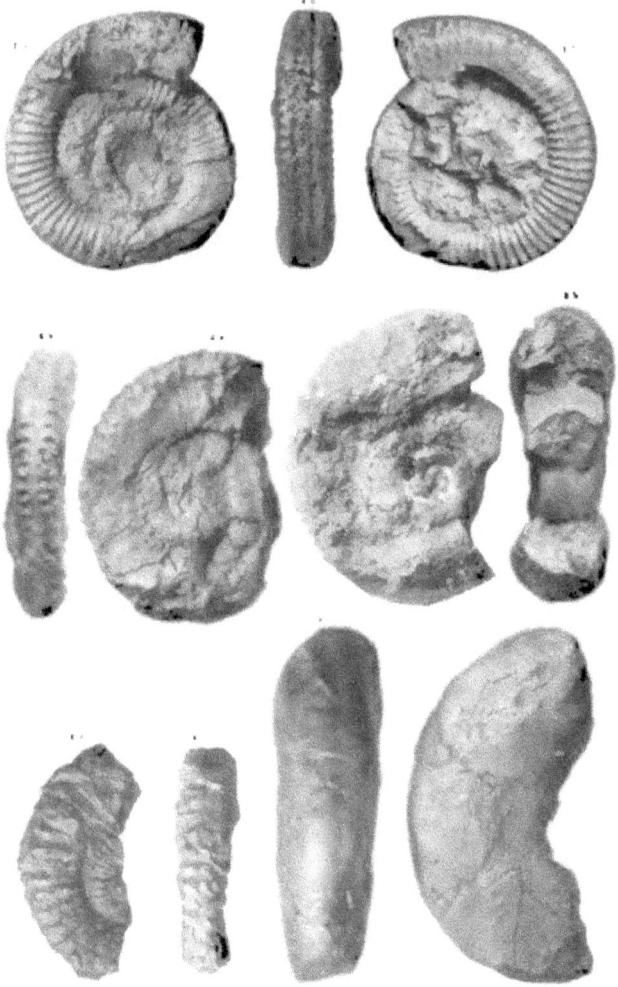

Pl. XVI.

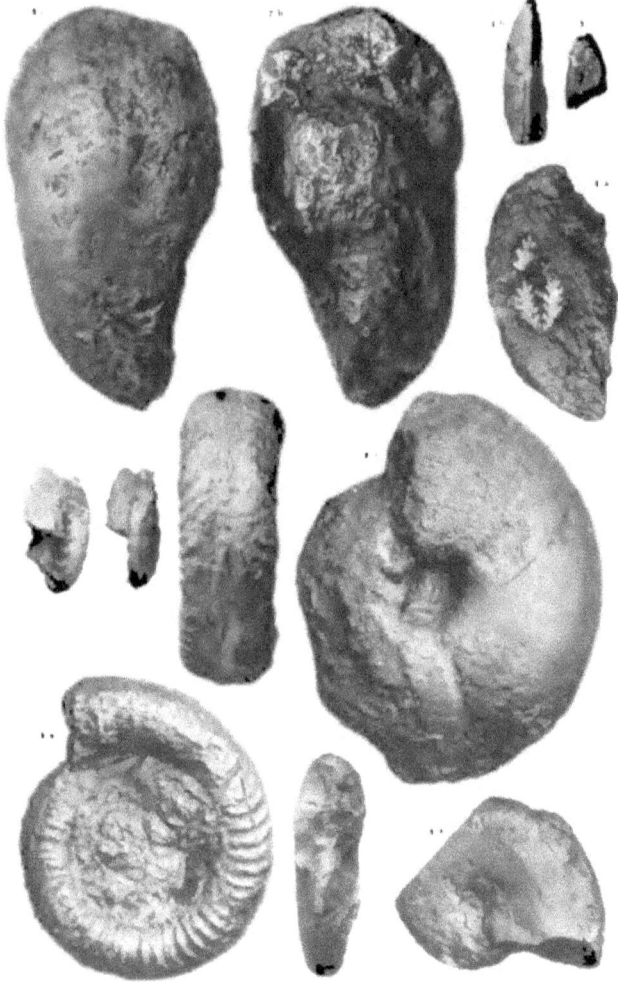

MEMOIRS

OF

THE GEOLOGICAL SURVEY OF INDIA.

Palæontologia Indica,

BEING

FIGURES AND DESCRIPTIONS OF THE ORGANIC REMAINS PROCURED DURING THE
PROGRESS OF THE GEOLOGICAL SURVEY OF INDIA.

PUBLISHED BY ORDER OF HIS EXCELLENCY THE GOVERNOR GENERAL OF INDIA IN COUNCIL.

Ser. XV.

HIMÁLAYAN FOSSILS.

Vol. I, Part 2.

ANTHRACOLITHIO FOSSILS OF KASHMIR AND SPITI.

By CARL DIENER, Ph.D.,

Professor of Geology at the University of Vienna.

PLATES I TO VIII.

CALCUTTA:
SOLD AT THE
GEOLOGICAL SURVEY OFFICE.
LONDON: KEGAN PAUL, TRENCH, TRÜBNER & CO.

MDCCCXCIX.

PRINTED BY THE SUPERINTENDENT OF GOVERNMENT PRINTING, INDIA, 8, HASTINGS STREET, CALCUTTA.

MEMOIRS OF THE GEOLOGICAL SURVEY OF INDIA.

HIMÁLAYAN FOSSILS.

VOL. I, PART 2.

ANTHRACOLITHIC FOSSILS OF KASHMIR AND SPITI.

HIMÁLAYAN FOSSILS.

VOLUME I, PART 2.

ANTHRACOLITHIC FOSSILS OF KASHMIR AND SPITI.

BY

C. DIENER, Ph.D.,

Professor of Geology at the University of Vienna.

WITH PLATES, I-VIII.

INTRODUCTION,

In the introduction to my monograph on the fauna of the permian Productus shales of Johár and Painkhánda it was proposed to devote a proper part of this volume to the description of the fossil contents of the Kuling shales of Spiti and of the Zewán or Barus beds of Kashmir, which have been considered to be of carboniferous age by previous authors.

Even then it seemed to me highly probable that representatives of both the carboniferous and permian systems were mixed together among the fossils contained in the Geological Survey's Himálayan collections from the upper palæozoic rocks of Kashmir and Spiti. This probability has been greatly strengthened by a closer examination of the fossil material entrusted to me for description. In my opinion the Kuling shales of Spiti, or more exactly, their lower portion underlying the triassic *Otoceras* stage, correspond stratigraphically to the Productus shales of Johár and Painkhánda, as has already been suggested by Griesbach, and must consequently be correlated with the permian system. In Kashmir this system seems to be likewise represented by black micaceous shales, observed by Lydekker on a ridge N.E. of Prnagam Trál, whereas the large bulk of fossils from the Zewán or Barus beds are of upper carboniferous age, occupying, as it seems, the very highest stratigraphical position within the carboniferous system.

In spite of this it is impossible to fix the exact stratigraphical zone of every fossil in the Geological Survey's collections from the upper palæozoic rocks of Kashmir and Spiti. Notwithstanding the rather limited number of species, regarding which their geological age cannot be settled definitively at present, I deemed it preferable to let this uncertainty find an expression in the title of the present monograph. I have used the term " *Anthracolithic* " under which I understand both the carboniferous and permian fossils. This term, which I find very convenient,

B

considering the intimate stratigraphical and faunistic connection of the carboniferous, and permian systems of the Himálayas, was originally introduced by Waagen in the "Geological Results" of his Salt Range Fossils (Vol. IV, 1891, p. 241). I am glad to avail myself of it as one of the best denominations in our stratigraphical nomenclature, and I sincerely wish that it may be used more generally in geological literature than has been done hitherto.

Before entering into a detailed description of the anthracolithic fossils of Kashmir and Spiti, a few notes on the previous geological literature on the subject may be found useful.

The first reference to the existence of carboniferous rocks in the Himálayas was made by Dr. Hugh Falconer in 1838, who proved the carboniferous age of a lime-stone in the Kashmir Valley.[1]

In 1850 W. King[2] described the first anthracolithic fossil from the North-Western Himálaya, *Strophalosia Gerardi*, which had been collected by Dr. Gerard on the crest of a pass, leading from Ladakh into Bisáhir, at a height of 17,000 feet.

Among the fossils, picked up by Dr. Gerard in Spiti and entrusted by him to the Asiatic Society of Bengal in Calcutta, a large number was subsequently proved to be of anthracolithic age. One of the most common shells in his collection has been described as *Spirifera Rajah* by J. W. Salter[3] in 1865.

Although he never lost sight of its being a true carboniferous form, closely allied to *Spirifer Keilhavii*, von Buch, he erroneously inferred that it had been derived from the triassic beds of the Spiti-Pass. In an appendix to the same work, however, Mr. H. F. Blanford[4] correctly observed that *Spirifer Rajah* did not occur in the same bed with triassic ammonites described by himself in 1863,[5] but decidedly below them—" in beds, which other evidence combines to show, must be referred to the same general relative age, as the carboniferous of Europe."

Blanford's view regarding the stratigraphical position of the beds with *Spirifer Rajah* was fully confirmed by Dr. Ferdinand Stoliczka, who in 1864 had examined a number of geological sections in Spiti and Rupshu. Among the palæozoic rocks of Spiti three different series, the Bábeh series, Muth series and Kuling series were distinguished by that learned author. The Bábeh and Muth series he correlated to the silurian, the Kuling series to the carboniferous system of Europe. The prevalent rocks of the latter series he found to consist of "a dark brown crumbling shale and a light coloured, mostly whitish quartzite, generally speaking very difficult to distinguish from the top beds of the Muth series." The total thickness of these beds, considered by him to be carboniferous, he estimated to be from 100 to 400

[1] Palæontological Memoirs of Hugh Falconer, in "Official Report of Expedition to Kashmir and Little Tibet in 1837-38." Vol. I, p. 667.

[2] W. King, "A monograph of the permian fossils of England," London, 1850, p. 96, Pl. XIV, figs. 6, 7.

[3] J. W. Salter and H. F. Blanford, "Palæontology of Niti in the Northern Himálayas," Calcutta, 1864, p. 59.

[4] H. F. Blanford, ibidem, p. 111.

[5] H. F. Blanford, "On Dr. Gerard's collection of fossils from the Spiti valley in the Asiatic Society's Museum, Journal Asiat. Soc. of Bengal, 1863, No. 2, pp. 124—128.

feet. He records the following list of fossils, collected by himself and Dr. Gerard[1]:—

Spirifer Moosakheelensis, Davids.
　　,,　　Keilhavis, v. Buch (= Sp. Rajah, Salt.)
　　,,　　tibeticus, Stoliczka.
　　,,　　altivagus, Stol.
Productus Purdoni, Davids.
　　,,　　semireticulatus, Mart.
　　,,　　longispinus, Sow.
Avicula, sp.
Cardiomorpha, sp.
Aviculopecten, sp.
Orthoceras,? sp.

Among the species of *Lamellibranchiata* not a single one permits of a specific determination. A single indistinct cast has been identified as *Productus Purdoni* but it does not warrant a decided determination. The species mistaken for *P. semireticulatus* and for *P. longispinus* belongs to the subgenus *Marginifera*, Waagen, and is one of the leading fossils of the Kuling shales of Spiti and of the corresponding beds in Kashmir.

Whether *Spirifer tibeticus* and *Sp. altivagus* are really of anthracolithic age is very doubtful. The original geological position of the specimens collected by Dr. Gerard is not known. Stoliczka himself found only one loose specimen of *Spirifer tibeticus* near Kibber. On the other hand, *Spirifer tibeticus* is so closely allied to *Sp. Griesbachi*, Bittner,[2] of upper triassic age, that a distinction between them is very difficult. This question will be more fully discussed by Dr. Bittner in his memoir on the triassic Brachiopoda and Lamellibranchiata of the Himálayas (Palæontologia Indica, ser. XV, Vol. III, Pt. 2).

Stoliczka's notes on the stratigraphical sequence in Spiti have been partly corrected by R. D. Oldham and C. L. Griesbach. Oldham[3] in his interpretation of the Spiti sections inferred that Stoliczka's "Muth-quartzite" should be rather correlated with the carboniferous quartzite of Kashmir, than with the silurian system. Griesbach[4] confirmed this view and distinguished the following sequence of beds in the anthracolithic series of Spiti.

The dark unfossiliferous limestones, which rest on the flesh coloured quartzite series of upper silurian age and which probably correspond to the devonian system, are conformably overlaid by earthy, grey, crinoid-limestones from 600 to 800 feet in thickness. The red crinoid-limestone is overlaid by a fine-grained, white

[1] F. Stoliczka, "Geological sections across the Himálayan Mountains from Wangtu bridge on the River Sutlej to Sungdo on the Indus, etc.," Memoirs Geol. Surv. of India, Vol. V. Pt. 1, pp. 33—79.

[2] C. Diener, "Ergebnisse einer Geologischen Expedition in den Central Himálaya, etc.," Denkschr. Kais. Akad. d. Wiss. Wien, math. nat. Cl., 1895, Bd. LXII, p. 658.

[3] R. D. Oldham, "Some notes on the geology of the N.W. Himálayas," Records, Geol. Surv. of India, Vol. XXI, 1888, pp. 151—163.

[4] C. L. Griesbach, Records of the Geol. Surv. of India, 1889, Vol. XXII, pp. 155—167, and "Geology of the Central Himálayas," Mem. Geol. Surv. of India, Vol. XXIII, 1891, pp. 313—323.

quartzite of about 500 feet in thickness, which Stoliczka originally included in his
Muth series. This sequence of beds is exactly the same as in the Central Himálayas of Kumaon and Gurhwál. As in the eastern sections, the entire series is
characterised by the scarcity of organic remains. In the Geological Survey's
Himálayan collections this series is not represented by a single fossil which
would permit of a specific determination. This remark, unfortunately, likewise
applies to the next rock-group, a grey limestone which, Griesbach states, overlies
the white quartzite conformably. It is a "hard, splintery, grey limestone, in flaggy
beds of a total thickness of about 70 feet, which has yielded numerous fossils,
though few in species. Amongst them are several *Producti, Athyris Royssii* and
Corals. Its evident connection with the white quartzite and the character of the
fossils define its upper carboniferous age."

It is very much to be regretted that Griesbach's collection does not contain
fossils from this grey limestone exposed in the Pin river section near Muth, more
especially so because this horizon seems to be absent in the Central Himálayas of
Kumaon and Gurhwál and might perhaps be a representative of the Zewán or
Barus beds of Kashmir.

The grey limestone near Muth is overlaid by Stoliczka's Kuling series. In
this series two groups of a geologically different age have been included by that
author. The upper portion has yielded the characteristic fossils of the *Otoceras*
stage and consequently belongs to the scythian' series of the triassic system. The
lower portion, consisting of dark, crumbling, often micaceous shales, alternating
irregularly with sandstone-partings, Griesbach considers to be equivalent of the
permian *Productus shales* of Johár and Painkhánda.

I am of opinion that the local denomination of "*Kuling shales*," given by
Stoliczka, might advantageously be retained for these beds, owing to the claim of
priority, although beds of lower triassic age had been originally included in Stoliczka's
"Kuling series." But a restriction of the original name to the well defined horizon,
included between the grey limestone of Muth and the *Otoceras* beds, is not contrary
to the laws of stratigraphical nomenclature. Among more recent instances I only
need mention the interpretation of the term "Partnach Schichten" by Skuphos,
who restricts this name to the lower portion of Gümbel's Partnach beds and
which has met with the unanimous approval of all Alpine geologists.

The existence of true anthracolithic rocks in the Kashmir Valley, which had
been first supposed by Dr. Hugh Falconer in 1838, was definitely proved in a most
important paper by Captain Godwin-Austen and Th. Davidson, an abstract of which
appeared in 1864 in the Quarterly Journal of the Geological Society of London.[2]
The original paper was, however, only published in 1866.[3] The sections of the
fossiliferous rocks near Wasterwan, Barus, Loodoo and Khoonmoo, on the eastern

[1] E. v. *Mojsisovics,* W. *Waagen,* and C. *Diener.* "Entwurf einer Gliederung der pelagischen Sedimente des
Trias System," Sitzungsber. kais. Akad d. Wissensch Wien. math. nat. Cl. Bd. CIV., 1895. p. 1278.

[2] *Godwin-Austen,* "Geological notes on part of the N.W. Himálayas," Quart. Journ. Geol. Soc. Vol. XX,
1864. pp. 383-387

[3] On the carboniferous rocks of the valley of Kashmir, with notes on the brachi-poda, collected by Captain
Godwin-Austen in Tibet and Kashmir, by Th. *Davidson,* ibidem, Vol. XXII, 1866, pp. 39-48.

side of the Kashmir Valley, S. of Srinagar, were described in detail by Captain Godwin-Austen. The brachiopods which were all obtained from the Kashmir Valley, and not partly from Little Tibet, as stated erroneously in his memoir, have been examined by Professor Davidson. The fauna described by this eminent author and considered as carboniferous, consists of the following forms, excluding those which were too badly preserved to permit of a specific determination :—

Terebratula (= *Dielasma* ?) *Austeniana*, Dav.
    „   (*Dielasma*) *sacculus* ?, Mart.
Athyris subtilita, Hall.
Spirifer Moosh, Salter.
    „   *Fritsana*, Davids.
    „   *Kashmericnsia*, Davids.
    „   *Moosakhylensis*, Davids.
    „   *sp. ind.* (= *Lydekkeri*, Diener).
Rhynchonella pleurodon var. Davreuxiana, de Kon.
    „   *Kashmericnsis*, Davids.
    „   *Barusicnsis* (misnamed : *Barusmensis*), Davids.
Streptorhynchus crenistria, Phill. (= *Derbya cf. senilis*, Phill.)
Productus semireticulatus, Mart.
    „   *Cora*, d'Orb.
    „   *scabriculus*, Mart.
    „   *Humboldti*, d'Orb. (?).
    „   *longispinus*, Sow. (?).
    „   *striatus*, Fischer, (?).
    „   *spinulosus*, Sow.
    „   (?) *laevis*, Davids.
Itieria Kashmeriensis, Davids.
Chonetes Hardrensis var. Tibetensis, Davids.   (recte *Kashmeriensis*, Lydekker).
Chonetes laevis, Davids.
    „   *Austeniana*, Davids.
    „   (*Spirifer*, Davids.) *Barusicnsis*, Dav.

Altogether 25 species, of which 13 only, or about one half of this number, are also contained in the Geological Survey's Himálayan collections from Kashmir.

Professor Davidson sums up his views regarding the geological age of the beds from which these fossils were obtained, in the following remarks, (l. c. p. 40) :—

"Here again we find many of our common and widely spread European and American species, along with a few, that had not yet been noticed from other parts of the world, and which indicate that the carboniferous rocks of Tibet, Kashmir and the Punjab belong to one great formation."

Another very important paper on the geology of the anthracolithic region to the East of the Kashmir Valley was published by Dr. A. Verobère in 1866 and 1867.[1] He stated that the fossiliferous series is underlaid by slates, in which there

* A. Verobère, "Kashmir, the Western Himálaya and the "Afghan Mountains," Journal Asiatic Soc. of Bengal, Calcutta, 1866, Vol. XXXV, pt. 2, pp. 89-134, 169-200, 1867, Vol. XXXVI, pt. 2, pp. 201-229.

is an abundance of contemporaneous volcanic rock. The limestones and shales which rest on these volcanic rocks have been divided by Verchère into three divisions, called by him (in ascending order) Zeeawán beds, Weean beds and Kothair beds. The term "Zeeawán beds" is, however, taken in a wider sense than in Godwin-Austen's memoir, the Zewán beds having been restricted by the latter author to a distinct horizon of the anthracolithic series only, characterised by its abundance of *Fenestella*. Both the Zeeawán beds and Weean beds are considered to be carboniferous by Verchère, whereas the Kothair beds were placed by him in the triassic system.

The numerous fossils collected by Dr. Verchère in the anthracolithic series of Kashmir Valley have been studied partly by himself and partly by E. de Verneuil, but the results of these studies have, unfortunately, been published in a form which renders them almost useless.

The following species are quoted by Verchère from the Zewán beds :—

Nautilus Pleniepianus, de Kon.
Terebratula (Dielasma) sacculus, Mart.
Spirifer Verchèrei, de Verneuil.
  ,,   *striatus*, Mart.
  ,,   *Moosakheylensis*, Davids.
  ,,   *Rajah*, Salter.
Spiriferina octoplicata var. *transversa*, Verch.
Athyris ambigua, Hall.
  ,,   *Buddhista*, Verch.
     cf. *Royssii*, Lev.
Retzia (Eumetria) grandicosta, Davids.
Orthis recmpinata, Mart.
  ,,   *sp. ind.*
Strophomena analoga, Phill.
Productus costatus, Sowerby.
  ,,   *semireticulatus*, Mart.
  ,,   *Cora*, d'Orb.
  ,,   *Humboldti*, d'Orb.
  ,,   *Purdoni*, Davids.
  ,,   *lingipianus*, Sow.
  ,,   *Boliviensis*, d'Orb.
  ,,   *scabratus*, Mart.
Strophalosia (?) arachnoidea, Verch.
Fenestella Sykesii, de Kon.
  ,,   *sp. ind.*
  ,,   *megastoma*, de Kon.
Vincularia mollangolaris, ? Portl.
Dictrichia ? sp. ind.
Acanthocladia sp. ind.
Retepora lepida, de Kon.
Aberolites acptora, ? Flem.

Both the descriptions and figures (Pl. I. to X. Journ. Asiat. Soc. of Bengal, Vol.

XXXVI, Pt. 2) are so unsatisfactory that I have only been able to identify such forms as are represented in Verchère's collection entrusted to me for examination by the Director of the Geological Survey of India. For the following species the identification seems to be pretty certain :—

> Spiriferina cf. Kra'ickeana, Shum.
> Spirifer Moosakhylensis, Dav.
>   ,,   Rajah, Salter.
> Athyris subtilita, Hall.
> Retzia (Eumetria) cf. grandicosta, Dav.
> Productus semireticulatus, Mart.
>   ,,   Cora, d' Orb.
>   ,,   aculeatus, Mart.

Orthis resupinata, Verch., is probably identical with O. indica, Waagen. Athyris Buddhista seems to be really a new species. Spirifer Vercherei, on the contrary, must be removed from the list of independent species, having been founded on strongly weathered waterworn specimens of Sp. Rajah only, which have lost the original details of their ornamentation.

There can be no doubt about the identity of Verchère's Zewán beds—I prefer to adopt the spelling of this name as given by Captain Godwin-Austen—with the anthracolithic rock-group, from which the brachiopods, collected by Godwin-Austen and studied by Davidson, were obtained. The case is different with the Weean beds of Verchère. Neither from Verchère's description of the Kashmir sections, nor from his list of the fossil contents of this rock-group is it possible to make out whether they actually belong to the anthracolithic system or to one of the younger horizons in the stratigraphical sequence (Lydekker's "Supra-Kuling series").

The following species are quoted by Verchère from the Weean beds of Kashmir :—

> Goniatites gange'icus, L. de Kon.
> Nautilus clivellatus, ? Sow.
> Solenopsis imbricata, de Kon.
>   ,,   nov. sp.
> Cardinia Himalayana, Verch.
>   ,,   ovalis, ? Mart.
> Cucullaea, sp.
> Pecten, sp.
> Aviculopecten dissimilis, Phon.
>   ,,   acatus, Verch.
>   ,,   ranus, Verch.
>   ,,   circularis, Verch.
>   ,,   sp. ind.
>   ,,   testudo, Verch.
>   ,,   gibbosus, Verch.
> Arcanus, nov. sp.
> Spiriferina Strackeyi, Salt.
> Productus lineris, Davids.
> Chonetes Burasicanus, Davids.

With the exception of the brachiopods, all the type specimens quoted in the preceding list are contained in the Geological Survey's collection. Among them a single one only, *Goniatites gangeticus* from Banda, can be safely identified. It is identical with *Daonbites nivalis*, Dion., one of the leading fossils of the Himálayan *Subrobustus* beds of lower triassic age. Attention has been drawn to this interesting fact in my memoir on the Cephalopoda of the lower trias (Vol. II, pt. I, of the present series). The rest of specimens are all undeterminable fragments, quite unfit for a specific determination. The Lamellibranchiata especially are so poorly preserved casts that it would be perfectly useless to have them figured. I can only say that it is absolutely impossible to derive from them any satisfactory conclusion as to the geological age of the strata in which they occur. Among the brachiopods, *Spiriferina Strackeyi* points to a triassic age, this species having been found in the Lilang series of Spiti by Stoliczka. Thus beds of triassic age have undoubtedly been included in the Wrean group by Dr. Verchère, but it is impossbile to decide whether this group represents the lower trias only or may also include the topmost portion of the anthracolithic system, as was found to be the case with Stoliczka's Kuling series in Spiti.

Both Godwin-Austen's and Verchère's views of the stratigraphical sequence in the Kashmir sections were partly modified by Lydekker[1] from whose reports on the geological survey of the Kashmir, Ladakh and Chamba territories much valuable information may be obtained, although he was not able to establish a safe classification of the upper palæozoic and mesozoic rocks (his "Zanskar system"), based on palæontological evidence. Neither did he succeed in separating the triassic and anthracolithic systems, nor did he recognise the lower trias and the typical Kuling shales, although both of these horizons are certainly present in Kashmir, as was clearly proved by an examination of the fossils in the Geological Survey's Himálayan collection.

The most important additions to our knowledge of the anthracolithic system in Kashmir, for which we are indebted to Mr. Lydekker, are the following:—

The fossiliferous Zewán or Barus beds, the total thickness of which varies from 30 to 280 feet, rest conformably on a compact white quartzite, which is considered to be the equivalent of the carboniferous white quartzite of the Central Himálayas of Oldham and Griesbach. This quartzite is generally underlaid by massive amygdaloidal and other traps, which frequently, when the bottom quartzitic bed is less strongly developed, pass insensibly upwards into the fossiliferous strata. Although Lydekker hints at the possibility of these traps with their associated slates being at least partly of carboniferous age, be preferred to class them with his Panjal system, which corresponds to the older palæozoic rocks in other parts of the world. These slates, characterised by the abundance of contemporaneous volcanic rock, were found to be underlaid by conglomeratic slate, very similar to the Blaini conglomerate of the Simla sections, composed of subangular fragments and rounded pebbles of slates and quartzites imbedded in a matrix of fine-grained slate.

[1] R. *Lydekker*, "The geology of the Kashmir and Chamba territories and the British district of Khágán," Memoirs Geol. surv. of India, Vol. XXII, 1883, Chapters VI, VII. I have not cited the previous papers of this author published in the Records of the Geol. Surv. of India, as their contents are embodied in the memoir quoted.

Whereas Lydekker considered this conglomerate, the glacial origin of which he advocated, to be of older palæozoic age, R. D. Oldham[1] compares the Kashmir conglomerate to the boulder bed of the Salt-Range and consequently refers it to the carboniferous system.

The following fossils from the anthracolithic rocks of Kashmir have been figured by Lydekker, but were not described in detail :—

Protoretepora ampla, Lonsdale.
Productus semireticulatus, Mart.
,, Humboldti, d'Orb.
Spirifer striatus, Mart.
Phillipsia cf. seminifera, Phill.

These determinations, which seem to have been chiefly quoted on the authority of Dr. Feistmantel, have been considerably modified by my revision of Lydekker's fossils. For the following three forms an identification can be established with certainty :—

Productus semireticulatus = Marginifera himalayensis, Diener.
,, Humboldti = P. Abichi, Waagen.
Spirifer striatus = Sp. Lydekkeri, Diener.

Mr. Lydekker was also the first to draw attention to the occurrence of the genus Lyttonia in the anthracolithic rocks of the Kashmir Valley.[2]

In 1891 Professor W. Waagen published the geological results of his examination of the Productus limestone fossils of the Salt-Range. In this memoir[3] he briefly discusses the brachiopoda of the Zewan or Barus beds of Kashmir described by Davidson. He remarks that the percentage of truly carboniferous forms, that is to say, of mountain limestone forms, is far larger among them than in the fauna of the Amb beds (lower Productus limestone) of the Salt-Range, and that in the meantime slight affinities to Australian forms were indicated by the presence of Spirifer Vihianus, Dav., and Sp. Kashmeriensis, Davids.

Among the species quoted by Davidson, Waagen found only two identical with Salt-Range forms: Athyris subtilita = Spirigerella Derbyi, Waagen, and Spirifer Musakheylensis, Dav. To these species a third one, Discinisca Kashmeriensis, might perhaps be added, as its affinity to D. Warthi, Waag., amounts almost to identity. On the strength of this evidence Waagen came to the conclusion that "the Kashmir carboniferous strata should either be placed on a level with the lower speckled sandstone of the Salt-Range, or else they should be considered as intermediate in age between the latter and the lower Productus limestone or upper speckled sandstone."

There are several points on which I differ from the views of that learned author; these differences will be noticed in the descriptions of the Kashmir fossils in their proper places.

[1] A Manual of the Geology of India, 2nd Edition, by R. D. Oldham, Calcutta, 1893, p. 134.
[2] Records Geol. Surv. of India, Vol. XVII, 1889, p. 87.
[3] W. Waagen, Salt-Range fossils, Palæont. Indica, ser. XIII, Vol. IV, Geological Results, pp. 165, 166.

The material for the present memoir consists of the fossils collected in the Kuling shales of Spiti by Dr. Gerard, Stoliczka and C. L. Griesbach, of parts of the collections made by Captain Godwin-Austen, Dr. Verchère and Major Collet in Kashmir, and, last but not least, in the rich collections, brought together from the anthracolithic rocks of Ladakh and Kashmir by R. Lydekker, with the type specimens, figured in Vol. XXII, of the Memoirs of the Geological Survey of India.

The palæontological literature, which I chiefly consulted, when working out Parts II, III, and IV of this volume, is given in the following list.

PALÆONTOLOGICAL LITERATURE.

1809. *Martin*, "Petrificata Derbyensia, or figures and descriptions of petrifactions, collected in Derbyshire," Wigan.

1814. *Sowerby*, "Mineral Conchology of Great Britain." Vol. I, London.

1817. *Schlotheim* "Beiträge zur, Versteinerungskunde," Denkschriften der Kgl. Akademie d. Wissensch. München, Bd. VI.

1820. *Parkinson*, "Organic remains of a former world," Vol. I, London.

1826-44. *Goldfuss*, Petrefacta Germaniæ, Abbildungen und Beschreibungen der Petrefacten Deutschlands und der angrenzenden Laender.

1834. *Phillips*, Illustrations of the Geology of Yorkshire, 2nd edition.

1837. *Fischer von Waldheim*, Oryctographie du Gouvernement de Moscou.

1839-43. *A. d'Orbigny*, Voyage dans l'Amérique Méridionale, Paléontologie, T. III.

1840. *L. v. Buch*, Essai d'une classification des Delthyris, Mém. Soc. Geol. de France, Vol. IV.

1843-44. *L. de Koninck*, Description des animaux fossiles, qui se trouvent dans le terrain carbonifère de Belgique. Mem. de la Soc. Royale des Sciences de Liége.

1843-44. *Keyserling*. Beitrag zur Palæontologie Russlands, Verhandlungen Kais. Russ. Mineral. Ges. St. Petersburg.

1844. *M'Coy*, Synopsis of the characters of the carboniferous fossils of Ireland.

1844. *Ch. Darwin*, Geological observations on the volcanic islands, visited during the voyage of H. M. S. Beagle. Appendix, Descriptions of fossil shells and corals from Van Diemensland, by G. B. Sowerby and W. Lonsdale.

1845. *Strzelecki*, Physical Description of New South Wales and Van Diemensland. Palæontology by Morris.

1845. *Murchison, E. de Verneuil et A. de Keyserling*, Géologie de la Russie d'Europe et des Montagnes de l'Oural, Vol. II, Paléontologie.

1846. *L. v. Buch*, Ueber *Spirifer Keilhavii*, ueber dessen Fundort und Verhaeltniss zu ahnliche Formen, Abhandlge. Koenigl. Akad. d. Wissensch. Berlin, p. 65.

1846. *A. Graf Keyserling*, Wissenschaftliche Beobachtungen auf einer Reise in das Petschoraland, St. Petersburg.

1847. *L. de Koninck*, Monographie des genres Productus et Chonetes, Mém. de la Soc. Royale des Sciences de Liége.

1847. *M'Coy*, On the fossil botany and zoology of the rocks associated with the coal of Australia, Annals and Magazine of Nat. History, New York, 1 ser. Vol. XX.

1848. *L. de Koninck*, Nouvelles notices sur les fossiles du Spitzbergen, Bull. Acad. Royale de Belgique, Vol. XVI, pt. 3.

1849. *Dana*, Descriptions of fossil shells obtained by the U. S. Exploring Expedition in Australia.

1850. *King*, A monograph of the permian fossils of England, Palæontographical Society, London.

1851-85. *Th. Davidson*. A monograph of the British fossil Brachiopoda, Palæontographical Society, London.

1852. *Owen*, Report of the Geological Survey of Wisconsin, Iowa and Minnesota.

1852. *J. Hall* in *Howard Stansbury's* Report of an Exploration of the Valley of the Great Salt Lake in Utah, Philadelphia.

1854. *Shumard* in *B. Marcy's* Exploration of the Red River of Louisiana.

1856. *Shumard* in 1st and 2nd Annual Report Geol. Survey of Missouri, p. 186.

1855. *Rückmall*, Lethaea Rossica. Ameurena Périoda, Vol. I.

1855. *Norwood and Pratten*, Notice of Productus and Chonetes as found in the Western States and Territories. Journ. Acad. of Nat. Sciences of Philadelphia, Vol. III, pp. 5.

1855. *Norwood and Pratten*, Fossils from the Carboniferous series of the Western States, ibidem, p. 71.

1857. *Lyon, Cox and Lesquereux*. Palaeontological Report of Kentucky.

1857-58. *Shumard and Swallow*, Descriptions of new fossils from the Coal Measures of Missouri and Kansas. Transactions, Acad. of Sciences, St. Louis, Vols. I and II.

1858. *Hall*, Report on the Geological Survey of Iowa, Vol. I. Palaeontology.

1858. *Marcou*, Geology of North America. Zürich.

1860. *Grunewaldt*, Beiträge zur Kenntniss der sedimentären Gebirgsformationen in des Berghaupt mannschaften Yekatherinburg, Slatoust, etc., Mém. Acad. Imper. des sciences de St. Petersbourg, ser. VII, Vol. II.

1861. *Geinitz*, Die Dyas I. Abth.

1861. *Salter*, On the fossils from the High Andes, collected by E. Forbes, Quart. Journ. Geol. Soc. Vol. XVII, p. 62.

1862. *Davidson*, On some carboniferous Brachiopoda, collected in India by A. Fleming and W. Purdon. Quart. Journ. Geol. Soc. London, Vol. XVIII, p. 25.

1862. *L. de Koninck et Th. Davidson*, "Memoirs sur les fossiles paléozoiques recueillis dans l'Inde."

1863. *Davidson*, On the lower carboniferous Brachiopoda of Nova Scotia, Quart. Journ. Geol. Soc., London, Vol. XIX, p. 158.

1863. *F. Roemer*, Ueber eine marine Conchylienfauna im productiven Steinkohlengebirge Oberschlesiens, Zeitschr. d. Deutschen Geol. Gesellschaft, Bd. XV, p. 567.

1863. *Geinitz*, Beiträge zur Kenntniss des organischen Ueberreste der Dyas, Neues Jahrb. f. Mineral, p. 385.

1864. *Meek and Hayden*, Palaeontology of the Upper Missouri, Smithsonian Contributions of Knowledge, Vol. XIV, p. 1.

1864. *Meek in Whitney*, Geological Survey of California, Vol. I. Palaeontology.

1865. *Salter and Blanford*, Palaeontology of Niti in the Northern Himalayas, Calcutta.

1865. *Beyrich*, Ueber eine Kohlenkalk-Fauna von Timor, Abhandlgn. Koenigl Akad. d. Wissensch. Berlin, 1864, p. 61.

1866. *Geinitz*, Carbonformation und Dyas in Nebraska, Nova Acta Academ. Caes. Leopaldimae Carolinae, Vol. XXXIII, p. 1.

1866. *Davidson*, Notes on the carboniferous brachiopoda, collected by Capt. Godwin-Austen in the Valley of Kashmere, etc., Quart. Journ. Geol. Soc., London, Vol. XXII, pp. 36—43.

1866-75. *Meek and Worthen*, Palaeontology of Illinois, Geol. Surv. of Illinois, Vols. I—VI.

1867. *Verchère*, Kashmir, the Western Himalaya and the Afghan Mountains, Journ. Asiat. Soc. of Bengal, Calcutta, Vol. XXXVI, Pt. 2.

1867. *Fal. v. Moeller*, Ueber die Trilobiten der Steinkohlenformation des Ural, Bull. de la Soc. imper. des Naturalistes de Moscou.

1867. *Trautschold*, Crinoides und andere Thierreste des jungeren Bergkalks im Gouvernement Moskau. Bull. Soc. Imper. des Nat. de Moscou, Vol. XI, No. 3, p. 1.

1867. *Davidson and Thomson*, Description of the Carboniferous Brachiopoda of Camphletown, Trans. Geol. Soc. of Glasgow, Vol. II, pp. 46, 149.

1857-60. *McChesney*, Description of new species of fossils from the palaeozoic rocks of the Western States. Transactions, Chicago Acad. of Sciences, Vol. I, p. 1.

1857-69. *White and St John*, Description of new subcarboniferous and coal-measure fossils, etc., ibidem. Vol. I, p. 115.

1869. *Toula*, Ueber einige Fossilien des Kohlenkalkes von Bolivia, Sitzungsber. Kais. Akad. d. Wissensch. Wien, math. nat. Cl. Bd. LIX, p. 433.

1870. *Meek*, Description of new species from the palaeozoic rocks of the Western States, Proceedings, Acad. of Science, Philadelphia, p. 22.

1870. *Roemer*, Geologie von Oberschlesien, Breslau.

1872. *Etheridge*, Description of the palaeozoic and mesozoic fossils of Queensland, Quart. Journ. Geol. Soc., London, Vol. XXVIII, p. 317.

1872. *Meek*, Palaeontology of Eastern Nebraska, in *Meek and Hayden*, Final Report upon the U. S. Geol. Surv. of Nebraska, Washington.

1873. *L. de Koninck*, Monographie des fossiles carbonifères de Bleiberg en Carinthia, Mém. de la Soc. Royale des Sciences de Liége.

1873. *Toula*, Kohlenkalk-Fossilien von der Südspitze von Spitzbergen, Sitzungsber. Kais. Akad. d. Wissensch. Wien, math. nat. Cl. Bd. LXVIII. p. 267.

1873. *Meek*, Report of the Geological Survey of Ohio, Vol. II, Palæontology.

1874. *Derby*, On the carboniferous brachiopoda of Itaitaba, Brasil. Bull. Cornell University, Ithaca, Vol. i, No. 2.

1874-79. *Tromschold*, Die Kalkbrüche von Mistschkowa, Mém. Soc. Imper. des Nat. de Moscou.

1875. *Toula*, Permocarbon-Fossilien von der West Küste von Spitzbergen, Neues Jahrb. f. Min. p. 225.

1875. *Toula*, Kohlenkalk und Zechstein-Fossilien aus dem Hornsund an der S. W. Küste von Spitzbergen Sitzgsber. Kais. Akad d. Wiss. Wien, math. nat. Cl. Bd. LXX. p. 123.

1875. *Toula*, Eine Kohlenkalkfauna von den Barents Inseln (Nowaja Semlja), Ibidem, Bd. LXX, p. 527.

1877. *White*, in *Wheeler's* Report upon the U. S. Geographical Surveys west of the one-hundredth Meridian, Vol. IV, Palæontology.

1877. *Meek, Hall* and *Whitfield* in Cl. *King's* Report of the Geograph. and Geol. Exploration of the Fortieth Parallel. Vol. IV, Palæontology, Pts. 1 and 2.

1877-78. *Stache*, Beiträge zur Kenntniss der Fauna der Bellerophonkalke Südtirols, Jahrb. K. K. Geolog. Reichs Anst. Wien, Bd. XXVII and XXVIII.

1878. *Abich*, Geologische Forschungen in den Kaukasischen Ländern, I. Th. Eine Bergkalk-fauna aus der Araxesenge bei Djoulfa, Wien.

1876-78. *L. de Koninck*, Recherches sur les fossiles paléozoiques de la Nouvelle Galles de Sud, Mém. Soc. Royale des Sciences de Liége, Vols. VI, VII.

1878-87. *L. de Koninck*, Faune du Calcaire carbonifère de la Belgique, Annales du Musee d'hist. nat. de Belgique, Bruxelles.

1879. *Val. v. Möller*, Ueber die bathrologische Stellung der jüngeren palæozoischen Schichten-systems von Djulfa in Armenien, Neues Jahrb. f. Min., p. 225.

1879-86. *Waagen*, Salt Range Fossils, Palæontologia Indica, ser. XIII, Vol. I, Productus Limestone Fossils.

1880. *Roemer*, Ueber eine Kohlenkalkfauna von der Westküste von Sumatra, Palæontographica, Vol. 27, p. 1.

1881. *K Martin*, Beiträge zur Geologie Ostasiens und Australiens. Sammlungen des Reichs-Museums in Leyden, Vol. I. Die tertiärvermögo führenden Sedimente Timors.

1882. *Whitfield*, On the fauna of the lower carboniferous limestone of Spergen Hill, Bull. Amer. Mus. of Nat. History, Vol. 1, p. 39.

1883. *Hyatt*, Genera of fossil cephalopods, Proceed. of the Boston Soc. of Nat. Hist., Vol. XXII.

1883. *R. Lydekker*, The Geology of the Kashmir and Chamba Territories and the British district of Khagan, Mem. Geol. Surv. of India, Vol. XXII.

1882. *Hayden*, XIIth Annual Report U. S. Geol. Surv. of Wyoming and Idaho, Pt. I.

1883. *Stache*, Fragments einer afrikanischen Kohlenkalk-fauna aus dem Gebiete der Westsahara, Denkschr. kais. Akad. d. Wissensch. Wien, math. nat. Cl. Vol. 46, p. 369.

1883. *Kayser* Ueber triassische Fauna von Lo-ping. Richthofen's "China" Vol. IV, p. 160.

1883-84. *H. Woodward*, A monograph of the British Carboniferous Trilobites, Palæontograph. Soc., London.

1884. *Walcott*, The Palæontology of the Eureka district, Monographs, U. S. Geol. Surv., Vol. VIII.

1884. *White*, XIIIth Annual Report of the Geology of Indiana, Pt. 2, Palæontology.

1885. *Tschernyschew*, Ueber permische Kalksteine im Gouvernement Kostroma, Verh. Kais. Russ. Mineral. Ges. St. Petersburg.

1887-88. *Gemmellaro*, La fauna dei calcari con Fusulina della Valle del fiume Sosio nella provincia di Palermo, Palermo.

1888. *M. Tsvetaew*, Cephalopodes de la section superieure du calcaire carbonifère de la Russie centrale Mém. Com. Géol. St. Pétersbourg, Vol. V, No. 3.

1888. *Krotow*, Geologische Forschungen am westlichen Ural-Abhange in den Gebieten von Tschordyn und Ssolikamsk, ibidem, Vol. VI

1888. *Sturkenberg*, Anthozoen und Bryozoen des oberen mittelrussischen Kohlenkalkes, Ibidem, Vol. V, No. 4.

1888. *Kirkby*, On the occurrence of marine fossils in the coal-measures of Fife, Quart. Journ. Geol. Soc., London, Vol. XLIV.

1889. *A. Karpinsky*, Ueber die Ammoneen der Artinsk-Stufe und einige mit denselben verwandte carbonische Formen, Mém. Acad. Impér. des Sciences de St. Pétersbourg, sér. VII. T. XXXVII. No. 2.

1889. *Tchernyschew*, Allgemeine geologische Karte von Russland, sheet. 139, Geologische Beschreibung des Central Urals und des Westabhanges, Mém. Comité Géologique St. Pétersbourg, Vol. III, No. 4.

1890. *Foord*, Notes on the Palaeontology of Western Australia, Geological Magazine, London, new series, Decade III. Vol. VII.

1890. *Walther*, Ueber eine Kohlenkalk-fauna aus der aegyptisch-arabischen Wüste, Zeitschr. d. Deutschen Geolog. Gesellsch. Vol. XLII. p. 419.

1890. *S. Nikitin*, Dépôts carbonifères et puits Artésiens dans la région de Moscou, Mém. Com. Géol. St. Pétersbourg, Vol. V, No. 5.

1890. *A. Karpinsky*, Zur Ammoneenfauna der Artinsk-Stufe, Mélanges Géologiques et paléont. thèse de Bull. de l'acad. Impér. des Sciences de St. Pétersbourg, T. I.

1891. *A. Hyatt*, Carboniferous Cephalopoda, Second Annual Report, Geol. Surv. of Texas for 1890, pp. 330-356.

1891. *White*, The Texas Permian and its mesozoic types of fossils, Bull. U. S. Geol. Survey, Washington, No 77.

1892. *Schellwien*, Die Fauna des Karnischen Fusulinenkalks, I. Theil, Palaeontographica, Bd. 39, pp. 1-56.

1892. *Rothpletz*, Die Perm, Trias und Jura-Formation auf Timor und Rotti im malayischen Archipel, ibidem Bd. 39, pp. 57-106.

1892. *Jack and Etheridge, jun.* Geology and Palaeontology of Queensland and New-Guinea.

1894. *Schellwien*, Ueber eine angebliche Kohlenkalk Fauna aus der aegyptisch-arabischen Wüste, Zeitschr. d. Deutschen Geol. Ges. Bd. XLVI.

1894. *E. Suess*, Beiträge zur Stratigraphie Central Asien, Denkschr. Kais. Akad. d. Wissensch. Wien, math. nat. Cl. Bd. LXI. pp. 433-456.

1895. *Salomon*, Geologische und palaeontologische Studien über die Marmolata, Palaeontographica, Bd. XLII.

1895. *Stuckenberg*, Korallen und Bryozoen der Steinkohlenablagerungen des Ural und des Timan, Mém. Com. Géol. St. Pétersbourg. Vol. X, No. 3.

1896. *Tornquist*, Das fossilführende Untercarbon am östlichen Rossberg-Massiv in den Süd-vogesen, I. Theil, Brachiopoda, Abhandl. zur Geol. Special-Karte von Elsass-Lothringen, Bd. V. Heft. 4.

1896. *Foord*, Ueber palaeozoische Faunen aus Asien und Nordafrika. Neues Jahrb. f. Min. p. 54.

1896. *Julien*, Le terrain carbonifère marine de la France Centrale, Paléontologie Française.

1896. *Pervin Smith*, Marine Fossils from the coal-measures of Arkansas, Proceed. Amer. Philosoph. Soc. Vol. XXXV. No. 152.

DESCRIPTION OF FOSSILS.

Class: CRUSTACEA.

Order: TRILOBITÆ.

Family: *PROETIDÆ*, Phillips.

Genus: PHILLIPSIA, Portlock.

1. PHILLIPSIA SP. IND. AFF. SEMINIFERA, Phillips. Pl. I, figs. 1, 2.

1883. *Phillipsia cf. seminifera.* Lydekker, Geology of the Kashmir and Chamba territories, etc. Mem. Geol. Surv. of India. Vol. XXII. Pl. II, figs. 5, 5a.

This species is represented in the Himálayan collection by two pygidia only, which are too fragmentary and too badly preserved to permit an exact determination. One of them (fig. 1) has been discovered and was compared with *Phillipsia seminifera*, Phill., by Lydekker; the other was not discovered, until the whole collection had been looked over several times, owing to the fact that it was crushed and partly covered by fragments of a *Spirifer*. After carefully cleaning it, however, it was found to be a *Phillipsia*, which I think may be safely referred to the same species as Lydekker's type specimen.

The pygidium figured by Lydekker, which is the better preserved of the two specimens available for description, is of a semi-elliptical shape, rather strongly convex and a little wider than long. The axis is considerably elevated above the lateral lobes. It is about one-third of the breadth of the entire tail-shield at its anterior border. Its posterior portion has been broken off, but from its preserved outlines we may judge that its extremity was rather prominent and distinctly obtuse.

In my second specimen (fig. 2) this character of the central axis may likewise be noticed. In the preserved portion of the axis five coalesced somites are shown. The lateral lobes consist of eight pleuræ, terminating within the narrow marginal space. Traces of tubercles may be seen both on the surface of the pleuræ and of the axial rings.

The measurements of this pygidium can only be given approximately, owing to its fragmentary and partly deformed condition. They are as follow :—

Length of the entire pygidium	10 mm.
Breadth	cca. 12 "
Length of the axial lobe	8¼ "
Breadth „ „ „ at its anterior margin	4 "
Breadth of the smooth marginal space	1½ "

The fragment of the second pygidium (fig. 2) shows eight somites in the axial, and six pleuræ in each of the corresponding lateral lobes. Traces of an indistinct granulation may likewise be noticed, both on the surface of the strongly marked lateral pleuræ and of the axial segments.

No measurements of this pygidium can be given on account of its imperfect state of preservation. The dimensions may be gathered in a general way from the figure. Its original outlines seem to have been altered considerably by crushing, especially so in a transverse direction.

Locality and geological position; number of specimens examined.—N. of Eishmakam, Lidar Valley, Kashmir, in a dark-blue limestone with numerous *Fenestellae*; Coll. Lydekker; 2.

Remarks.—Among the congeneric species from carboniferous rocks of European districts, *Phillipsia seminifera*, Phillips, has been correctly compared with the present form by Lydekker. The two species are certainly closely allied, though probably not identical. It is especially the figure of *Ph. seminifera* given by Woodward on Pl. V, fig. 5, of his monograph of the British carboniferous Trilobites (Palæontographical Society, London, 1883, Part I) which strongly resembles Lydekker's specimen. It is, however, not possible to fix the affinities between the European species and our Himâlayan trilobite in a more positive way, as the state of preservation of the latter is too indifferent to warrant an exact determination.

Phillipsia seminifera has been mentioned by L. de Koninck (Recherches sur les fossiles paléozoiques de la Nouvelle Galles du Sud, 1876, p. 848) among the carboniferous fossils of Colooolo in New-South-Wales, but I am far from convinced of the correctness of this identification, especially regarding the pygidium figured by this author on Pl. XXIV, fig. 8a of his memoir.

An Australian species, which is probably closely allied to the present one, is *Griffithides dubius*, Etheridge (Quart. Journ. 1872, Vol. XXVIII, p. 338, Pl. XVIII, fig. 7) from the carboniferous rocks of the Don River in Queensland.

MOLLUSCA.

Class: LAMELLIBRANCHIATA.

There is no more difficult group among the anthracolithic fossils of Kashmir and Spiti than the Lamellibranchiata. Although this class is well represented, I have not met with one single complete specimen. Thus even the few identifications must necessarily remain uncertain. This will, I hope, explain why I deemed it preferable to indicate the described species which are probably new, as *sp. ind.* only, without adding a particular denomination. All the numerous specimens collected by Dr. Verchère have been purposely excluded from a special description, their fragmentary state of preservation rendering them absolutely unfit for determination.

I need hardly say that with such materials no idea of the real character of the fauna can be formed. No doubt, systematic researches at the proper localities will greatly add to our knowledge.

Order : ANISOMYARIA, Neumayr.

Family : *MYTILIDÆ*, Lam.

Genus : MODIOLA, Lamarck.

MODIOLA ? SP. IND. Pl. I fig. 5.

The umbonal region having been broken off entirely, the fragment of a right valve is provisionally and with much hesitation referred to the genus *Modiola* on account of its external similarity with some types, figured by L. de Koninck on Pl. 28 of his "Faune du Calcaire Carbonifère de Belgique" (Annales du Musée Roy. d'hist. nat T. XI. 5 ème ptie.). It especially reminds of *M. princeps* or of *M. fusiformis* by its obliquely elongated, slightly inflated shape, and by its numerous concentric striæ of growth, which are of somewhat irregular strength as in *M. Cordoliana*.

Locality and geological position ; number of specimens examined.—Dark-blue micaceous shales with numerous *Penestella*, near Eishmakam, Kashmir Valley ; Coll. Lydekker ; 1.

Family : *PECTINIDÆ*, Lamarck.

Genus : AVICULOPECTEN, M'Coy.

AVICULOPECTEN SP. IND. Pl. I. fig. 3.

The only known specimen of this species is a right valve with a nearly circular outline and a very indifferent sculpture. The valve is strongly inequilateral, flat, and about as long as high. The apex is anterior in its position, slightly prominent, limited on both sides by wings of unequal size, the posterior ear being considerably larger than the anterior one. It is on the strength of this character that I have considered this species as belonging to the genus *Aviculopecten*.

Both wings are marked off distinctly from the remainder of the shell. The marginal edges, which separate the apical region of the valve from the wings, slope very steeply towards the surface of the latter, although their general elevation above them is not considerable. The anterior margin projects far in front of the anterior wing and passes gradually into the ventral margin. The posterior margin is continuous, not sinuated at the commencement of the posterior wing. The hinge-line is only one-half the length of the antero-posterior diameter of the shell.

The whole valve is very flatly arched, its greatest thickness being situated about the middle of its height.

The surface is smooth with the exception of numerous and delicate concentric lines of growth. Traces of a few radiating costæ may be seen on the surface of the posterior wing.

The measurements are as follow :—

Entire length of the shell	21 mm.
„ height	21 „
Length of the hinge-line	9 „
Thickness of the right valve	1·3 „
Apical angle without the wings	ca. 90°

Locality and geological position; number of specimens examined.—Dark-blue shales and limestone partings with *Fenestellæ*, near Eishmakam; Coll. Lydekker; 1.

Remarks.—A specific determination is impossible. In ornamentation it resembles *A. squamula*, Waagen (Salt Range Fossils, Pal. Ind., ser. XIII, Vol. I, Prod. Limest. Foss., p. 315, Pl. XXIV, fig. 5), but it differs from it radically owing to its strongly inequilateral outline. From *A. sibirica*, Verneuil (Géologie de la Russie d'Europe, Vol. II. Paléontologie, p. 329, Pl. XXI, fig. 7) and *A. ellipticus*, Phill., it is also readily distinguished by this latter conspicuous character. There is also a distant similarity between it and *Avicula circularis*, Hall (Palæontology of Iowa, p. 522, Pl. VII, fig. 9) or *Streblopteria cellensis*, L. de Koninck (Faune du calcaire carb. de Belgique, p. 209, Pl. 30, fig. 14), although the Kashmir shell is certainly not a *Streblopteria*.

Genus : PECTEN, Klein.

PECTEN SP. IND. Pl. I. fig. 4.

This species is represented by a single left valve only.

Outline transversely oval and very inequilateral. Apex anterior not very prominent, pointed. Two wings, of which the anterior is by far the larger. I based the determination as *Pecten* on this character, no other generic features being seen. Anterior wing flat, almost rectangular, separated from the anterior marginal edge of the valve by a furrow, bounded by a comparatively high perpendicular wall. Towards the posterior wing the shell is less steeply inclined, but the edge separating them is also distinctly defined. This wing is rather small and cut off obliquely, its posterior margin meeting the hinge-line at an obtuse angle. Hinge-line rather short, barely one-half the entire length of the shell.

The anterior margin forms a broad curve and projects only slightly in front of the anterior wing. It is continuous with the broadly arched ventral margin. The posterior margin passes into the ventral one in a kind of an obtusely rounded-off angle, from where it ascends in a nearly straight, oblique line, passing into the furrow which separates the posterior wing from the remainder of the shell.

The valve is moderately inflated, its greatest thickness being situated about the middle of its height.

D

The ornamentation consists of numerous extremely thin concentric striæ of growth, which are crossed by a system of radiating costæ. The latter are augmented in number towards the margins by intercalation. They are unequal in strength, roof-shaped and separated by valleys of unequal width, but stronger and thinner costæ do not alternate regularly. On the anterior wing a delicate radiating sculpture is likewise noticed. The posterior wing is entirely smooth.

The measurements are as follow :—

Entire length of the shell	14·5 mm.
„ height „ ... . ;	19 „
Length of the hinge-line	7 „
Thickness of the left valve	4 „
Apical angle without the wings	89°

Locality and geological position ; number of specimens examined.—Dark-blue micaceous shales, associated with *Sperifer Lydekkeri* and *Fenestella*, Rishmakam, Kashmir Valley ; Coll. Lydekker ; 1.

Remarks.—The specimen is too fragmentary to determine its relationship to other congeneric forms. In the character of ornamentation it is not unlike *Pecten præcox*, Waagen (Salt Range Fossils, l. c., p. 318, Pl. XXIII, fig. 3), from the top beds of the upper Productus limestone, or some of the *Pectines* from the Bellerophon-limestone of the Comelico district described by Stache, but it radically differs from them by its strongly inequilateral shape.

Class : PTEROPODA ?

Order : CONULARIDA, Waagen.

Family : *CONULARIDÆ*, Walcott.

Genus : CONULARIA, Miller.

CONULARIA TENUISTRIATA, M'Coy. Pl. VII, fig. 6.

1847. *Conularia tenuistriata*, M'Coy, On the fossil botany and zoology of the rocks associated with the coal of Australia. Annals and Mag. of Nat. History, Vol. XX, p. 307, Pl. XVII, figs. 7, 8.

1877. *C. tenuistriata*, L. de Koninck, Recherches sur les foss. paléozoïques de la Nouvelle Galles du Sud., p. 310, Pl. XXIII, fig. 2.

1884. *C. tenuistriata*, Waagen, Records Geol. Surv. of India, Vol. XIX, Pt. I., p. 26 ; Pl. I, fig. 2.

1891. *C. tenuistriata*, Waagen, Salt Range Fossils, Pal. Ind., ser. XIII, Vol. IV, Geological Results, p. 129, Pl. V, figs. 2, 3.

A single fragment of this species was quite accidentally discovered when chiselling out a specimen of *Spirifer Lydekkeri* from a block of quartz-sandstone. I only succeeded in clearing two of the perfectly well preserved faces from the tough matrix and in developing the transverse section. Although incomplete, the specimen is, I think, sufficiently well preserved to permit of identification.

The fragment belongs to a young individual of a total length of about 60 mm., of which 39 mm. is preserved, whilst the apical portion was broken off. Apical angle very small, a little less than ten degrees, imparting to the shell a strongly elongated shape. Outline of the transverse section somewhat rhomboidal and inequilateral. The narrower side is nearly two-thirds the breadth of the broader one. Both the longer and narrower faces are distinctly impressed in the middle. The four corners of the pyramid are marked by narrow furrows, in which the ribs from both sides meet, being slightly bent backwards and alternating with each other. These transverse ribs, which ornament the four faces, meet in the middle of the latter under an obtuse angle and there mostly alternate. Very few only unite directly with each other and then form simply broken, upward curved lines. The ribs are very thin, smooth on their crests, and rather regularly distributed. From seventeen to twenty ribs are counted within a space of ten millimetres.

I have not observed any striation on the surface between the ribs.

I do not think that there can be any doubt about the identity of this fragment with the specimens from the *Conularia*-nodules of the Salt Range boulder-group, which have been described as *C. tenuistriata* by Waagen. Whether the specimens from the Salt Range ought to be identified with M'Coy's Australian species appears to me less certain. The latter seems to be distinguished by its strongly inequilateral transverse section, the narrow sides of which attain scarcely one half the length of the broader ones. The difference in the number of ribs in the Indian and Australian types has been satisfactorily explained by Waagen. In the small fragment figured by L. de Koninck (l. c. Pl. XXIII, fig. 2a) exactly the same number of ribs is counted within a space of 10 mm. as in the present specimen. If the difference in the shape of the transverse section should be thought a sufficient reason for distinguishing the Salt Range and Kashmir types from the Australian ones by a varietal denomination, the name *Conularia tenuistriata var. Indica* might be applied to them.

It cannot be identified with *C. laevigata*, Morris (in Strzelecki's Physical description of New South Wales, etc., p. 290, Pl. XVIII, fig. 9), on account of its considerably smaller apical angle, its rhomboidal, not rectangular, laterally impressed transverse section, and its more numerous ribs. From *C. Warthi*, Waagen, and its allies it is at once distinguished by its different sculpture.

Locality and geological position; number of specimens examined.—Quartz sandstone with *Spirifer Lydekkeri*, Ladakh Valley, Kashmir; Coll. Lydekker; 1.

Remarks.—The occurrence of this species in the quartzitic sandstones of the Ladakh Valley is of no small interest. Its importance is, however, diminished by the fact that the exact stratigraphical position of these sandstones in the anthracolithic system of the Himálayas is as little known as that of the *Conularia*-nodules in the Salt Range boulder bed, discovered by Dr. Warth in 1885.

Conularia tenuistriata has been quoted from the carboniferous sandstone of Murree by L. de Koninck, from the boulder bed of the Eastern Salt Range by Waagen and from the Gympie beds of Queensland with some hesitation by Etheridge, jun.

MOLLUSCOIDEA.

Class: BRACHIOPODA.

Order: TESTICARDINES, Bronn.

Suborder: APHANEROPEGMATA, Waagen.

Family: *PRODUCTIDÆ*, Grey.

Subfamily: PRODUCTINÆ, Waagen.

Genus: PRODUCTUS, Sowerby.

The *Productidæ* are the most numerous fossils from the anthracolithic rocks of Kashmir and Spiti. Four genera (*vis.* subgenera) of this family are represented among them. They are: *Productus, Marginifera, Strophalosia, Chonetes.* Among them the genus *Productus* takes the most important part, at least regarding the number of species, of which I count not less than eleven altogether. Three of them are to be attributed to the section of *Fimbriati*, two to the *Semireticulati* and *Spinosi*, one to the *Lineati*, *Undati*, *Caperati*, and *Irregulares*.

A classified list of the *Producti*, from Kashmir and Spiti, which I have been able to determine specifically, is drawn up in the following scheme:—

I. Section, LINEATI.
1. *Productus Cora*, d'Orbigny.

II. Section, UNDATI.
2. *Productus undatus*, Defrance.

III. Section, SEMIRETICULATI.
3. *Productus semireticulatus*, Martin.
4. *P. cf. longispinus*, Sowerby.

IV. Section, SPINOSI.
5. *Productus cf. scabriculus*, Martin.
6. *P. cf. spinulosus*, Sowerby.

V. Section, FIMBRIATI.
7. *Productus Abichi*, Waagen.
8. *P. pustulosus*, Phillips.
9. *P. punctatus*, Martin.

VI. Section, CAPERATI.
10. *Productus aculeatus*, Martin.

VII. Section, IRREGULARES.
11. *Productus mongolicus*, Diener.

In addition to these species a very remarkable one may be mentioned, which has been described and figured by Professor Davidson (Quart. Journ. Geol. Soc., London, Vol. XXII, 1866, p. 45, Pl. II, fig. 16) as *Productus laevis, nov. sp.* It was founded on a single ventral valve, "occurring in a coarse limestone in the Zéwan beds, valley of Kashmir, but less compact and of a lighter grey, than the bed, from which most of the species, found at Khoonmoo, were obtained."

Davidson's description of this species is as follows:—"Shell small, nearly circular; hinge-line slightly shorter than the greatest width of the shell; ventral valve evenly convex; ears small, surface smooth (?); length 3 lines, the width slightly exceeding the length."

This species is not represented in our collection. Without venturing a decided opinion on this subject, I do not think it superfluous to draw attention to the striking resemblance of *Productus laevis* to the triassic *Koninckina Leonhardi*, Wissm.,[1] from St. Cassian. As Professor Davidson himself asserts his specimen to have been found in a rock lithologically different from the limestone, from which the majority of brachiopoda collected by Captain Godwin-Austen had been obtained, the identity of *P. laevis* with a species of *Koninckina* is not impossible. In the upper triassic beds of the Bambanag Range (Kumaon) the presence of *Amphiclina*, a typical representative of the *Koninckinidæ*, has been proved by Dr. Bittner.[2]

On the other hand an identity of *Prod. laevis* with *Leptaena indica*, Waagen (Salt Range Fossils, Pal. Ind. ser. XIII, Vol. I, Prod. Limest. Foss., p. 609, Pl. LVIII, figs. 7—9) from the Katta beds of the middle Productus limestone is likewise possible as has been suggested by Waagen himself, although to me this solution of the question does not seem very probable. For Mr. Davidson should scarcely have placed, I think, his Kashmir species in the genus *Productus* had he been able to demonstrate the presence of an area in his specimen of a similar kind, as it is developed in *Leptaena indica*.

Besides these species *Productus Humboldti*, d'Orb., *P. Boliviensis*, d'Orb., *P. costatus*, Sow., *P. Purdoni*, Dav., have been quoted from the carboniferous rocks of Kashmir by Dr. Verchère. The last mentioned species has also been described from the Kuling shales of Spiti by Stoliczka, but it is absolutely impossible to venture on an identification of these forms, as the interpretation of species, as applied by Verchère, widely differs from that adopted in the present memoir.

[1] A. Bittner, Brachiopoden der Alpinen Trias, Abhandlgn. K. K. Geol. Reichs-Anst. Wien, Bd. XIV, Pl. XXX, figs. 45-50.

[2] C. Diener, Ergebnisse einer Geologischen Expedition in den Central Himalaya, Denkschr. Kais. Akad. d. wissenschaften, Wien. math. nat. Classe, Bd. LXII, 1896, p. 568.

L Section : LINEATI.

1. Productus Cora, d'Orbigny, Pl. I, fig. 12.

1872. *Productus Cora*, d'Orbigny, Voyage dans l'Amérique Méridionale, T. III., 4 ème ptie., Paléontologie, p. 55, Pl. V, figs. 8, 9.

1866. *P. Cora*, Davidson, Quart. Journ. Geol. Soc., Vol. XXII, p. 73.

1867. *P. Cora*, Verchère, Kashmir, the Western Himalaya and the Afghan Mountains, Journ. Asiat. Soc. of Bengal, Vol. XXXV, Pt. II, pp. 216, 212.

For further synonyms my memoir on the Chitichun-fossils (Pt. III of the present volume) ought to be consulted.

Two ventral valves referable to this well-known species have been obtained by Lydekker from the carboniferous limestone of Barus in Kashmir. One of them is fairly complete and agrees very well with the figures and description of *P. Cora* given by Waagen in his monograph of the Salt Range Fossils (Pal. Ind., ser. XIII, Vol. I, Productus Limestone Fossils, p. 677, Pl. LXV, fig. 3; Pl. LXVII, figs. 1, 2). It is larger than any specimens from the permo-carboniferous limestone of Chitichun, No. I, and nearly equals in size the example figured by Waagen on Pl. LXVII, fig. 2.

It is considerably broader than long and provided with tolerably large asymmetrical wings. The curve of the valve is rather irregular, being somewhat flattened in the apical region. The apex is strongly curled inwards. The hinge-line corresponds to the greatest breadth of the shell. The trail is not preserved.

Not the slightest trace of a sinus is exhibited in the specimens, which consequently must be separated from the group of *Productus Neffedievi* and placed in the group of *P. corrugatus*, M'Coy. The latter species and *P. Cora* are very closely related to each other. According to Waagen their only difference consists in the general absence of spines on the surface of *P. corrugatus*. As a few irregularly scattered spines are exhibited on the surface of my type specimen, I deemed it preferable to identify it with d'Orbigny's species.

Apart from these few, irregularly scattered spines, the ornamentation consists of numerous, delicate, radiating striae, which are descending straight across the frontal region of the valve. A few indistinct concentric folds or wrinkles are developed on the wings.

Neither the dorsal valve, nor the internal characters are preserved.

The approximate measurements are as follow :—

Length of the shell in a straight line	14 ½ mm.
„ „ „ along the curve	29 „
Breadth of the shell	25 „
Thickness of the ventral valve	13 „

Locality and geological position ; number of specimens examined.—Barus, Kashmir Valley, in a dark blue limestone; Coll. Lydekker; 2. From the same locality three or four specimens were obtained by Capt. Godwin-Austen, which have been referred to this species by Professor Davidson (*vide antea*).

Remarks.—In identifying these specimens with *Productus Cora* I am taking this species in the circumscription attributed to it chiefly by Russian geologists. Further remarks on this subject as well as on the geological range of *P. Cora* will be found in the third part of this volume (pp. 16 and 17).

II. Section : UNDATI.

2. PRODUCTUS UNDATUS, Defrance. Pl. I, figs. 9, 10.

1826. *Productus undatus*, Defrance, Dictionnaire des sciences nat., Vol. XLIII, p. 364.

1843. *P. undatus*, L. de Koninck, Description des animaux foss. du terrain carbonif. de Belgique, p. 156, Pl. 11, fig. 3.

1844. *P. tortilis*, M'Coy, Synopsis of the characters of the carbon. fossils of Ireland, p. 116, Pl. XX, fig. 14.

1845. *P. undatus*, E. de Verneuil, Géologie de la Russie d'Europe, Vol. II., Paléontologie, p. 261, Pl. XV, fig. 15.

1847. *P. undatus*, L. de Koninck, Monographie des genres Productus et Chonetes, p. 69, Pl. V, fig. 3.

1860. *P. undatus*, Davidson, Monogr. of the Scottish Carb. Brachiopoda, p. 61, Pl. IV, figs. 16-17.

1861. *P. undatus*, Davidson, Monogr. British Carbonif. Brachiopoda, p. 161, Pl. XXXIV, figs. 7-13.

1874. *P. undatus* (?) Toula, Kohlenkalk- und Zechstein-Fossilien aus dem Hornsund an der Südwestküste von Spitzbergen, Sitzgsber. Kais. Akad. d. Wissensch. Wien math nat. Cl. Bd., LXX, p. 9.

1875. *P. undatus*, Trautschold, Die Kalkbrüche von Mjatschkowa, p. 65, Pl. V, fig. 3.

1876. *P. undatus*, L. de Koninck, Recherches sur les familles Paléozoïques de la Nouvelle Galles de Sud, p. 190, Pl. IX, fig. 4.

1883. *P. cf. undatus*, Stache, Fragmente einer afrikanischen Kohlenkalkfauna aus dem Gebiete des West-Sahara, Denkschr. Kais Akad. d. Wissensch, Wien. math. nat. Cl. Bd. 46, p. 404, Taf. VII, 2L.

1883. *P. undatus*, Kayser, Oberearbonische Fauna von Loping, Richthofen's China, Bd., IV, p. 188, Taf. XXVI, figs. 12, 13.

1890. *P. undatus*, Foord, Notes on the Palaeontology of Western Australia, Geological Magazine, London, new. ser., Decade III, Vol. VII, p. 182, Pl. VII, fig. 8.

1892. *P. undatus*, Etheridge jun. in Jack and Etheridge, Geology and Palaeontology of Queensland and New-Guinea, p. 254, Pl. 13, fig. 16.

1896. *P. undatus*, Tornquist, Das fossilführende Untercarbon am nördlichen Rossbergmassiv in den Süd-Vogesen, Th., I. Abhandlungen zur Geologischen Spezial-Karte von Elsass-Lothringen, Bd. V, Hft. 4, p. 70, Taf. XIV, figs. 9, 11.

Two casts of dorsal valves which have been collected by Lydekker in the carboniferous rocks near Eishmakam are referable to this characteristic and easily recognised species.

One of them is of an unusually large size, larger even than Kayser's type-specimens from Loping, whereas the second is an average-sized example, of nearly the same dimensions as L. de Koninck's type specimen from the Belgian Calcaire de Visé. Both of them are broader than long, making however exception of the trail, which is very well developed in the larger specimen (fig. 9).

Valve slightly concave, with a regularly excavated apex and with indistinctly defined auriculate expansions. No trace of a median fold is developed. The hinge-line is shorter than the greatest width of the shell.

The ornamentation is very characteristic and agrees in every respect with the figures and descriptions given by L. de Koninck and Davidson of the sculpture in their European type-specimens. The surface is covered by very numerous and

delicate radiating striæ, which are not continuous but interrupted by an equally
delicate, concentric sculpture. The number of these radiating thread-like striæ is
augmented by intercalation. The most prominent feature in the ornamentation of
the shell is, however, the numerous concentric lamellæ or wrinkles of growth, which
sharply imbricate and occur at irregular distances, but all over the surface
of the valve, imparting to the latter a terrace-shaped appearance. The
steeper, occasionally perpendicular slope of these undulating lamellæ is directed
towards the apex. Only very few lamellæ can be traced across the entire valve.
As a rule, either two separate wrinkles unite during their passage or are absorbed,
whilst a new one is rising at some distance. In the trail this concentric, crumpled-
like sculpture is but very indistinctly developed. In my two specimens it is altogether less strongly marked than in the Chinese specimens from Loping described
by Professor Kayser.

No traces of spines or tubercles have been discovered.

On account of the variability of the outline of my larger type-specimen with
its partially preserved trail, it appears barely practicable to give exact measurements. The drawing (fig. 9) will however give a sufficiently clear idea of its
dimensions and features. The second specimen is not complete. Nevertheless the
indication of its length, breadth and thickness may perhaps be of interest.

The measurements are as follow :—

Length of the dorsal valve	. . . . mm. 25 m.m.
Breadth „ „ „	. . . . 19 „
Depth „ „ „	. . . . 7 „
Length of the hinge-line	. . . . ca. 24 „

Locality and geological position ; number of specimens examined.—Dark blue
limestone with *Fenestella*, North of Eishmakam, Kashmir Valley ; Coll. Lydekker ; 2.

Remarks.— Productus undatus is a comparatively rare but wide spread species,
both in lower and upper carboniferous rocks of the Eastern hemisphere. In
Europe it is known from Belgium, where it occurs chiefly in the Calcaire
de Visé, from Great Britain, from Central France, from Alsatia and from Russia,
where it occurs in the Moscovian stage of Miatchkowa. It has been described by
Stache from the lower carboniferous beds of the Western Sahara, by Kayser from
the upper carboniferous deposits of Loping in China, by L. de Koninck from a
carboniferous sandstone near the Paterson River in New South Wales, by Etheridge
jun. from the Gympie beds of Queensland. Toula quotes the species as doubtful
from the permo-carboniferous rocks of the Hornsund in Spitzbergen.

A. H. Foord figures a *Productus* from the carboniferous rocks of the Irwin
river (Victoria district of Western Australia) which he considers to be identical
with *P. undatus.* I am however not convinced of the absolute identity of the two
forms. The Australian specimen differs from the typical shape of *P. undatus,* as it
is described by L. de Koninck, by the larger number of its spines and reminds very
strongly of *P. cancriniformis,* Tschernyschew (Mém. du Comité Géol. de la Russie,
Vol. III, No. 4, St. Pétersbourg, 1889, p. 373, Pl. VII, figs. 32, 33). This spinose
variety of *P. undatus* can barely be distinguished from *P. cancriniformis,* if one

has to deal with ventral valves only, the chief difference consisting in the shape of the dorsal valve, which is regularly concave in *P. undatus*, but distinctly geniculate in Tschernyschew's species.

The fragment from Djulfa in Armenia, considered first as identical with *P. undatus* by Abich (Geologische Forschungen in den Kaukasischen Lœndern I. Theil, Eine Bergkalk-Fauna aus der Araxes-Enge bei Djulfa, p. 31, Taf. V, fig. 10), but referred later to the permian *Productus hemisphærium*, Kutorga, both by Abich himself and by Val. von Moeller, has certainly nothing to do with the carboniferous species.

III. Section: SEMIRETICULATI.

3. Productus semireticulatus, Martin. Pl. I, fig. 14; Pl. II, fig. 10.

1809. *Anomites semireticulatus*, Martin. Petrificata Derbiensia, Pl. XXXII, figs. 1, 2; Pl. XXXIII, fig. 4.
1858. *Productus semireticulatus*. Davidson, Quart. Journ. Geol. Soc., London, Vol. XXII, p. 36, Pl. I, figs. 6, p. 43, Pl. II, fig. 13.
1867. *Prod. semireticulatus*, Verchère, Journ. Asiatic Soc. of Bengal, Vol. XXXV, Pt. 2, pp. 201, 212.

For a more complete list of synonyms *vide* my monograph on the permo-carboniferous fauna of Chitichun No. I (Pt. 3 of this volume). To these synonyms the following ought to be added:—

1863. *P. semireticulatus*, Davidson, On the lower carboniferous brachiopoda of Nova Scotia. Quart. Journ. Geol. Soc., Vol. XIX, p. 174, Pl. IX, figs. 20, 21.
1896. *P. semireticulatus*, Julien, Le terrain carbonifère marin de la France Centrale, p. 65, Pl. I, figs. 1-4, 13, Pl. VII, figs. 4-6, Pl. XI, fig. 6, Pl. X II, fig. 8.
Non P. semireticulatus, Stoliczka, Mem. Geol. Surv. of India, Vol. V, Pt. I, (1865) p. 29, *sen* Lydekker, ibid, Vol. XXII, Pl. II, fig. 1.

This well-known and far-spread species is not at all rare in the carboniferous rocks of Kashmir, but most of the specimens are in a rather inferior state of preservation. A small number of specimens is however sufficiently complete for identification.

Most of the specimens are of considerable size, some of them reaching 90 mm. in width. The ventral valves, which as a rule are entirely crushed or show no shelly substance, are provided with a distinctly developed sinus, which in some specimens is quite as deeply indented as in the Carinthian variety, described as *P. semireticulatus var. bathykolpos* by Schellwien (Palæontographica, Bd. 39, 1892, p. 22). Curiously enough, dorsal valves of this species are generally much better preserved than the ventral ones. A tolerably complete specimen of a dorsal valve of *P. semireticulatus* from Barus is represented on Pl. I, fig. 14. This specimen is provided with a long hinge-line, distinctly developed wings, and with a shallow median fold, becoming more prominent in the vicinity of the front only. The flat proximal portion, which is covered by the characteristic reticulate sculpture, meets the frontal portion at a right angle.

In one of the specimens (Pl. II, fig. 10) the internal structure of the ventral valve is clearly exhibited. In the deep valley between the coarsely striated divaricator muscular scars the phylloid petal-shaped adductor, or occlusor impressions are situated. The proximal portion of the cast is separated from the hinge margin by deep furrows. The apex is smooth, crossed only by a low, distinctly produced ridge. The internal surface of the shell is covered by numerous coarse granulations.

The measurements of this cast are as follow :—

Length of the shell in a straight line	50	mm.
„ „ „ „ along the curve	83	„
Breadth of the shell	ca. 60	„
Thickness of the ventral valve	30	„

Locality and geological position ; number of specimens examined.—Shaly limestone with *Fenestella*, Barus, Kashmir Valley ; Coll. Godwin-Austen ; 6 ; dark blue limestone with *Fenestella*, N. of Khúmmu, near Pampur ; Coll. Lydekker ; 5 (including the two figured type-specimens) ; Tangar, N.W. of Avantipur ; Coll. Lydekker ; 1.

The specimens described by Davidson were obtained from Barus and from Loodoo, W. of Westerwan.

Remarks.—The fossils from Kashmir and Spiti which have been quoted as *Productus semireticulatus* by Stoliczka and Lydekker are different from Martin's species and must be classed among the subgenus *Marginifera*. They will be described hereafter as *Marg. himalayensis.*

4. PRODUCTUS CF. LONGISPINUS, Sowerby. Pl. I, fig. 11.

1814. *Productus longispinus*, Sowerby, Mineral Conchol., Vol. I, p. 154, Pl. LXVIII, fig. 1.

1814. *P. Flemingii*, Sowerby, ibidem, fig. 2.

1814. *P. spinosus*, Sowerby, ibid. Pl. LXIX, fig. 3.

1822. *P. lobatus*, Sowerby, Ibd. Vol. IV p. 16, Pl. 318, figs. 2-6.

1836. *P. setosa*, Phillips, Geology of Yorkshire, Vol. II, Pl. VIII, figs. 9, 17.

1841. *P. lobatus*, L. von Buch, Abhandlgn. Königl. Akad. d Wissensch. Berlin, I. Th. p. 32, Pl. II, fig. 17.

1843. *P. longispinus*, L. de Koninck, Déscr. des animaux fossiles du Terrain carbon. de Belgique, p. 187, Pl. XII, fig. 11, Pl. XII. bis, fig. 2.

1845. *P. lobatus*, E. de Verneuil, Géologie de la Russie d'Europe, Vol. II, Paléontologie, p. 266, Pl. XVI, fig. 3, Pl. XVIII, fig. 6.

1846. *P. tuberatus* (?) Graf Keyserling, Reise in das Petschoraland, p. 208, Pl. IV, fig. 6.

1847. *P. Flemingii*, L. de Koninck, Monographie des genres *Productus* et *Chonetes*, p. 96, Pl. X, figs. 2, 3.

1860. *P. Flemingii*, Grœnewaldt, Beitrage zur Kenntniss der sedimentären Gebirgsformationen etc., Mém. de l'acad. impér. des sciences de St. Pétersbourg sér. VII, T. II, p. 125, Taf. III, fig. 4.

1861. *P. longispinus*, Davidson, Monograph British Carbon. Brachiopoda, p. 154, Pl. XXV, figs. 5—17.

1865. *P. Orbignyanus*, Meinitz, Carboniformation and Dyas in Nebraska, p. 64, Taf. IV, figs. 8—11.

1867. *P. lobatus var. pauriaicatus*, Trautschold, Bull. soc. impér. des natur. de Moscou, T. XI. p. 87, Taf. V, fig. 3.

1870. *P. longispinus*, Rœmer, Geologie von Oberschlesien, p. 69, Taf. VIII, fig. 2.

1872. *P. longispinus* ? Meek, in Meek and Hayden, Final Report of the U. S. Geological Survey of Nebraska, p. 161, Pl. VIII fig. 6 (non Pl. VI, fig. 7).

1873. *P. Flemingii*, L. de Koninck, Monographie des fossiles carbonifères de Bleiberg, p. 94, Pl. I, fig. 14.

1874. *P. longispinus* (?) Toula, Kohlenkalk und Zechstein Fossilien aus dem Eiswasser an der Südwest Küste von Spitzbergen, Sitzgsber. Kais. Akad. d. Wiss. Wien, LXX, Bd. math. nat. Cl. I. Abth. p. 44, fig. 7.

1874. *P. longispinus*, Trautschold, Die Kalkbrüche von Mniszchkowo, p. 67, Taf. 1. fig. 4.

1876. *P. rimeatus*, Trautschold, ibid. p. 61, Taf. V, fig. 5.

1880. (?) *P. longispinus*, Römer, Über eine Kohlenkalk Fauna der Westküste von Sumatra, Palaeontographica, XXVII. Bd. p. 5.

1883. *P. longispinus*, Kayser, Obercarbonische Fauna von Loping, Richthofen's China, Bd. IV, p. 183, Taf. XXVII, fig. 1, nos 3-4.

1890. *P. longispinus*, Nikitin, Mém. Com. Géol. de la Russie, Vol. V, No. 5, p. 159, Pl. 1, figs. 7-13.

1892. *P. longispinus*, Schellwien, Die Fauna des Karnischen Fusulinenkalks, Palaeontographica, Bd. 39, p. 24, Taf. III, fig. 4-5, Taf. VIII, fig. 36.

The only specimen referable to this species consists of an incomplete ventral valve, which, however, agrees in every respect so perfectly with some British specimens from Yorkshire that I do not hesitate to identify it with the latter.

In my specimen the apical region is partly broken, but the rest corresponds exactly with Davidson's type-specimen from Yorkshire, figured on Pl. XXXV, fig. 7, of his monograph. The shell is slightly transverse, a little wider than long, evenly convex in a longitudinal direction, but in the transverse direction divided by a broad sinus, which flattens gradually towards the front. The lateral parts appear strongly depressed and descend in a very steep curve to the margin. The small auricular expansions are slightly curled.

There is not the least trace of any marginal ridge, which forms the distinguishing character of Waagen's subgenus *Marginifera*. This specimen certainly cannot therefore belong to the latter, the shell margin having been broken off in such a manner that the absence of a prominent shelly ridge within the wings warrants its separation from *Marginifera*. Having a large number of true *Marginiferæ* from Chitichun No. 1 and from Kashmir at hand for comparison, I am fully convinced that this specimen belongs to the genus *Productus*, s. s.

The surface of the ventral valve is covered by numerous rounded longitudinal ribs which are of about equal width for their entire length and slightly converge towards the mesial sinus. This radiating sculpture is crossed by delicate concentric ribs in the visceral and apical portions. Some of the longitudinal ribs are dichotomous. Two points of attachment of broken-off spines may be observed in the vicinity of the mesial sinus.

Exact measurements of this specimen can barely be given on account of its incomplete state of preservation.

Locality and geological position ; number of specimens examined.—Dark blue limestone with *Fenestella*, Barus, Kashmir ; Coll. Lydekker ; 1.

Remarks.—It strongly resembles the delicately ribbed variety of *Productus grationsus*, Waagen. This is also the case with the Yorkshire specimens of *P. longispinus*, as has already been remarked by Rothpletz (Palaeontographica, Bd. 39, 1892, p. 76). This strong resemblance even induced L. de Koninck (Monogr. des foss. carb. de Bleiberg, p. 25) to class a typical representative of *P. grationsus*, *P. semireticulatus*, Boyrich, from Timor (Abhandl. Koenigl. Akad. d. Wiss, Berlin, 1865, Taf. II, fig. 4) among the synonyms of *P. longispinus*. The differences, enumerated by Rothpletz, especially the more delicate ornamentation in the apical region and the less strongly developed wing in Sowerby's species appear however sufficient for a distinction of the two forms.

E 2

Davidson (Quart. Journ. Geol. Soc., Vol. XXII, p. 43) quotes *Productus longi-spinus* (?) from the grey limestone of Khoonmoo. It is however impossible to say whether he really had a true *P. longispinus* at hand, or rather a species of the subgenus *Marginifera*. He asserts that his specimen from Kashmir is iden-tical with some specimens found in the Punjab. But the majority of the Punjab specimens considered as identical with Sowerby's species have been proved by Waagen to belong to *Marginifera* (especially to *M. typica*), although the true *P. longispinus* is certainly not altogether absent in the Salt Range.

Productus longispinus is a wide-spread species of a considerable horizontal and vertical range. It occurs in strata of lower and upper carboniferous age in Western Europe, in the Moscovian and Gshelian stage of Russia, in the upper car-boniferous rocks of Loping and Sumatra (?) and in the Productus limestone of the Salt Range. Its occurrence in the coal-measures of North America is not yet beyond every doubt. The specimen considered as identical with *P. longispinus* by Meek and with *P. Orbignyanus* by Geinitz, is the only one which resembles some of Sowerby's types so very closely that their identity is probable. The case is different with the specimen figured by Meek on Pl. VI, fig. 7, of the final report on the Palæontology of Nebraska, which differs from *P. longispinus* by having an almost smooth shell. *P. longispinus*, Meek (Report of the geological ex-ploration of the 40th parallel, Vol. IV, Palæont., p. 78, Pl. VIII, fig. 4) from Nevada is more gibbous and provided with larger costæ than the true *P. longispinus* from the British and Belgian mountain-limestone. *P. splendens*, Norwood and Pratten (Journ. Acad. of nat. sciences of Philadelphia, 2nd ser., Vol. III, p. 11, Pl. I, fig. 5), and *P. Wabashensis*, Norw. and Pratt. (ibid. p. 13, Pl. I, fig. 6) must according to Waagen be classed among the subgenus *Marginifera* although the latter species is considered to be directly identical with *P. longispinus* by Meek, who insists that the figures given by the two above-mentioned authors are quite defective and misleading. *Productus scitulus*, Meek and Worthen (Geological Survey of Illinois, 1868, Vol. II, Palæont., p. 280, Pl. XX, fig. 5) and *P. parvus*, Meek and Worthen (ibidem, p. 297, Pl. XXIII, fig. 4), which have been placed among the synonyms of *P. longispinus* by L. de Koninck (Fossiles carbonifères de Bleiberg, p. 25), ought to be maintained as proper species, especially *P. scitulus*, which is distinguished by its very delicate ornamentation. A similar remark applies to *P. capacii*, d'Orbigny (Voyage dans l'Amérique Méridionale, T. III, 4 ème ptie., Paléontologie, Pl. III, figs. 24—26) from Yarbichambi in South America. *P. longispinus*, White (Report upon the U. S. geograph. surv., West of the 100th Meridian, Vol. IV, Paleontology, 1877, Pt. II, p. 116, Pl. VIII, fig. 6) from Santa Fé (New Mexico) is probably a representative of Waagen's sub-genus *Marginifera*, judging at least from the figure given by Prof. White.

Nor am I convinced of the identity of *P. longispinus* with the species described by Toula from the Hornsund in Spitzbergen; Toula's description, it is true, agrees very well with his identification of the two forms, but the figure does not strengthen this view. In this figure the concentric ornamentation is represented as quite indistinct in the apical portion of the ventral valve, whereas it continues all over its visceral part in the shape of broad stripes, thus differing considerably from the delicate reticulation which is so conspicuous in Davidson's type specimens.

The specimen from the Nordfjord of Spitsbergen described as *P. longispinus var. setosa* by Toula (Permocarbon Fossilien von der Westkuste von Spitsbergen, Neues Jahrb. f. Mineral. 1875, p. 252, Taf. VIII, fig. 4) must, according to my humble opinion, be kept separate from the British species on account of its considerably less numerous and coarse ribs. Thus the presence of *P. longispinus* in the permo-carboniferous deposits of Spitsbergen has not yet been established with certainty.

Both L. de Koninck (Recherches sur les fossiles paléozoiques de la Nouvelle Galles du Sud, p. 191, Pl. XI, fig. 3) and R. Etheridge (Quart. Journ. Geol. Soc., 1872, Vol. XXVIII, p. 333, Pl. 18, fig. 9) mention the occurrence of *P. longispinus* in carboniferous beds of Australia. I think however that L. de Koninck's specimen from New South Wales can only be identified with Sowerby's species, if such wide interpretation of the latter is permitted. The specimen from the Don River in Queensland is a rather indifferently preserved cast of a dorsal valve. Its identification with *P. longispinus* has been questioned by Etheridge, jun. (Geology and Palæontology of Queensland and New Guinea, p. 255).

IV. Section: SPINOSI.

5. Productus cf. scabriculus, Martin. Pl. II, figs. 8, 9.

1809. *Anomites scabriculus*, Martin, Petrificata Derbiensia. p. 8, Pl. XXXVI, fig. 6.

1836. *Productus scabriculus*, Phillips, Geology of Yorkshire, Vol. II, Pl. VIII, fig. 2.

1836. *P. quincuncialis*, Phillips, ibidem, Pl. VIII, fig. 5.

1843. *P. scabriculus*, L. de Koninck, Description des animaux fossiles du terrain carbon. de Belgique, p. 190. Pl. XI, fig. 3.

1845. *P. scabriculus*, E. de Verneuil, Géologie de la Russie d'Europe, Vol. II, Paléontologie, p. 271. Pl. XVI, fig. 5, Pl. XVIII. fig. 5.

1847. *P. scabriculus*, L. de Koninck, Monographie des genres Productus et Chonetes, p. 111, Pl. XI, fig. 6.

1862. *P. scabriculus*, Davidson, Monogr. British Carb. Brachiopoda, p. 169, Pl. XLII, figs. 3-8.

1866. *P. scabriculus*, Davidson, Quart. Journ. Geol. Soc., London, Vol. XXII, p. 63, Pl. 11, fig. 13.

1873. *P. scabriculus*, L. de Koninck, Monographie des fossiles carbonifères de Bleiberg, p. 37, Taf. I, fig. 16.

1876. *P. scabriculus*, Trautschold, Die Kalkbrueche von Mistschkowa, p. 69, Taf. VI, fig. 1.

1876. *P. scabriculus* (?), L. de Koninck, Recherches sur les foss. paléozoique de la Nouvelle Galles du Sud, p. 196.

1889. *P. scabriculus*, Tscherny schew, Mem. Com. Géol. St. Pétersbourg, Vol. III, No. 4, p. 271, Taf. VI, fig. 12.

This species is probably represented by two specimens. One of them is a ventral valve, whereas the second, smaller one represents a dorsal valve. Neither of them is sufficiently well preserved to permit identification.

The ventral valve (fig. 8) is medium sized, and in its dimensions and general shape agrees pretty well with Davidson's type-specimen of *P. scabriculus var. quincuncialis*, Phill., from Yorkshire (Pl. XLII, fig. 6). Much wider than long, tolerably inflated, provided with a hinge-line, which is inferior in length to the greatest width of the shell, and with a broad, but not deeply excavated mesial sinus. The auriculated expansions are small and pointed.

Although this specimen is much crushed and its surface weathered, its ornamentation is still partially visible. It consists of numerous, radiating ribs, swelling out at short intervals into elongated, protracted tubercles, which exhibit an indistinctly quincuncial arrangement. Both the longitudinal ribs and the numerous tubercles covering them appear to be rather more delicate than in the majority of the British and Russian types of *P. scabriculus*, but in this respect agree well with the ornamentation exhibited by Davidson's specimens from Kashmir.

Traces of concentric wrinkles are very indistinctly marked.

The dorsal valve (fig. 9) is also much wider than long, slightly concave, provided with a short hinge-line and with a low median elevation, corresponding to the sinus in the opposite valve. Its surface is covered by numerous, elongated, radiating grooves, which, like the tubercles in the ventral valve, are arranged in a sort of irregular quincunx. This radial ornamentation is crossed by a concentric sculpture, which is most distinctly marked in the apical region and on the wings.

The measurements of the larger specimen (fig. 8) are as follow :—

Length of the ventral valve in a straight line	. . . . . .	37 mm.
„ „ „ „ along the curve	. . . . . .	ca. 64 „
Length of the hinge-line	. . . . . .	ca. 34 „
Breadth of the ventral valve	. . . . . .	46 „
Thickness	. . . . . .	19 „

The measurements of my second specimen (fig. 9) are as follow :—

Length of the dorsal valve	. . . . . .	22·5 mm.
Breadth „ „	. . . . . .	34 „
Length of the hinge-line	. . . . . .	22 „

Locality and geological position ; number of specimens examined.—Dark limestone with mica, Barus, Kashmir Valley ; Coll. Lydekker ; 2. The specimens referred to this species by Prof. Davidson were found at Barus and Khoonmoo.

Remarks.—On the whole the determination of the present specimens as *Productus scabriculus* cannot be far wrong and accords with the description of Prof. Davidson's specimen. *P. scabriculus* is a widespread species, which ranges from lower carboniferous into permo-carboniferous strata, but seems to be most common in beds of middle and upper carboniferous age. In Europe it has been found in the mountain-limestone of Great Britain and Belgium and in the lower carboniferous Noetscher Schichten of Bleiberg in Carinthia. In Russia it was found in the different stages of the carboniferous system and in the Artinskian strata of the Ural. L. de Koninck mentions the species from the carboniferous beds of New South Wales, but without giving any figure. Among the fossils collected by Drasche near the Norilfjord of Spitzbergen and described by Toula, there is a specimen of *Productus* which has been identified as *P. cf. scabriculus* (Permo-Carbon-Fossilien von der Westkueste von Spitzbergen, Neues Jahrb. f. Mineral., 1875, p. 252, Taf. VIII, fig. 6). This identification seems to me extremely doubtful, judging by the figure, which represents a *Productus* distinguished by the presence of very coarse, irregular and mostly dichotomous ribs, and by the rarity of tubercles.

Some American shells, as *Productus asperus*, M'Chesney (Description of new species of fossils from the palaeozoic rocks of the Western States, 1868, p. 34, Pl. I, fig. 7), *P. Wilberanus*, M'Chesney (ibidem, p. 36, Pl. I, fig. 8), *P. Rogersii*, Norwood and Pratten (Journ. Acad. of Natural Sciences, Philadelphia, Vol. III, 1854, p. 9, Pl. I, fig. 3), *P. symmetricus*, Meek (Final Report of the U. S. Geol. Surv. of Nebraska, p. 167, Pl. V, fig. 6), *P. Nebrascensis*, Owen (ibid., p. 165, Pl. II, fig. 2; Pl. IV, fig. 6; Pl. V, fig. 11), have been classed among the synonyms of the British species by L. de Koninck and looked upon as local varieties only. Leaving the question undecided whether these forms should be regarded as distinct species, which I consider preferable, or merely as variations of *P. scabriculus*, still the fact remains that the typical *P. scabriculus* is certainly absent in the coal-measures of North-America. The specimen from Pecos Village described and figured by Marcou (Geology of N. America, Zurich, 1858, p. 47, Pl. V, fig. 5) is certainly different from Martin's species. The same remark applies to *P. scabriculus*, Abich (Geologische Forschungen in den Kaukasischen Ländern, 1, Theil, Eine Bergkalk Fauna aus der Araxes-Enge bei Djoulfa, p. 33, Taf. V, fig. 3), which this author himself has excluded from the synonyms of the true *P. scabriculus* in his additional remarks and which has been made the prototype of another species, *P. Abichi*, by Waagen.

6. PRODUCTUS CF. SPINULOSUS, Sowerby. Pl. II, fig. 12.

1814. *Productus spinulosus*, Sowerby, Mineral Conch., Pl. LXVIII, figs. 5, 6.
1836. *P. granulosa*, Phillips, Geology of Yorkshire, Pl. VII, fig. 14.
1843. *P. Cancrini*, L. de Koninck, (non de Verneuil), Déscription des animaux foss. du terrain carbonifère de Belgique, Pl. IX, fig. 3.
1843. *P. papillatus*, L. de Koninck, ibidem, Pl. X, fig. 6, p. 9-1.
1847. *P. granulosus*, L. de Koninck, Monographie des genres Productus et Chonetes, p. 135, Pl. XVI, fig. 7.
1862. *P. spinulosus*, Davidson, Monogr. British Carb. Brachiopoda, p. 176, Pl. XXXIV, figs. 18-21.
1866. *P. spinulosus, ?* Davidson, Quart. Journ Geol. Soc., London. Vol. XXII, Pl. II, fig. 16.
1868. *P. granulosus*, Krotow, Geologische Forschungen am Westlichen Ural-Abhange in den Gebieten von Tschardyn and Knolskansch, Mém. Com. Géol. St. Pétersbourg, Vol. VI, p. 4-4, Taf. I, figs. 14, 16
1862. *P. spinulosus*, Tschernyschew, Mém. Com. Géol. St. Pétersbourg, Vol. III, No. 4, p. 281.

Among the carboniferous fossils obtained by Captain Godwin-Austen at Barus there is a single ventral valve, bearing much resemblance to this species. It is however too imperfectly preserved to permit of a decided identification.

The shell is small, transversely semicircular and regularly curved in both directions. Hinge-line shorter than the greatest width of the shell. The beak is not very involute and scarcely overhangs the hinge-margin. Ears very small. There is no indication of a distinct sinus, only a flat depression in the middle part of the front.

The ornamentation is indifferently preserved and can only be made out in the vicinity of the lateral and frontal margins. It seems to consist exclusively of small but numerous, subregular tubercles, without any regular quincuncial arrangement, as in the variety, to which the term "*granulosus*" was originally applied.

Neither longitudinal ribs nor concentric striæ can be made out, but the specimen is too much weathered to state this with any degree of certainty.

In general shape and sculpture the specimen seems to agree better with *P. spinulosus* than with any other form of this genus. It also strongly resembles *P. opuntia*, Waagen (Salt Range Fossils, Pal. Ind., Ser. XIII, Vol. I, Prod. Limest. Foss., p. 707, Pl. LXXIX, figs. 1, 2) from the Cephalopoda (Jabi) beds of the upper division of the Productus Limestone, but I should not like to identify it with this species, which, according to Waagen's description, is always provided with a very strongly elevated median and apical part of the ventral valve. In this character, however, my specimen does not agree with the Indian shell. *Productus Wallacianus*, Derby (Bull. of the Cornell University, Ithaca, 1874, Vol. I, No. 2, p. 57, Pl. III, figs. 46—49, Pl. VI, fig. 5) from the coal-measures of Itaituba may also be compared with our specimen. Taking into consideration Derby's statement that *P. Wallacianus* chiefly differs from *P. spinulosus* by the absence of concentric wrinkles on the ears, this Brazilian species may perhaps be very closely related to our *Productus* from Kashmir, but the figures given by Derby are too bad to allow of a closer comparison.

The measurements of my specimen are as follow :—

Length of the shell in a straight line	18·6 mm.
" " " along the curve	18 "
Greatest breadth of the shell	16 "
Length of the hinge-line	12 "
Thickness of the ventral valve	5 "
Distance of the apex from the frontal margin	12 "

The proportion between the last dimension and the entire length of the shell most clearly shows the difference between this specimen and Waagen's *P. opuntia*, in which the apical region is always highly elevated above the proper beak. The measurements of my specimen agree almost exactly with those of Sowerby's type-specimen of *P. spinulosus*, as figured in Davidson's monograph (Pl. XXXIV, fig 18).

Locality and geological position; number of specimens examined.—Coarse grey, semi-crystalline limestone, Barus, Kashmir Valley ; Coll. Godwin-Austen ; 1. Two specimens referred with some doubt to *P. spinulosus* by Davidson were obtained from Khoonmoo.

Remarks.—*Productus spinulosus* has been quoted from the mountain limestone of Great Britain and Belgium by Davidson and L. de Koninck, from the upper-carboniferous rocks of the Ural Mountains by Krotow and from the Artinskian stage by Tschernyschew. The specimens considered as identical with *P. spinulosus* by Abich (Geologische Forschungen in den Kaukasischen Laendern, 1. Theil, Eine Bergkalk Fauna aus der Araxes-Enge bei Djulfa, p. 51, Taf. V, fig. 9, Taf. IX, fig. 22) have been referred to *P. horridus* by Val. von Moeller.

V. Section : FIMBRIATI.

7. Prodcotus Abiohi, Waagen. Pl. I, fig. 8.

1863. *Productus Humboldti*, Lydekker. Geology of the Kashmir and Chamba territories, Memoirs Geol. Surv. of India, Vol. XXII, Pl. II, fig. 3.

1884. *P. Abichi*, Waagen, Salt Range Fossils, Palæont. Indica, Ser. XIII, Vol. I, Prod. Limest. Foss., p. 697, Pl. LXXIV, figs. 1—17.

For a complete list of synonyms I refer the reader to my memoir on the permo-carboniferous fauna of Chitichun No. 1, (Pt. III of this vol.).

Of this beautifully sculptured *Productus* an excellently preserved ventral valve has been figured by Lydekker and considered identical with *P. Humboldti*, d'Orbigny. But its coarse sculpture and its less numerous, elongated tubercles which are mostly arranged in a rather regular quincunx, distinguish the present specimen from the true *P. Humboldti*. In general shape this specimen bears much resemblance to my type specimen of *Prod. Abichi* from the permo-carboniferous limestone of Chitichun No. I. (Pt. III of this volume, Pl. III, fig. 8), especially as regards the attenuated character of its apical region. It is however less inflated and provided with a very shallow median sinus only. In this respect it agrees best with the example figured by Waagen on Pl. LXXIX, fig. 4, of his monograph. The hinge-line is considerably shorter than the greatest breadth of the shell.

The ornamentation is exactly the same as in Waagen's type-specimen from the upper Productus limestone. The coarse, elongated, quincuncially arranged tubercles are crossed by concentric, delicate lines of growth. A narrow zone of roundish pustules is restricted to the immediate vicinity of the lateral and front margins only.

The measurements of this specimen are as follow :—

Length of the shell in a straight line	29·5 mm.
„ „ „ along the curve	33 „
Greatest breadth of the shell	28·5 „
Thickness of the ventral valve	8 „
Length of the hinge-line	20 „

The small size of this specimen makes its distinction from the closely allied *P. gangeticus*, Diener (Pt. IV of this volume, Pl. I, figs. 1—3, Pl. II, fig. 3), one of the leading species of the permian Productus shales of Painkhanda, an easy matter.

Locality and geological position ; number of specimens examined.—Black limestone, summit of ridge N. E. of Prongam Trál, Kashmir; Coll. Lydekker; 1.

Remarks.— *Productus Abichi* is one of the most characteristic fossils of permo-carboniferous and permian rocks in Armenia, India and Timor. It has as yet never been found in beds of an older than permo-carboniferous age. At the locality, where it has been obtained by Lydekker, it is associated with *Marginifera himalayensis*, Diener, *Strophomena analoga*, Phill., and *Chonetes grandicosta*, Waagen.

The specimen from Khoonuwo, which has been referred to *P. Humboldti* by Davidson (Quart. Journ. Geol. Soc., Vol. XXII, 1866, p. 43, Pl. II, fig. 14), does not permit an exact determination, on account of its fragmentary condition. But it seems at all events to be specifically different from *P. Abichi*.

8. PRODUCTUS PUSTULOSUS, Phillips, Pl. I, fig. 13.

1834. *Productus pustulosus*, Phillips, Geology of Yorkshire, Vol. II, p. 316, Pl. VII, fig. 15.

1836. *P. ovalis*, Phillips, ibidem, p. 216, Pl. VIII, fig. 14.

1843. *P. punctatus*, L. de Koninck (non Martin), Déscription des animaux foss. du terrain carbon. de Belgique, p. 196, Pl. IX, fig. 6, Pl. XII, bis fig. 3.

1847. *P. pustulosus*, L. de Koninck, Monographie des genres Productus et Chonetes, p. 118, Pl. XII, fig. 6, Pl. XIII, fig. 1, Pl. XVI, figs. 6—9.

1862. *P. pustulosus*, Davidson, Monogr. British Carb. Brachiopoda, p. 168, Pl. XLI, figs. 1—8, Pl. XLII, figs. 1—4.

1863. *P. pustulosus*, Roemer, Zeitschr. d. Deutsch. Geol. Ges. p. 591, Taf. XVI. fig. 8.

1870. *P. pustulosus*, Roemer, Geologie von Oberschlesien, p. 60, Taf. VIII, fig. 5.

1878. *P. pustulosus*, L. de Koninck, Monographie des fossiles carboniferes de Bleiberg en Carinthie, p. 26. The reference to Pl. I, fig. 31 is erroneous, no figure of this species being actually given in L. de Koninck's memoir.

1883. *P. pustulosus var. palliata*, Kayser, Obercarbonische Fauna von Loping, Richthofen's "China," Vol. IV, p. 166, Taf. XXVII, figs. 9—13.

1883. *P. cf. pustulosus* (?), Krotow, Geologische Forschungen am Westlichen Ural-Abhange, etc., Mém. Com. Géol. St. Pétersbourg, Vol. VI, p. 406.

1895. *P. pustulosus*, Tornquist, Das fossilführende Untercarbon am östlichen Rossberg-Massiv in de Südvogesen, Abhandl. Geol. Spec. Karte von Elsass-Lothringen, Bd. V, Heft 4, p. 72, Taf. XIV, fig. 3.

1896. *P. pustulosus*, Jalom, Le terrain carbonifère marin de la France Centrale, p. 67, Pl. VII, figs. 1—3, Pl. IX, fig. 1, Pl. X, fig. 3, Pl. XII, fig. 7, Pl. XIII, figs. 5, 6.

A very well preserved cast of a dorsal valve exhibits the characteristic shape and sculpture of this *Productus*, which, in its typical forms at least, may be easily distinguished from congeneric species by its peculiar ornamentation.

The shell is rotundate-square, nearly as long as wide, very flatly concave in the apical and visceral portions, but is distinctly geniculated, where the trail commences, exactly as in the Chinese variety, described as *"palliata"* by Kayser. It is divided by a broad but rather shallow median elevation which originates at some distance from the apex. The hinge line is a little shorter than the greatest width of the shell. The ears are broad and perfectly flat.

The surface of the cast is covered with numerous, tolerably regular, concentric wrinkles, crossing the radial ornamentation, which consists of coarse elongated pustules, distinctly arranged in quincunx. The immediate vicinity of the very apex is covered with closely packed granulations. At a distance of 2 mm. from the apex the concentric wrinkles make their first appearance and gradually increase in strength towards the shell-margin.

The specimen does not seem to me distinguishable from the *var. palliata* of *P. pustulosus*, as it has been described by Kayser. It can scarcely be mistaken for a species of the group of *P. Humboldti*, d'Orb., on account of its much coarser ornamentation. From *P. punctatus*, Martin, it distinctly differs by the character and distribution of its tubercles. It is distinguished from *P. Buchianus*, de Kon., by its much larger size, from *Prod. Leuchtenbergensis*, de Kon., by the concave shape of the dorsal valve, from *P. fimbriatus*, Sow., by the larger number of its concentric wrinkles.

Measurements of the specimen :—

Length of the dorsal valve	34 mm.
Breadth „ „ „	27 „
Depth „ „ „	7 8 „
Length of the hinge line	30 „

Locality and geological position; number of specimens examined.—Coarse grained, dark limestone, with a reddish weathered surface, containing remains of *Fenestellæ* ; Barus, Kashmir Valley ; Coll. Godwin-Austen ; 1.

Remarks.— *Productus pustulosus* is known from beds of lower carboniferous age in England, Ireland, Belgium, Central France, Alsatia, Silesia, Carinthia and Central Russia (?). A variety of this species has been described from Loping by Kayser. It differs from the typical form by the presence of a broad, distinctly marked off, geniculate trail. It is this upper carboniferous variety with which the type specimen from Kashmir ought to be identified.

The specimen from New Mexico considered as identical with *P. pustulosus* by Marcou (Geology of N. America, p. 48, Pl. VII, fig. 1), is distinguished from this species by the presence of a finely tuberculated zone in the vicinity of the anterior and lateral shell-margins.

P. pustulosus var. minutus, Abich (l. c. p. 37, Taf V, fig. 10) from the permian rocks of Djulfa has been referred to *Strophalosia horrescens* by Val. von Moeller.

9. PRODUCTUS PUNCTATUS, Martin, Pl. II, fig. 11.

1809. *Anomites punctatus*, Martin, Petrificata Derbiensia, p. 8, Pl. XXXVII, fig. 6 (figs. 7, 8 exclusis).

1814. *Terquexia rossus*, Parkinson, Organic Remains, etc., Vol. III, p. 177, Pl. XII, fig. 11.

1823. *Anomites thecarius*, Schlotheim, Nachträge zur Versteinerungskunde, Vol. 1, p. 63, Pl. XIV, fig. 1.

1823. *Productus punctatus*, Sowerby, Min. Conch., Vol. IV, p. 23, Pl. 323.

1836. *Producta punctata*, Phillips, Geology of Yorkshire, Vol. 11, p 216 ; Pl. VIII, fig. 14.

1837. *Leptæna subrufa*, Fischer von Waldheim, Oryctographie du Gouvernement de Moscou, p. 143, Pl. 22, fig. 2 (non Sowerby).

1841. *Productus punctatus*, L. v. Buch, Abhandlungen, herausg. Akademie d. Wiss. Berlin 1. Th. p. 34, Taf. 11, figs. 10, 11.

1843. *P. punctatus*, L. de Koninck, Description des animaux foss. du terrain carbon. de Belgique, p. 198, Pl. VIII, fig. 4 ; Pl. X, fig. 3

1845. *P. punctatus*, E. de Verneuil, Géologie de la Russie d'Europe, Vol. 11. Paléontologie, p. 276, Pl. XVI, fig. 11.

1847. *P. punctatus*, L. de Koninck, Monographie des genres Productus et Chonetes, p. 133, Pl. XII, fig. 2

1854. *P. punctatus*, Shumard in R. Marcy's Exploration of the Red River of Louisiana, p. 188, Pl. 1, fig. 3 ; Pl. 11, fig. 1.

1858. *P. punctatus* (?), Marcou, Geology of North America, p. 48, Pl. VI, fig. 3.

1865. *P. subulispina*, M'Chesney, Description of new species of fossils from the palæozoic rocks of the Western States, p. 37. Illustrations of the same : 1865, Pl. 1, figs. 10, 11.

1862. *P. punctatus*, Davidson, Monogr. British carboniferous Brachiopoda, p. 172, Pl. XLIV, figs. 9-16.

1863. *P. punctatus*, Roemer, Geologie von Oberschlesien, p. 65, Pl. VII, fig. 2.

1872. *P. punctatus*, Meek, Final Report of the U. S. Geological Surv. of Nebraska, p. 169, Pl. II, fig. 4 Pl. IV, fig. 8.

1873. *P. punctatus*, L. de Koninck, Monographie des familles carbonifères de Bleiberg, p. 90, Pl. I, fig. 1v.

1873. *P. punctatus*, Meek and Worthen, Geological Surv. of Illinois, Vol. V, p. 566, Pl. 25, fig. 12.

1875. *P. punctatus* (?), Toula, Eine Kohlenkalk-Fauna von den Baren's Inseln. Sitzgber. Kais. Akad d. Wiss. Wien. math. nat. Cl. Bd. LXXI, I. Abth. p. 56.

1876. *P. punctatus* (?), L. de Koninck. Recherches sur les foss. paléozoïques de la Nouvelle Galles du Sud, p. 193, Pl. X, fig. 3.

1877. *P. punctatus*, White, Report upon the U. S. Geogr. Surveys W. of the one hundredth Meridian. Vol. IV, Palæontology, p. 114, Pl. VII, fig. 2.

1880. *P. punctatus*, Krotow, Mém. Com. Géol. St. Pétersbourg, Vol. VI, p. 6 4.

1889. *P. punctatus*, Tschernyschew, Mém. Com. Géol. St. Pétersbourg, Vol. III, No. 4, p. 273.

1890. *P. punctatus*, Nikitin, ibid Vol. V, No. 5, p. 68.

1892. *P. punctatus*, Schellwien, Die Fauna des Karnischen Fusulinenkalks. Palæontographica, Bd. 39, p. 25, Taf. V, fig. 1.

1896. *P. punctatus*, Julien, Le terrain carbonifère marin de la France Centrale, p. 89, Pl. VII, fig. 9 ; Pl. XI, figs. 3, 4 ; Pl. XII, fig. 8, Pl. XIV, figs. 4, 5.

This beautiful and easily recognised species which in Western Europe is most frequently associated with *Productus pustulosus* is represented in the Geological Survey's Himálayan collection from Kashmir by two ventral valves, which, although partly injured by weathering, are sufficiently well preserved to warrant a certain identification.

My specimens are somewhat broader than long, moderately arched and provided with a distinct medial sinus, originating a short distance from the very apex and extending to the front. The hinge-line is shorter than the greatest width of the shell. The auriculate expansions are barely defined from the swell of the umbo.

The most characteristic feature of this species is its peculiar sculpture. I am entirely unable to detect any differences in the ornamentation of my specimens and Davidson's or L. de Koninck's type-specimens. My two specimens clearly exhibit the numerous concentric and regular bands, separated from each other by smoother interstices and thickly set with very numerous and delicate spines. The concentric ridges increase in size from the apex towards the front, but in my second specimen, which attains 38 mm. in length, they become again smaller in the vicinity of the margins.

The measurements of the figured specimen, which can only be given approximately, are as follow :—

Length of the ventral valve in a straight line	77 mm.
„ „ „ „ along the curve	87 „
Breadth of the ventral valve	80 „
Thickness „ „ „	45 „
Length of the hinge-line	77 „

Locality and geological position ; number of specimens examined.—Dark blue shale with limestone partings, containing numerous *Fenestella* and indeterminable casts of *Strophalosia*, Barus, Kashmir Valley ; 2.

Remarks.—*Productus punctatus* is a very characteristic species, which ranges through the entire carboniferous system into permo-carboniferous strata. In Europe it has been described from the mountain-limestone of Great Britain, Belgium,

Central France and Silesia, from the Nootscher Schichten of Bleiberg in Carinthia of lower carboniferous age, from the upper carboniferous *Fusulina* limestone of the Krone (Carnian Alps), from the Moscovian, Gshelian and Artinskian stage of Russia. In North America it likewise ranges through all the strata of the carboniferous period, from the sub-carboniferous beds of Iowa, Illinois and Missouri into the upper carboniferous coal measures of Nebraska.

A dorsal valve figured by L. de Koninck from the carboniferous rocks of New South Wales has been attributed to *P. punctatus* by this learned author. But judging from the figure the specimen is too incomplete to allow a safe identification.

VI. Section : CAPERATI.

10. Productus aculeatus, Martin. Pl. I, figs. 6, 7.

1809. *Anomites aculeatus.* Martin, Petrificata Derb. p. 8, Pl. XXXVII, figs. 9, 10.

1814. *Productus aculeatus.* Sowerby, Miner. Conch. Vol. I, p. 156, Pl. LXVIII, fig. 4.

1836. *Producta tessypina*, Phillips, Geology of Yorkshire. Pl. VIII, fig. 13.

1836. *P. spinulosa*, Phillips (nec Sowerby), ibidem, Vol. II, Pl. VII, fig. 14.

1843. *P. gryphoides.* L. de Koninck (ex parte) Description des animaux foss. du terrain carbonif. de Belgique, p. 182, Pl. IX, fig. 1, Pl. XII, fig. 14.

1847. *P. aculeatus*, L. de Koninck ; Monographie des genres *Productus et Chonetes*, p 144, Pl. XVI, fig. 6.

1861. *P. aculeatus*, Davidson, Monogr. British Carb. Brachiopoda, p. 166, Pl. XXXIII, figs. 16-20.

1867. *P. aculeatus*, Verchère, Kashmir, the Western Himalaya and the Afghan Mts. Journ. Asiatic Soc. of Bengal, Vol. XXXV, Pt. 2, pp. 202, 213.

1873. *P. aculeatus*, L. de Koninck. Monographie des fossiles carbonifères de Bleiberg en Carinthie, p. 53, Pl. I, fig. 30.

1875. *P. aculeatus*, Toula, Eine Kohlenkalk Fauna von den Barents-Inseln, Sitzungsber. Kais. Akad. d. Wiss Wien, math. nat. Cl. Bd. LXXI, I Abth. p. 56, Taf. II, fig. 10.

1883. *P. aculeatus*, Kayser, Obercarbonische Fauna von Loping, Richthofen's China, Bd. IV, p. 166, Taf. XXVI figs. 1-8.

1885. *P. aculeatus*, Krotow, Mém. Com. Géol. St. Pétersbourg, Vol. VI, p. 479, Taf. I. figs. 16, 17.

1892. *P. aculeatus*, Schellwien, Die Fauna des Karnischen Fusulinenkalks, Palaeontographica, Bd. 39, p. 25, Taf. III, figs. 10, 11.

Among the few fossils of undoubtedly anthracolithic age which have been collected in the Kashmir valley by Dr. Verchère, a small species of *Productus* is most numerously represented, which agrees very well with *P. aculeatus*, Martin.

All my specimens are of moderate dimensions, nearly as broad as long, or of slightly elongated outline. The greatest width of the shell is nearly always situated nearer to the front margin than to the beak, thus imparting to my forms a somewhat trapezoid shape. The ventral valve is very strongly convex and provided with high, steeply curved, lateral parts. The hinge-line is considerably shorter than the greatest width of the shell and is overlaid by the attenuated, strongly involute apex. The ears are very small and not distinctly defined. No trace of a median sinus is developed in any of my specimens.

In the majority of my forms the sculpture consists of irregularly scattered tubercles, which in the anterior portion of the valve are occasionally transformed into longitudinal ribs. These radiating ribs are however never as numerous and

regular as in the Chinese variety described by Kayser, or in the specimen figured
by Davidson on Pl. XXIII, fig. 19, of his monograph of the British carboniferous
Brachiopoda.

Besides this peculiar sculpture a much more delicate concentric ornamentation
is exhibited in the less weathered specimens.

No dorsal valve of this species is known to me. The measurements of one of
my type specimens (fig. 7) are as follow :—

Length of the ventral valve in a straight line	.	.	.	.	.	.	16 mm.
„ „ „ „ along the curve	.	.	.	.	.	.	20 „
Breadth . . .	„	.	.	.	.	.	.	18 „
Thickness „ „ „	.	.	.	.	.	.	8 „
Length of the hinge-line	.	.	.	.	.	.	.	7 „

Locality and geological position ; number of specimens examined.—Coarse, dark
grey limestone, Kashmir ; Coll. Verchère ; 8.

Remarks.—*Productus aculeatus* is a very variable species, under which many
shapes differing considerably from Martin's type specimen have been united by
various authors. My forms seem to hold an intermediate position between
Martin's type, which almost entirely agrees with Toula's specimen from Hoefer
Island, and the Chinese variety described by Kayser. The Carinthian variety
figured by Schellwien is characterised by yet stronger ribs in the frontal portion
of the ventral valve. The specimen from New South Wales, considered as identical with *P. aculeatus* by L. de Koninck (Recherches sur les foss. paléozoiques
de la Nouvelle Galles du Sud. p. 204, Pl. XI, fig. 6) seems to differ considerably
from Martin's species by its uncommonly broad and strongly developed concentric
laminæ.

Productus aculeatus is a characteristic species of the carboniferous period, being
known from the mountain limestone of Belgium, Great Britain and Bleiberg in
Carinthia, and from the upper carboniferous beds of the Carnian Alps, of Russia,
Hoefer Island and China.

VII. Section : IRREGULARES.

11. Productus mongolicus, Diener. Pl. VI, figs. 7, 8.

1883. *Productus cf. Cora.* Kayser, Obercarbonische Fauna von Loping. Richthofen's China. Bd. IV, p. 184,
Taf. XXVII, fig. 8.

1899. *P. mongolicus,* Diener. Himálayan Fossils, Palæont. Indica, ser. XV, Vol I, Pt. 3. The permocarboniferous fauna of Chitichun No. 1. Pl. IV, figs. 8-10.

Two specimens from the anthracolithic rocks of Kashmir are referable to this
characteristic and easily recognised species. One of them is a tolerably well preserved ventral valve, which, although somewhat deformed by crushing, exhibits all
the leading features peculiar to the present species. It is of an elongately triangular
shape, provided with an acuminated, involute apex, and, as far as I am able to
judge, with very small, strongly depressed wings. The ornamentation is of exactly
the same pattern as in my type specimens from the permo-carboniferous lime-
stone-crag of Chitichun No. 1.

I think this specimen may be safely identified with *P. mongolicus*, not with
the closely allied *P. compressus*, Waagen (Salt Range Fossils, Pal. Ind., ser. XIII,
Vol. I, Prod. Limest. Foss., p. 710, Pl. LXXXI, figs. 1, 2) on account of its more
strongly developed, concentric sculpture, which equally affects the median and
lateral portions of the valve and on account of its lateral margins not being concealed
below the strongly compressed lateral parts of the shell, as is the case with the Salt
Range species, according to Waagen's description.

The measurements are approximately as follow:—

Length of the shell in a straight line . . . . 34 mm.
     "   "   "   along the curve . . . . 41 "
Breadth of the shell . . . . ca. 24 "
Thickness of the ventral valve . . . . 8 "

My second specimen is the cast of a large ventral valve, with a few fragments
of the dorsal valve adhering to it. The distance between the two valves must have
been very small, especially in the apical region. This specimen only differs from
the previously described one by its larger size and greater flatness. I think there
can be little doubt as to its identity with the present species.

Locality and geological position; number of specimens examined.—Dark, micaceous shales with limestone partings, containing numerous remains of *Bryozoa*,
Barus, Kashmir valley; Coll. Godwin-Austen; 2.

Remarks.—*Productus mongolicus* has been described from the upper carboniferous beds of Loping in China by Kayser, and from the permo-carboniferous limestone
of Chitichun No. 1 in Tibet by myself.

Subgenus : MARGINIFERA, Waagen.

1884. *Marginifera*, Waagen, Salt Range Fossils, Palæontologia Indica, ser. XIII, Vol. I. Prod. Limest. Foss.,
p. 713.

For a discussion of the subgeneric value of this group of *Producti*, *vide* my
monograph of the Chitichun fossils (Pt. 3 of this volume, pp. 30 to 82).

1. MARGINIFERA HIMALAYENSIS, nov. sp.   Pl. II, figs. 1-7; Pl. VI, figs. 1, 2.

1865. *P. semireticulatus*, Stoliczka, Geological sections across the Himalayan Mountains from
Wangtu bridge on the river Sutlej to Sungdo on the Indus, Mem. Geol. Surv. of India, Vol. V, Pt.
I, p. 70.
1865. *P. longispinus*, Stoliczka, ibidem, p. 69.
1883. *P. semireticulatus*, Lydekker, The Geology of the Kashmir and Chamba Territories, Mem. Geol.
Survey of India, Vol. XXII, Pl. II, fig. 2.
1896. *Marginifera cf. typica*, Diener, Triassische einer Geologischen Expedition in den Central Himalaya
von Johar, Hundes und Painkhánda, Lamkahr, Kais. Akad. d. Wiss. math. nat. Cl. Bd. LXII,
p. 596.
1898. *M. himalayensis*, Diener, The permocarb. fauna of Chitichun No. 1, Pal. Ind. ser. XV, Himálayan
Foss. Vol. I, Pt. 3, p. 35.

This is the most common species of Stoliczka's " Kuling Shales " in Spiti and is
met with in corresponding beds in Kashmir. The majority of specimens contained
in the Geological Survey's Himálayan collection have been mistaken for *Productus*

semireticulatus by Stolicska and Lydekker. With this species the present form has however scarcely anything in common but an external similarity in the sculpture of the dorsal valve. Its internal characters, on the contrary, prove it to be a typical representative of Waagen's subgenus *Marginifera* and to be most closely allied to *B. typica*, from which indeed it is distinguished by very subordinate characters only.

The present species varies considerably in its general shape and outlines. It is, as a rule, transversely oval or even transversely rectangular, but specimens of a subquadrate outline are also occasionally met with, and a few exceptional forms even exhibit an elongately oval shape.

The ventral valve is always very strongly inflated. Its visceral part is distinctly prominent above the hinge-line. It is either regularly curved in the longitudinal direction throughout its entire length from the apex to the front margin, or its apical region appears to be more or less flattened. In some specimens even a blunt geniculation sets in where the visceral and anterior portions of the valve unite. Close to the apex, almost at its point, a mesial sinus takes its origin and extends down to the front line. It is of variable width and depth, but is always distinctly developed. In some of my specimens it is extraordinarily deep and narrow, exactly of the same character as in the type specimen of *Productus gratiosus*, figured on Pl. III, fig. 7, of my monograph on the Chitichun brachiopoda (Pt. 3 of this volume). As a rule the mesial sinus is most deeply impressed in the visceral portion of the valve, but becomes more shallow towards the front. The beak is not much bent over, pointed, but not passing far within the hinge-line. The latter is straight and marks the greatest breadth of the shell. The wings are rather large and prominent, pointed at their extremities, somewhat triangular and strongly arched. They are defined by a sinuosity or furrow from each lateral margin. The lateral portions of the ventral valve are bent down in a very steep curve towards the wings. The lateral margins are rounded anteriorly from the sinuosity or furrow in advance of each wing, to the front, which is distinctly, though, as a rule, but slightly sinuous in the middle.

If the shell has been entirely preserved, the ornamentation consists of very numerous longitudinal striæ or costæ. The majority of the ribs remain quite regular for their entire length. Occasionally, however, two or three are united into a coarser one. The striæ or costæ are parallel and do not exhibit any tendency to converge towards the centre line of the mesial sinus, as it is the case in the typical form of *Marginifera typica*. In the apical region the delicate, radial plication is crossed by an equally delicate concentric sculpture, imparting to this portion of the shell an indistinctly reticulate appearance. The concentric striæ are always less numerous and separated by larger intervals than the radiating ones. Traces of spines are almost entirely absent.

Generally on somewhat worn-out examples the concentric striæ are so nearly obsolete that the radial plication only is exhibited. Specimens of this kind strongly remind of the figure of *Productus longispinus var. setosus*, given by Krotow in his memoir on the geological structure of the districts of Tschérdyn and

Ssolikamsk (Mém. Com. Géol. de la Russie, Vol. VI, St. Pétersbourg, 1588, Taf. I, figs. 12, 13). In specimens which are still more weathered the surface, at a first glance, presents the appearance of being entirely smooth. But even these specimens, like the one figured by Lydekker, almost always show the remains of either the concentric or radiating striæ on the more protected parts.

A small number of specimens which have been collected in the Kuling shales near Muth in Spiti by Dr. Ferdinand Stoliczka are distinguished by a slightly stronger ornamentation of their ventral valves than the rest of my forms. In these specimens (Pl. VI, figs. 1, 2), which clearly show the peculiar internal characters of the subgenus *Marginifera*, both the concentric and radiating plications are rather strongly developed. In the specimen, figured on Pl. VI, fig. 1, moreover, the radiating costæ standing next to the mesial sinus show a slight tendency to converge towards the latter. Regarding the small importance of these characters of difference, I do not think that these specimens ought to be considered as more than variations of the present species.

The dorsal valve is deeply concave and follows very closely the curve of the opposite one, thus leaving but very little room for the animal within. The visceral and anterior portions are very often separated by a strong geniculation. A broad, strongly elevated mesial ridge, corresponding to the sinus in the ventral valve, extends from the beak towards the front. The wings are flattened, but not distinctly marked off from the remainder of the valve. The sculpture is much more prominent than in the ventral valve. Both the apical and visceral portions are strongly reticulated, imparting to the casts of this valve an external similarity to small forms of *Productus semireticulatus*. In the trail this reticulated ornamentation is replaced by a delicate radial plication. The sculpture of the wings is more delicate than in the remainder of this valve. Traces of spines have but seldom been noticed.

The shelly substance of both valves is rather thin.

The measurements of one of my largest and most complete specimens (Pl. II, fig. 1) are as follow :—

Length of the shell in a straight line .	.	.	.	.	.	.	.	72½	mm.
„ „ „ along the curve	.	.	.	.	.	.	.	40	„
„ „ „ dorsal valve	.	.	.	.	.	.	.	19	„
Entire breadth of the shell	.	.	.	.	.	.	.	43	„
Thickness of the ventral valve	.	.	.	.	.	.	.	14	„
Distance of the two valves from each other .	.	.	.	.	.	.	7	„
Breadth of the shell without the wings	.	.	.	.	.	.	30	„

The characteristic feature of the subgenus *Marginifera*, viz., the prominent shelly ridge within the wings of the ventral valve, is clearly seen in Lydekker's type specimen of this species from Kashmir (Pl. II, fig. 5). The wing having been broken off on one side of this specimen, the peculiar crenulated ridge, as described and figured by Waagen, is exhibited. This internal ridge has also been noticed in several forms from the Kuling shales of Spiti collected by Dr. Stoliczka. In a large number of my specimens, moreover, the trail of the ventral valve is marked

u

off from the remainder of the shell by a distinct band or furrow, which corresponds
to the internal ridges along the margins of the dorsal valve.

Of other internal characters of this species nothing is known to me.

Locality and geological position; number of specimens examined.—The
Kuling shales of Spiti are very rich in specimens belonging to this species.
Among the Geological Survey's Himálayan collection from Spiti the following
localities are represented :—Kuling, Coll. Stoliczka, 35 ; Coll. Griesbach, 14 ; Khar,
Coll. Stoliczka, 4 ; Coll. Griesbach, 18 ; Muth, Coll. Stoliczka, 8 ; Lilang, Coll.
Stoliczka, 8. Six specimens have been collected by Dr. Gerard, but the exact
locality from which they have been obtained is not known.

In Kashmir our species has been collected by Lydekker in shales with limestone partings of the same lithological character as the Kuling shales of Spiti. The
specimens, 16 in number, were found on the western summit of a ridge N.E. of
Prongam Tral, associated with *P. Abichi,* Wang., *Strophomena analoga,* Phil., and
Chonetes grandicosta, Waagen.

It is not improbable that Dr. Verchère (l. c. p. 213) had this species in view
when describing his specimens of *Productus longispinus* from Kashmir as differing
from Davidson's types by " their well-defined, enrolled and horn-like ears."

Remarks.—The present species is very closely allied to *Marginifera typica,*
Waagen (Salt Range Fossils, Pal. Ind., ser. XIII., Vol. I, Prod. Liment. Foss., p.
717, Pl. LXXVI, figs. 4-7, Pl. LXXVIII, fig. 1), and it was only after much
consideration that I did not feel justified in identifying them altogether. If one
had to deal with ventral valves only, it should, in some cases at least, be difficult to
distinguish between them. This remark chiefly applies to the variety from Muth,
which, by its coarser sculpture, approaches very closely the typical form of the
Salt-Range species. The almost entire absence of spines, however, is a remarkable
feature in all my specimens of *M. himalayensis.*

The most striking proof of the distinctness of the two species rests on the
character of the dorsal valve which is much more strongly reticulated and provided
with a very prominent median fold in *M. himalayensis.* Nor are the two
remarkable rows of grooves present in the latter species, which in *M. typica* extend
inside along the ridges, separating the wings from the remainder of the shell.

A closer comparison of the present species with other forms of the subgenus
Marginifera seems hardly necessary, as the distinguishing characters are almost the
same as have been indicated by Waagen as separating *M. typica* from the rest of
congeneric species.

Genus : STROPHALOSIA, King.

The genus *Strophalosia* is not at all rare in the anthracolithic system of the
Western Himálayas, but nearly all the specimens which I have examined are
too fragmentary for specific determination.

The only Himálayan representative of this genus, the characters of which are tolerably well known, is *Strophalosia Gerardi*, King (A monograph of the permian fossils of England, London, 1850, p. 96, Pl. XIX. figs 6, 7). The forms on which this species has been founded by Prof. King were picked up by Dr. Gerard on the crest of a pass, leading from Ladakh into Biskhir, at a height of 17,000 feet. This species, which is not represented among my fossil material, has been compared by Waagen to *Strophalosia plicosa* from the lower Productus limestone of the Salt-Range and has been identified recently with an Australian shell from the Bowen river coalfield in Queensland by R. Etheridge, jun. (Geology and Palæontology of Queensland and New Guinea, London, 1892, p. 260, Pl. XIII, fig. 18, Pl. XIV, fig. 18; Pl. XL, figs. 7, 8). Whether the two dorsal valves from the Zewan beds of Kashmir, described and figured by Dr. Verchère as *Strophalosia* (?) *arachæoides* (Journ. Asiat. Soc. of Bengal, Vol. XXXV, Pt. 2, p. 213, Pl. IV, fig. 1a, 1b), really belong to this genus, is at least very doubtful.

Among the specimens of *Strophalosia*, contained in the Geological Survey's Himálayan collection, two species can be distinguished. One of them appears to be related to *St. costata*, Waagen, the other to *St. tenuispina*, Waagen.

1. STROPHALOSIA SP. IND. AFF. ST. COSTATA, Waagen. Pl. I, figs. 15, 16.

Besides a considerable number of very poorly preserved fragments, two incomplete ventral valves with partly preserved shell are referable to a species, which to me seems rather closely allied to *St. costata*, Waagen (Salt-Range Fossils, Pal. Ind., ser. XIII, Vol. I. Prod. Limest. Foss., p. 655; Pl. LXIII, figs. 7, 8; Pl. LXIV, fig. 1) from the lower Productus limestone (Amb beds) of the Salt-Range.

Both specimens are of transversely oval outline, but so much crushed as to appear strongly asymmetrical. The apex is too badly preserved to show the presence of point of attachment. Nor has it been possible to state the presence of an area with full certainty. A medial sinus is but slightly indicated.

The most characteristic feature of these two ventral valves is their ornamentation. It consists of tolerably sharp, elevated, radiating ribs, which are not quite regular for their entire length and increase in number towards the front either by bifurcation or by intercalation of new costæ. At irregular intervals the ribs are strongly nodose and ornamented with spines. These spines are most numerous in the vicinity of the lateral margins near both extremities of the hinge-line. This coarse radiating sculpture is crossed by a very delicate concentric ornamentation.

The measurements of the smaller specimen (fig. 15) are approximately as follow :—

Length of the shell	.	.	.	.	.	.	.	.	.	.	21 mm.
Breadth „ „ „	.	.	.	.	.	.	.	.	.	.	22 „
Length of the hinge-line	.	.	.	.	.	.	.	.	.	15 „

Locality and geological position; number of specimens examined.—Dark blue, shaly limestone, with *Fenestella*, N. of Eishmakam, Kashmir Valley; Coll. Lydekker; 2.

Remarks.—Among the Indian representatives of this genus the present

G 2

species has probably its nearest ally in *Strophalosia costata*, Waagen. It is, however, distinguished from the latter by its considerably larger dimensions, by the absence of a strongly developed mesial sinus and by its more numerous spines.

2. STROPHALOSIA CF. (?) TENUISPINA, Waagen. Pl. I, fig. 17.

1887. *Strophalosia tenuispina*, Waagen, Salt-Range Fossils, Pal. Ind., ser. XIII, Vol. I. Productus Limestone Foss., p. 664, Pl. LXIV. figs. 2—7.

A single incomplete ventral valve of a *Strophalosia* so closely resembles this species from the lower Productus limestone of the Salt-Range that I can scarcely be far wrong in referring it to the latter as its nearest ally.

Its general outline is almost circular. The ventral valve, which alone is accessible to observation, is but moderately inflated, as in Waagen's type specimen, figured on Pl. LXIV, fig. 2 of his monograph and is very regularly curved in either direction. The apex being broken off, one of the most important characters for the identification of a *Strophalosia* is unfortunately missing.

The ornamentation is nearly the same as in the specimen figured by Waagen on Pl LXIV, fig. 6 of his monograph. It consists of a comparatively small number of very thin, elongated spines, which are directed forward, firmly appressed to the surface of the valve, and arranged into approximately concentric rows. A few imbricating, concentric striæ of growth may be noticed besides this radiating sculpture.

I have not succeeded in cleaning the dorsal valve from the adhering matrix.

Exact measurements of this species cannot be given, as the materials at hand are too fragmentary. The figure will give, I hope, a sufficiently clear idea of its features.

Locality and Geological position; number of specimens examined.—Shaly, micaceous limestone with *Productus semireticulatus*, Barus, Kashmir Valley ; Coll. Godwin-Austen ; 1.

Remarks.—As this specimen is too imperfect to warrant a decided identification, its direct reference to *Strophalosia tenuispina* may have elements of doubt, but, on the other hand, I cannot satisfactorily compare it to any other among the congeneric species. Had it not been for the moderate inflation of the ventral valve and for the less numerous and more delicate spines, *S. plicata*, Waagen, might have put in a claim for a closer comparison.

In the Salt-Range *Strophalosia tenuispina* is restricted to the Chonetes-bed of the lower Productus limestone at Amb. The Barus beds of Kashmir with *P. semireticulatus* seem to hold a similar geological position.

Subfamily : CHONETINÆ, Waagen.

Genus : CHONETES, Fischer v. Waldh.

In the Himálayan collections of the Geological Survey from Kashmir and Spiti the genus *Chonetes* is represented by altogether four species. Among the subdivisions or sections of this genus established by L. de Koninck and partly emendated by Waagen, they can be grouped in the following manner:—

I. Section: STRIATÆ.

1. *Chonetes cf. Lissarensis*, Diener.
2. *Chonetes Austriana*, Davidson.

II. Section: GRANDICOSTATÆ.

3. *Chonetes grandicosta*, Waagen.
4. *Chonetes Barusiensis*, Davidson.

One of these species, *Chonetes cf. Lissarensis*, from the Kuling shales of Spiti, is probably identical with one of the most common leading fossils of the permian *Productus* shales of Johar, described by myself in the fourth part of this volume. *Chonetes grandicosta* had been previously discovered in the upper Productus-limestone by Waagen. The two remaining species are peculiar to the anthracolithic system of Kashmir. One of them, *Ch. Barusiensis*, had been originally classed among the genus *Spirifer* by Davidson. In placing it in the genus *Chonetes* I am following Waagen's opinion.

Besides *Chonetes Austriana* and *Ch. Barusiensis* two more species of this genus have been described by Davidson in his memoir on the carboniferous fossils, collected by Captain Godwin-Austen in Kashmir. These are *Ch. lævis*, Davidson (Quart. Journ. Geol. Soc., London, Vol. XXII, 1866, p. 44, Pl. II, fig. 17) and *Ch. Hardrensis var. Tibetensis*, Davidson (ibid, p. 36, Pl. I, fig. 7). The latter denomination ought to be changed into " *Kashmeriensis*," according to Lydekker, because the specimens in question were obtained from Kashmir, not from Skardú in Little Tibet, as it had been first erroneously supposed.

Chonetes lævis is a small and rather indifferent species, with a nearly smooth surface. Waagen has compared it to *Ch. Ambiensis*, Waag., *Ch. rotundata*, Toula, and *Ch. planumbona*, Meek and Worthen. He even suggested its probable identity with *Ch. rotundata* from Heefor Island. I think, however, that all these forms are specifically different from *Ch. lævis*. *Ch. rotundata*, Toula (Eine Kohlenkalk Fauna von den Barents-Inseln, Sitzungsber Kais. Akad. d. Wiss. Wien, 1875, math. nat. Cl. Bd. LXXI., p. 28, Taf. II. fig. 12, non 11) is much more strongly convex and not flattened towards the ears, which according to Davidson's description, is a prominent feature in the Himálayan form. *Ch. Ambiensis* Waagen (Salt-Range Fossils, Pal. Ind., ser. XIII., Vol. I. Prod., Limest. Foss., p. 618, Pl. LVIII, figs. 1-6) is also more strongly inflated and provided with a shorter hinge-line and with a prominent apex. From *Ch. planumbona*, Meek and Worthen (Geol. Surv. of Illinois, Vol. II. Palæont. p. 253, Pl. 18, fig. 1), *Ch. lævis* differs by similar characters.

The second species, *Chonetes Hardrensis var. Kashmeriensis*, belongs to a group of the carboniferous descendants of the devonian *Ch. Hardrensis*, Phill., of which *Ch. Laguessiana*, L. de Kon., is the prototype.

I. Section: STRIATÆ.

1. Chonetes cf. Lissarensis, Diener.   Pl. VI, fig. 3.

1897. *Chonetes Lissarensis*, Diener, Himálayan Fossils. Pal. Ind. ser. XV, Vol. I, Pt. 4. The fauna of the
       permian Productus shales of Johár and Painkhánda. Pl. II, figs. 4, 5, p.

Among the fossils collected by Stoliczka in the Kuling shales of Spiti
there is a single ventral valve of a *Chonetes*, which seemed to me indis-
tinguishable from *Ch. Lissarensis*, one of the most typical species of the permian
Productus shales of Johár.

Its outline is transversely trapezoidal, with rounded-off margins. It is
very gently curved in the longitudinal direction. The hinge-line corresponds to
the greatest width of the shell. The lateral parts are distinctly flattened and form
slight triangular wings. The apex is very slightly developed and pointed. No
spines have been noticed on the margins, which are limiting the area above.
The mesial sinus is cut out very sharply and bordered by steep marginal walls. It
is, however, neither deep nor wide, and disappears almost completely in the
vicinity of the front. A narrow, sharp, median fold rises from the bottom of the
sinus and is distinctly developed almost throughout its entire length. It is exactly
of the same character as the majority of the specimens from the Productus shales
of the Lissar Valley in Johár.

The specimen is a cast only. Its central portion is ornamented with very
numerous and delicate striæ, covered with numerous puncta. In the marginal
region the ornamentation is partly obscured by weathering, but in some places,
especially in the vicinity of the wings, the presence of deep grooves is distinctly
seen, which are directed towards the shell margin.

The measurements of this specimen are as follow :—

Entire length of the shell	. . . .	9 mm.
„ breadth „ „ along the hinge line	. . . . . .	17 „
Thickness of the ventral valve	. . . .	3 „
Apical angle, without the wings	. . . . ca.	120°

Locality and geological position ; number of specimens examined.—Kuling
shales, associated with *Marginifera Himalayensis*, Kuling, Spiti ; Coll. Stoliczka ; 1.

*Remarks —*This specimen can hardly be otherwise determined, I think, than
as *Ch. Lissarensis*. If I do not venture on a direct identification, it is because
the example is not complete and differences might perhaps be noticed, if the
dorsal valve were preserved. A closer comparison with other forms of this group,
Ch. Vishnu, Salter, *Ch. lobata*, Schellwien, *Ch. mesoloba*, Norw. and Pratt., seems
hardly necessary.

Chonetes Lissarensis is a very common shell in the permian Productus shales
of the Lissar Valley in Johár. Its presence in the Kuling shales of Spiti would
therefore be of no slight importance, with regard to a determination of their exact
stratigraphical position. The Kuling shales of Spiti, or, more exactly, their lower
portion, which contains a palæozoic fauna and is overlaid by the triassic *Otoceras*

stage, have been correlated with the Productus shales of Johár and Painkhanda
by Griesbach. The presence of *Ch. Lissareusis* in these beds would be strongly
in favour of this view.

2. CHONETES AUSTENIANA, Davidson. Pl. II, fig. 13.

1866. *Chonetes Austeniana*, Davidson, Quart. Journ. Geol. Soc., London, Vol. XXII, p. 44, Pl. II, fig. 18.
Ibid. *Ch. Austeniana*, Waagen, Salt-Range Fossils, Palæont. Indica, ser. XIII., Vol. 1, Prod. Limest. Foss., p. 643.

This species was introduced by Davidson from a single ventral valve, collected
by Captain Godwin-Austen in the anthracolithic rocks of Kashmir. My materials
for its description are unfortunately in no way more complete.

The only specimen, available for description, is a ventral valve of a transversely
oval outline. It is moderately inflated and very regularly curved both in the
longitudinal and in the transverse directions. There are no auricular expansions, although the lateral parts are slightly flattened near the extremities
of the hinge-line, which corresponds to the greatest width of the shell. The apex
is barely at all prominent, pointed, but not or only very little involute. At its
very extremity a narrow, concave, median sinus originates and extends to the
front, increasing gradually but very slowly in width and depth. On the lateral
portion of the valve two obscure, rounded depressions of very small dimensions
may be traced on each side of the medial sinus. These depressions are more
conspicuous than the broad, barely elevated folds, into which the lateral portions
are divided by the former. Otherwise the surface is almost smooth. But this
feature, I think, is only due to the weathering of the cast, because in some places
a large number of rounded punctures is indistinctly visible, which are arranged
into radiating striæ.

Neither the dorsal valve nor the internal characters of this species are known.
The measurements of my type specimen are as follow:—

Entire length of the shell	. . . .	19.5 mm.
„ breadth „ „	. . . .	16 „
„ thickness „ „	. . . .	24 „
Apical angle of the ventral valve without the wings	. . . .	130°

Locality and geological position; number of specimens examined.—Coarse,
grey limestone, made up almost entirely of undeterminable fragments of brachiopoda, Zewan beds, Barus, Kashmir Valley; Coll. Godwin-Austen; 1.

Remarks.—The present species has been classed by Waagen among the
section of *grandicostatæ*, and I must state my reasons for not concurring in
this view with that learned author. In its general characters *Ch. Austeniana*
approaches more nearly the "*striatæ*" than any form of the "*grandicostatæ*," being
characterised by a delicate, radiating striation, which is made up of a large number
of regularly arranged punctures, whereas the traces of larger plications are rather
obscure and restricted to the vicinity of the margins only. In this respect *Ch.
Austeniana* cannot advantageously be compared to *Ch. semiovalis*, Waagen (l. c.
p. 633, Pl. LXI, fig. 5) as has been suggested by Waagen. The latter species

undoubtedly belongs to the section of *grandicostatæ*, being distinguished by the presence of strong and moderately high, radiating ribs. I therefore should not think it advisable to unite *Ch. semioralis* and *Ch. Austeniana* in the same group of forms.

Among the "*striatae*" *Chonetes Austeniana* may be considered as the proto-type of a proper group, distinguished from the rest of congeneric species by the presence of a few, low depressions, affecting the lateral portions in the vicinity of the shell-margin.

II. Section : GRANDICOSTATÆ.

3. Chonetes grandicosta, Waagen. Pl. II, fig. 14.

1884. *Chonetes grandicosta*, Waagen, Salt Range Fossils, Palæont. Ind., ser. XIII, Vol. I, Prod. Limest. Foss., p. 688, Pl. LXL, figs. 4, 7.

The materials of this species in the Himálayan collection are very small, but the specimen is so characteristic that I think the determination can be made with sufficient accuracy.

The specimen serving for description is slightly inferior in size to Waagen's types, but perfectly agrees with them in all its characters. It is easily distin-guished from all the rest of congeneric forms by its strongly inflated ventral valve, its deep mesial sinus, and its well developed, radiating folds.

The ventral valve, which alone is accessible to observation, is a little wider than long, strongly but rather regularly convex. The hinge-line corresponds to the greatest width of the shell. The apex is slightly prominent, pointed and incurved, thus concealing the area almost entirely. In the specimen at my disposal the largest portion of the area is covered up by the rocky matrix.

Transversely the valve appears impressed in the middle by the presence of a very deep and broad, mesial sinus, which originates in the apex and is limited on both sides by prominent, rounded folds. These folds descend rather abruptly towards the flattened, pointed wings, from which they are marked off very sharply.

The entire valve is covered by a radiating sculpture, which is most prominent on the two elevated folds. The sinus is ornamented in its bottom by delicate, longi-tudinal striæ only, whereas on the elevated folds regular costæ make their appear-ance, the highest among them forming the very crest of each fold. The wings are apparently devoid of a radial sculpture, but are ornamented with imbricating, transverse striæ of growth. This transverse sculpture is but very faintly indicated in the remaining portions of the valve.

Neither the dorsal valve nor the internal characters of this species are known to me.

The measurements of the present specimen are as follow :—

Entire length of the shell	. . . .	8 mm.
„ breadth „ „ along the hinge-line	. . . .	9 „
„ thickness „ „	. . . .	4 „
Apical angle of the ventral valve, without the wings	. . .	ca. 80°

Locality and geological position ; number of specimen examined.—Micaceous, dark shales, with *Productus Abichi* and *Marginifera himalayensis*, western summit of a ridge N.E. of Prongam Trál, Kashmir ; Coll. Lydekker ; 1.

Remarks.—The only differences between this form and Waagen's type-specimens of *Chonetes grandicosta* are the somewhat more longitudinal shape and the smaller size of the former. In these characters it approaches more nearly *Ch. aequicosta*, Waagen (l. c. p. 639, Pl. LX, fig. 7), from which it differs however by its ornamentation and by its deeply impressed mesial sinus. The two above-mentioned points of difference are certainly too insignificant to forbid its identification with *Ch. grandicosta.*

In the Productus limestone of the Salt Range the present species is restricted to the Cephalopoda beds of Jabi.

4. CHONETES BARUSIENSIS, Davidson. Pl. VI, fig. 4.

1865. *Spirifer Barusiensis*, Davidson, Quart. Journ. Geol. Soc., London, Vol. XXII, p. 43, Pl. II, fig. 7.
1884. *Chonetes Barusiensis*, Waagen, Salt Range Foss., Pal. Ind. ser. XIII, Vol. I. Prod. Limest. Foss., p. 618.

Professor Davidson had only a single ventral valve of this species at his disposal, for which he introduced the present denomination. Nor are my materials in any way more complete. Thus our knowledge of the species necessarily must yet remain rather imperfect.

Davidson himself states his description of this shell, which he provisionally classed among the genus *Spirifer*, to be very incomplete. Waagen was the first to discover the close resemblance of *Spirifer Barusiensis* to the largely costate species of *Chonetes* from the *Productus* limestone of the Salt Range, and he consequently deemed it preferable to consider it as a ventral valve of a *Chonetes*. I fully agree with this opinion of that learned author.

Chonetes Barusiensis is of very small size and is easily distinguished from the rest of largely costate *Chonetes* by its transversely trapezoidal outline, being twice as broad as it is long. The greatest width of the shell corresponds to the hinge-line. The ventral valve is but very little inflated and equally curved in the longitudinal direction. A broad, rounded sinus originates in the very apex and extends to the front, increasing gradually in width and depth. The wings are large, flattened and pointed. The apex is not prominent. There are three folds within the mesial sinus and five on each of the two elevated parts of the valve on both sides of the sinus. The costæ within the sinus are less strongly developed than those on the elevated parts of the shell. The wings are devoid of any ornamentation.

Neither the dorsal valve nor the internal characters of this species are known to me.

The measurements of my type specimen are as follow :—

Entire length of the ventral valve	.	4 mm.
" breadth " " "	.	8 "
" thickness " " "	.	cca 1 "
Apical angle of the ventral valve, without the wings	.	ccs. 100°

Locality and geological position ; number of specimens examined.—Dark shales, weathering in greenish and reddish colours, with *Protoretepora*, Barus, Kashmir Valley ; Coll. Godwin-Austen ; 1.

Remarks.—Chonetes squamulifera, Waagen(l. c. p. 634, Pl. LX, figs. 1 to 4) from the middle and upper Productus limestone of the Salt Range has been considered as the nearest ally to the present species by Waagen. The two forms are, however, readily distinguished by the smaller size and greater width of the Himálayan shell, by its flatness, its larger wings and its distinctly impressed sinus, which is limited off more sharply from the elevated parts of the ventral valve.

On account of its great flatness and of its large, pointed wings *Ch. deplanata*, Waagen, might put in a claim for a closer comparison.

Family : *LYTTONIDÆ*, Zittel.

Sub-family : LYTTONIINÆ, Waagen.

Genus : LYTTONIA, Waagen.

LYTTONIA sp. ind.   Pl. II, figs. 15, 16.

Among the fossil materials, collected in the anthracolithic rocks of Kashmir by Lydekker, there are several fragmentary casts of a *Lyttonia*, which is probably very closely related to *L. nobilis*, Waagen, or to *L. tenuis*, Waagen, but does, unfortunately, not allow a specific determination, on account of the fragmentary character of the forms available for observation.

The smaller of the two figured casts, corresponding to the cardinal portion of a ventral valve, exhibits the peculiar triangular outline, which is common to all the species of this genus. The shelly substance has been entirely destroyed, but its characteristic, distinctly porous structure has been partly preserved on the cast. Both the strongly developed median septum and the numerous lateral septa are marked by deep furrows, which are corresponding to them in the casts. The surface of the latter is partly covered by the branches of *Protoretepora*.

The specimen, figured on Pl. II, fig. 16, seems to have attained very considerable dimensions, its width at the anterior border measuring about 60 mm.

The present species is perhaps identical with one of the congeneric forms from the Productus limestone of the Salt Range described by Waagen. With *Lyttonia Richthofeni*, Kayser (Obercarbonische Fauna von Loping, Richthofen's "China," IV. Bd., p. 161, Taf. XXI, figs. 9-11) it cannot be identified, because the median septum, which is very distinctly developed in my casts, seems to be entirely absent in the Chinese shell.

*Locality and geological position ; number of specimens examined.—*Dark shales with limestone partings, made up almost entirely of *Bryozoa*, especially of *Protoretepora*, Marble Pass, Kashmir ; Coll. Lydekker ; 4.

*Remarks.—*Lydekker (Records Geol. Surv. of India, 1884, Vol. XVII, p. 37) was the first to draw attention to the occurrence of *Lyttonia* in the anthracolithic deposits of the Kashmir Valley. The presence of this genus is of considerable importance for a safe correlation of these beds, as it has hitherto never been discovered in older than upper carboniferous strata.

Family : STROPHOMENIDÆ, King.

Sub-family : STROPHOMENINÆ, Waagen.

Genus : STROPHOMENA, Blainville.

1. STROPHOMENA ANALOGA, Phillips. Pl. II, fig. 17.

1836. *Producta analoga*, Phillips, Geology of Yorkshire, Vol. II, p. 215, Pl. II, fig. 10.
1840. *Leptæna dubata*, Sowerby, Miner. Conch., Vol. VII, Pl. 615, fig. 2.
1843. *Leptæna depressa*, L. de Koninck, Description des animaux fossiles qui se trouvent dans le terrain carbonifère de Belgique, p. 215, Pl. XII, fig. 3.
1844. *Leptagonia multirugata*, M'Coy, Synopsis of the characters of the carb. foss. of Ireland, p. 118, Pl. XVIII, fig. 12.
1861. *Strophomena rhomboidalis var. analoga*, Davidson, Monogr. British Carb. Brachiopoda, p. 119, Pl. XXVIII, figs. —15.
1862. *Strophomena analoga*, Davidson, On the lower carboniferous brachiopoda of Nova Scotia, Quart. Journ. Geol. Soc., London, Vol. XIX, p. 178, Pl. IX, fig. 18.
1867. *Str. analoga*, Verchère, Kashmir, the Western Himalaya and the Afghan Mts. Journ. Asiat. soc. of Bengal, Vol. 36, Pl. 2, p. 312, Pl. 11, fig. 4.
1872. *St. rhomboidalis var. analoga*, Etheridge, Quart. Journ. Geol. Soc., Vol. XXVIII, pp. 88? and 338, Pl. XV, figs. 3 and ? 5, 11. XVI, fig. 7, Pl. XVIII, fig. 1.
1874. *St. depressa*, Toula, Eine Kohlenkalk Fauna v. d. den Baren-In Inseln Spitzgsber. Kais. Akad. d. Wiss. Wien. LXXI, Bd. math. nat. Cl. I, Abth p. 22 Taf. II, fig. 8.
1876. *Strophomena analoga*, L. de Koninck, Nouvelles Recherches sur les fossiles paléozoïques de la Nouvelle Galles du Sud, p. 204, Pl. IX, fig. 3, Pl. XI, fig. 7.
1876. *St. rhomboidalis var analoga*, R. Etheridge jun. in Jack and Etheridge, Geology and Palæontology of Queensland and New Guinea, p. 245 Pl. XII, figs. 8, 9, Pl. LX, fig. 6.
1884. *St analoga*, Julien, L'terrain carbonifère moten de la France Centrale p. 84, Pl. III, figs. 8, 9.

This species which by the majority of palæontologists has only been admitted as a variety of the silurian *Strophomena rhomboidalis*, Wahlenberg, is represented among the fossils collected by Lydekker, in the anthracolithic rocks of Kashmir, by a tolerably well preserved, though incomplete, ventral valve. Exactly at the place where we ought to expect the sudden geniculation in the convexity of the shell, the latter has been broken off in my specimen.

The ventral valve is of a subtrapizoidal outline, considerably wider than long, moderately curved in both directions, and provided with large, slightly depressed auricular expansions, which join the lateral borders in a very regular curve. The hinge-line corresponds to the greatest width of the shell. The apical region strongly reminds one of a *Productus*, but exhibits the trace of the characteristic perforation in the vicinity of the beak, peculiar to the genus *Strophomena*.

The geniculated portion of the valve having been broken off entirely, the surface of my specimen shows the reticulate ornamentation only, restricted to the posterior portion of complete forms. The longitudinal striæ are very numerous and somewhat irregular. The concentric wrinkles are more strongly marked.

Among them those wrinkles, which are situated in the vicinity of the apex, meet the cardinal border at right angles or are even slightly converging towards the beak, whilst those, which are situated on the visceral portion of the valve follow the marginal curves in their direction and are consequently turned outwardly towards the cardinal angles.

My specimen is too incomplete to allow exact measurements to be given.

Locality and geological position; number of specimens examined.—Shales with limestone partings containing *Productus Abichi, Chonetes grandicosta* and *Marginifera himalayensis,* western summit of a ridge North-East Prongam Trál, Kashmir ; Coll. Lydekker ; 1.

Remarks.—This species belongs to a series of forms, which ranging from silurian into permo-carboniferous deposits, are changing so slightly in their shape and sculpture that the majority of palæontologists are inclined to consider them as variations of a single species only. L. de Koninck (Recherches sur les foss. pal de la Nouvelle Galles du Sud, p. 210), however, does not share in this view, but believes the carboniferous *St. analoga* to differ by a few though very subordinate constant characters from its silurian and devonian allies.

St. analoga, in L. de Koninck's circumscription of this species, is a form of a tolerably large geographical and geological distribution. It is known from the lower carboniferous deposits of western Europe and of Nova Scotia, from the upper carboniferous rocks of Queensland and of New South-Wales, and from the permo-carboniferous limestone of Barents' Island (North-West Nowaja Semla). With the specimens which have been collected from the last mentioned locality by Harler and described by Toula, Lydekker's example from Kashmir resembles most closely by the comparatively strong convexity of its ventral valve.

A very large form of a *Strophomena* has been figured by Dr. Verobère. But the author does not state whether it was obtained from the Zewan beds of Kashmir or from the Productus limestone of the Salt Range.

Sub-family : ORTHOTHETINÆ, Waagen.

Genus : DERBYIA, Waagen.

DERBYIA CF. SENILIS, Phillips. Pl. VI, figs. 5, 6.

1856. *Spirifer senilis*, Philips. Geology of Yorkshire. Vol. II, Pl. IX, fig. 5.

1861. *Streptorhynchus crenistria var. senilis*, Davidson, Monograph British carboniferous Brachiopoda, p. 154, Pl. XXVII, figs. 2, 3, 4, Pl. XXX, figs. 12, 14 (varietas senistria).

1865. *Streptorhynchus crenistria*, Davidson, Quart. Journ. Geol. Soc., London, Vol. XXII, p. 42, Pl. II, fig. 10.

1880. *Streptorhynchus crenistria var. senilis* (?) Roemer Ueber eine Kohlenkalk-fauna von den Westhuste von Sumatra, Palæontographica, 27 § 4., p. 6.

1880. *Orthothetes crenistria var. senilis*, Etheridge jun., Proceed. Royal Pro. Soc. of Edinburgh, p. 383, Pl. VII, figs. 12–16.

1883. *Orthothetes crenistria var. senilis.* Kayser, Obermshon-sche Fauna von Loping, Richthofen's "China," IV. Bd., p. 178, Pl. XXIII, figs. 1—7.

1884. *Derbyia senilis,* Waagen. Salt Range Fossils, Pal. Ind., ser. XIII, Vol. I, Prod. Limest. Fase., p. 583.

1892. *Derbyia senilis,* Etheridge jun., Jack and Etheridge, Geology and Palæontology of Queensland and New Guinea, p. 245, Pl 12, figs. 1 - 4.

It is with great reserve only that I refer some of the numerous fragments from the carboniferous rocks of Kashmir which have been identified with *Orthothetes crenistria* by Davidson to the present species. As has been stated by Professor Waagen himself, a distinction of the different species which he introduced in his genus Derbyia is extremely difficult "as the forms are very variable and seemed to be linked together by more or less numerous transitional shapes." A specific distinction therefore becomes impossible, if one has to do with fragmentary specimens.

Among my materials of *Derbyia*, accepting Waagen's separation of this genus and *Orthothetes*, there is only one single, fairly complete dorsal valve. It is on account of its strong similarity to the British and Chinese types of *Derbyia senilis* that I refer a few fragments of ventral valves to the same genus, although no trace of the characters, which led Waagen to a generic distinction between *Streptorhynchus*, *Derbyia* and *Orthothetes*, has been preserved in any of my specimens.

The dorsal valve, which alone, among all the rest of fragments provisionally referred to this species, is fairly complete, appears to me to have a stronger affinity with *Derbyia senilis*, than with any of the species described from the Productus Limestone of the Salt Range by Waagen, or from the upper carboniferous limestone of the Krone by Schellwien. It agrees almost perfectly with the British specimen figured by Davidson on Pl. XXVII, fig. 3 of his monograph. It is moderately convex, but the regularity of the curve is slightly deformed by pressure. No distinct median furrow has been noticed. The hinge-line is considerably shorter than the greatest width of the shell. In this respect the present specimen is also most closely allied to the British form from Bolland and to the Chinese types from Loping, described by Kayser, but distinctly differs from all the Salt Range species of the group of *D. senilis*. Etheridge's type specimens of *D. senilis* from Queensland are likewise provided with a short hinge-line and therefore agree best with the Himálayan and Chinese types.

Towards both ends of the hinge-line the valve is somewhat flattened. Thus two small flattened wings are formed on each side. The frontal line seems to be nearly straight, not depressed in the middle.

The surface of this valve is ornamented with numerous radiating striæ of unequal strength. These striæ, the edges of which appear locally crenelated by the intersection of concentric wrinkles, are somewhat irregular, often flexuous and dichotomising.

The measurements of this specimen are as follow :—

Length of the dorsal valve	. . . .	. 21·6 mm.
breadth „ „ „	. . . .	. 26 „
Thickness „ „ „	. . . .	. 4 „
Length of the hinge-line	. . . .	. 16 „

Whereas this specimen may be referred to *Derbyia senilis* with great probability, although I am not convinced of its identity with the latter species, the appurtenance of several other fragments to the genus *Derbyia* is much more doubtful. One of the more complete, which has been figured on Pl. VI, fig. 6 of this memoir, exhibits the peculiar semiconic shape of the valve and the step-like interruptions in the sculpture " produced by two or three very large and irregular concentric undulations," which are among the leading features of *D. senilis*. These are however the only claims my fragment can put in for a comparison with the present species.

Locality and geological position ; number of specimens examined.—Micaceous shales with *Spirifer Musakheylensis* and numerous *Bryozoa*, Barus ; Coll. Godwin-Austen ; 4.

Remarks.—The large list of synonyms given for *Streptorhynchus crenistria* by L. de Koninck cannot be accepted for the present species, if the latter is taken in the narrow circumscription, which has been proposed by Waagen. Following the interpretation of that learned author, none of the American or Spitzbergen forms of the genus can be united with *Derbyia senilis.*

The true *D. senilis* has hitherto only been described from the mountain lime-stone of Western Europe and from the upper carboniferous deposits of China, Australia and (?) Sumatra. Its presence in the permian rocks of Timor is as yet very doubtful.

Sub-order : HELICOPEGMATA, Waagen.

Family : *NUCLEOSPIRIDÆ*, Davidson.

Sub-family : RETZIINÆ, Waagen.

Sub-genus : EUMETRIA, Hall.

1. EUMETRIA CF. GRANDICOSTA, Davidson.   Pl. VI, fig. 10.

1857. *Retzia radialis var. grandicosta*, Davidson, Quart. Journ. Geol. Soc., London, Vol. XVIII, p. 28, Pl. I, fig. 5.

1863. *Retzia radialis*, Phill., *var. grandicosta*, Davidson, in L. de Koninck, Mémoire sur les fossiles paléozoiques, etc. recueillis dans l'Inde, p. 39, Pl. IX, fig. 5.

1887. *Retzia radialis var. grandicosta*, Verchère, Kaelmir, the Western Himalaya and the Afghan Mts. Journ. Asiat. Soc. of Bengal, XXXVI, Pt. 3, No. 3, p. 311.

1862. *R. grandicosta*, Kayser, Oberschlesische Fauna von. Loping, Richthofen's "China." IV Bd., p. 176.

1884. *Eumetria grandicosta*, Waagen, Salt Range Foss., Pal. Ind., ser. XIII, Vol. I. Productus Limestone Foss., p. 491, Pl. XXXIV. figs. 6-12.

1890. *Retzia grandicosta*, Nikitin. Mém. Comité Géol. St. Pétersbourg, Vol. V. No. 5, p. 86, Taf. III, figs. 9-11.

1902. *Retzia grandicosta*, Rothplatz, Die Perm-Trias und Juraformation auf Timor und Rotti Paläontographica, 39 Bd., p. 83, Taf. X, fig. 11.

Two incomplete forms of a small brachiopod with a punctate shell struc-ture and with radiating costæ have been obtained from the carboniferous rocks near Eishmakam by R. Lydekker. I think there can be but little doubt that the species is a representative of the genus *Retzia* or of its sub-genus *Eumetria*. It is not impossible that we are dealing here indeed with the widely distributed

Eumetria grandicosta, but the state of preservation of my specimens does not warrant a certain identification.

The more complete one of my two forms is a little longer than large, and provided with moderately and evenly convex valves. The beak of the ventral valve has unfortunately been broken off. Neither a sinus nor a median fold is developed in any of the two valves. Each valve bears about eight or ten prominent costæ, which are distinctly rounded on their tops. Their exact number cannot be made out, because the lateral portions of the shell are partly covered by the tough, adhering matrix, which I have not been able to remove.

In my second specimen consisting of a dorsal valve, nine ribs are counted. The median rib does not surpass the others in strength.

The shell is punctate.

The measurements of the figured specimen are, approximately, as follow :—

Length of the shell	cm.	7 mm.
„ „ „ dorsal valve		6 „
Breadth of the shell		5·5 „
Thickness of the two valves		3·5 „

The reference of this shell to *Eumetria grandicosta* is provisional only, although it seems to exhibit a stronger resemblance to this species than to the rest of congeneric forms. From *E. ulotrix*, de Koninck, or from *E. indica*, Waagen, it differs by the greater number and by the less prominent character of its costæ. In *Retzia radialis*, Phill., the number of ribs is, as a rule, larger, and in typical shapes at least a median sinus or corresponding fold is developed. With *Retzia pseudocardium*, Nikitin, a closer comparison is scarcely necessary. From *Retzia Mormonii*, Marcou, (Geology of North America, Pl. VI, fig. 11), it is distinguished by its less globose shell.

A similar remark applies to *Retzia compressa*, Meek (Palæontology of California, Vol. I, p. 14, Pl. II, fig. 7), which is, however, very closely allied to my Kashmir specimens and specially agrees with them perfectly well in size and number of ribs. It is therefore not impossible that a larger number of better preserved types of the present species might prove the latter to be rather a variety of *Retzia compressa* than of *Eumetria grandicosta*.

Locality and geological position; number of specimens examined.—Dark limestone, crowded with young individuals of indeterminable brachiopods, Eishmakam, Kashmir Valley ; Coll. Lydekker ; 2.

Remarks.—In the Salt Range *Eumetria grandicosta* is equally distributed throughout the entire thickness of the Productus limestone, with the only exception of the Chidru beds. It has been, moreover, described from the Gshelian stage of the carboniferous system in Central Russia by Nikitin and from the permian rocks of Timor by Rothpletz.

Retzia compressa is quoted from the anthracolithic rocks of California by Meek and from the upper carboniferous *Fusulina* limestone of Loping in China by Kayser.

Family: *ATHYRIDÆ*, Phillips.

Sub-family: ATHYRINÆ, Waagen.

Genus: ATHYRIS, M'Coy.

(SPIRIGERA, d'Orbigny.)

The family *Athyridæ* is represented among the brachiopoda from the anthracolithic rocks of Kashmir and Spiti by the genus *Athyris* only.

The sub-genus *Spirigerella*, Waagen, which is so largely represented in the Salt Range, is entirely absent from my fossil materials from the North-Western Himálayas. I must lay a special stress on this fact because Waagen himself supposed Davidson's and Verchère's *Athyris subtilita* from Kashmir to be identical with *Spirigerella Derbyi*, an identification in which this learned author has however been mistaken.

Among the representatives of the genus *Athyris* in the Geological Survey's Himálayan collection from Kashmir and Spiti the three following species can be distinguished :—

> 1. *Athyris Gerardi*, nov. sp.
> 2. *A. cf. expansa*, Phillips.
> 3. *A. subtilita*, Hall.

The first is restricted to the Kuling shales of Spiti, the two others have been obtained from the upper carboniferous beds of Kashmir.

Besides these three species *Athyris cf. Royssii*, Lev., and *A. Budhista* have been quoted from the Zewán beds of Kashmir by Dr. Verchère.

1. ATHYRIS GERARDI, nov. sp. Pl. VI, figs. 12, 13, 14.

1897. *Athyris Royssii*, Diener, ex parte, Himálayan Fossils, Palæontologia Indica, ser. XV, Vol. I, Pt. 4. The fauna of the permian Productus shales of Johár and Painkhánda, Pl. V, fig. 5 (non 7).

In my monograph of the brachiopoda obtained by Griesbach from the permian Productus shales of Johár and Painkhánda, I identified the present shell with *Athyris Royssii*, Leveillé, the materials then at my command not being sufficient to enable me to introduce a new species. A number of better preserved forms having subsequently turned up from Dr. Gerard's collections from the Kuling shales of Spiti, I was able to assure myself that they differ from *A. Royssii* by some constant features. The differences, which might be established between them, are of a similar character to those which induced Count A. de Keyserling (Wissenschaftliche Beobachtungen auf einer Reise in das Petschoraland, St. Petersburg, 1846, p. 237) to separate *A. Royssiana* from *A. Royssii*.

As leading features of this species the following ought to be considered. The shell attains a larger size than in any of the hitherto described forms of *Athyris*. The ventral valve is almost perfectly flat and the difference in the inflation of the

two valves is much more remarkable than in *A. Royssii* or in *A. Royssiana*. With
the latter species the present one agrees in the presence of an uncommonly large
apical angle, which in full-grown individuals attains about 150°, and in the small
size of the beak, which barely overhangs the hinge-margin. A remarkable difference between the Himálayan and the Russian species exists, however, in the shape
of the mesial sinus. In *A. Royssiana* the sinus is rather strongly developed and
shaped into a highly prominent tongue, which is bordered by parallel margins. In
A. Gerardi the sinus is but slightly impressed, in adolescent types of 15 mm. in
length barely if at all perceptible, and in full grown individuals is always bordered by distinctly converging margins.

The only two specimens, which I have been able to secure from the Productus
shales of Kiunglung near the Niti Pass, were in a rather imperfect state of preservation. I consequently failed in making out their characters of distinction with
sufficient certainty so as to feel justified in separating them from *Athyris Boyssii* (in
a wider interpretation of this latter species than has been admitted by L. de Koninck). The specimen, figured on Pl. V, Fig. 7 of my above-quoted memoir, ought,
however, to remain with *A. Boyssii*. Its sub-pentagonal outline, the small size of
its apical angle (100° only), and the presence of a well developed sinus are in favour
of an identification with the latter species. I first thought this type to be linked
to the true *A. Gerardi* by intermediate shapes, but this view I find no more tenable
since numerous young forms of *A. Gerardi* from the Kuling Shales of Spiti have
come to my knowledge, which, although approaching in their outline and in the
size of the apical angle the specimen from Kiunglung are constantly differing
from the latter by the greater flatness of their ventral valves and by the absence of
any distinctly marked sinus. I consequently deemed it preferable to leave this
specimen with *Athyris Boyssii*.

The specimen from the Productus shales of the Chor Hoti, figured by Salter in
the Palæontology of Niti (p. 68, Pl. V, fig. 13) ought, according to my humble opinion, likewise to be identified with *A. Boyssii* rather than with the present species.

The measurements of a tolerably well preserved ventral valve are approximately as follow :—

Entire length of the shell	60 mm.
„ breadth „ „	52 „
Thickness of the ventral valve	14 „
Apical angle of „ „	145°

My largest specimen attains an approximate length of 45 mm., corresponding to
a width of 60 mm.

Locality and Geological position ; number of specimens examined.—Sandstone
partings in the Kuling shales N. W. of Po, Spiti ; Coll. Gerard ; 7.

Remarks.—*Athyris Gerardi* must be added to the numerous elements peculiar
to the fauna of the Productus shales, the large number of which imparts to the
latter its characteristic aspect. The sandstone partings, in which this species occurs
in the Kuling shales of Spiti, do not differ lithologically from the layers in which
it has been found imbedded in Painkhánda, associated with *Spirifer Moosakheylensis, Sp. Nitiensis* and *Productus gangeticus*.

2. ATHYRIS CF. EXPANSA, Phillips.   Pl. VI, fig. 11.

1836. *Spirigera expansa*, Phillips, Geology of Yorkshire, Vol. II, p. 220, Pl. X, fig. 18.

1847. *Atrypa expansa*, Sowerby, Mineral Conchology, Pl. DCXVII, fig. 1.

1851. *Atrypa Ambrицata*, Sowerby (non Phillips), ibidem, fig. 4.

1857. *Athyris expansa*, Davidson, Monograph British carbon. Brachiopoda, p. 82, Pl. XVI, figs. 14, 16-19,
        Pl. XVII, figs. 1-5.

1858. *Athyris expansa*, Krotow, Mém comité géol. St. Pétersbourg, Vol. VI, p. 421.

A single, rather poorly preserved specimen may be provisionally referred to this species. It is very transversely elliptical, nearly twice as broad as long. The ventral valve, which alone is accessible to observation, is very evenly convex, but not strongly inflated. No trace of a mesial sinus has been noticed. The ornamentation of the shell consists of numerous, concentric lines of growth, which are occasionally intersected by an indistinct radiating sculpture.

The present specimen appears by its general shape and outline to be most closely allied to *Athyris expansa*. From *A. subexpansa*, Waagen, it is distinguished by the absence of any mesial sinus and of fringed lateral expansions.

The measurements of this specimen are approximately as follow :—

Entire length of the ventral valve	23 mm.
„ breadth „ „ „	40 „
„ thickness „ „ „	6 „
Apical angle of the ventral valve	ca. 140°

Locality and Geological position ; number of specimens examined.—Dark, micaceous shales N. of Khummú, near Pampur; Coll. Lydekker; 1.

Remarks.—*Athyris expansa*, Phill., has been quoted from the mountain limestone of England and Ireland by Davidson, and from the lower portion of the carboniferous limestone of Tscherdyn (Russia) by Krotow.

3. ATHYRIS SUBTILITA, Hall.   Pl. VII, figs. 1-8.

1843. *Spirifer Royssii*, d'Orbigny, Voyage dans l'Amérique Méridionale, T. III, 4 ème ptie, Paléontologie,
        p. 98, Pl. III, figs. 17-19.

1852. *Terebratula subtilita*, Hall, in Howard Stansbury's Report of an exploration of the valley of the Great
        Salt Lake of Utah, Philadelphia, p. 409, Pl. IV, figs. 1, 2.

1855. *Terebratula subtilita*, Schiel, Pacific Railroad Report, Vol. II, p. 108, Pl. I, fig. 3.

1856. *T. subtilita*, Hall, ibidem, Vol. III, p. 101, Pl. II, fig. 4.

1857. *Terebratula subtilita*, Davidson, Monograph British Carb. Brachiopoda, p. 16, Pl. I, figs. 21, 22.

1857. *Athyris subtilita*, Davidson, ibidem, p. 96, Pl. XVII, figs. 8-10.

1858. *Terebratula subtilita*, Marcou, Geology of North America, p. 48, Pl. VI, fig. 9.

1858. *Terebratula subtilita*, Hall, Report on the Geological Survey of Iowa, Vol. I, Pl. 2, Palaeont., p. 716.

1861. *Athyris subtilita*, Salter, Quart. Journ. Geol. Soc. London, Vol. XVII, p. 64, Pl. IV, fig. 4.

1863. *A. subtilita*, Davidson, ibidem, Vol. XIX, p. 170, Pl. IX, figs. 4, 5.

1866. *A. subtilita*, Davidson, ibidem, Vol. XXII, p. 40, Pl. II, fig. 2.

1866. *A. subtilita*, Geinitz, Carboniformation und Dyas in Nebraska, p. 60, Tab. III, figs. 7-9.

1867. *A. subtilita*, Verchère, Kashmir, the Western Himalaya and the Afghan Mts., Journ. Asiat. Soc. of
        Bengal, Calcutta, Vol. XXXVI, Pt. II, No. 3, pp. 303, 310, Pl. II, figs. 1, 1a.

1869. *Spirigera (Athyris) subtilita*, Toula, Ueber einige Fossilien der Kohlenkalkes von Bolivia, Sitzungsber.
        Kais. Akad. d. Wiss. Wien, math. nat., Cl. LIX, Bd. I, Abth. p. 6, fig. 4.

1873 *Athyris subtilita.* Meek, Final Report of the U. S. Geol. Surv. of Nebraska. Pt. II, p. 160, Pl. 1 fig. 13; Pl. V, fig. 9, Pl. VII, fig. 4.

1874 *A. subtilita,* Derby (as parte), on the carboniferous brachiopods of Itaituba, Bull. Cornell University, Ithaca. Vol. 1, No. 2, p. 7.

1875 *A. subtilita,* Toula. Eine Kohlenkalk-fauna von den Baren-Inseln, Sitzungsber. Kais. Akad. d Wiss. Wien, math. nat. Cl. Bd. LXXI, I. Abth. p. 20.

1877 *A. subtilita.* Meek, in Cl. King's Report of the U. S. Geological Exploration of the 40th parallel, Vol. IV, Palæontology, p. 82, Pl. VIII, fig. 6.

1877 *Spirigera subtilita,* White, in Wheeler's Report upon the U. S. Geological Surv., West of the one hundredth Meridian, Vol. IV, Palæontology, p. 141, Pl. X, fig. 6.

1884 *Spirigera subtilita,* White, XIII, Annual Report of the Geol. of Indiana, Pt. II, p. 136, Pl. XXXV, fig. 6-9.

1887 *Athyris subtilita,* L. de Koninck, Faune du calcaire carbonifère de la Belgique, 6me ptie, Annales du Musée Royal d' hist nat. de Belgique, Bruxelles, T. XIV, p. 73, Pl. XVIII, figs. 1-4, 7-10, 12-20, Pl. XII, figs. 47-54.

1887 *A. subtilita,* Prvin Smith, Marine Fossils from the coal-measures of Arkansas, Proceed. Amer. Philos. Soc., Vol. XXXV, No. 162, p. 31.

This species, which is one of the most abundant and widespread carboniferous brachiopods, has been quoted from the Barus or Zewán beds of Kashmir by Prof. Davidson in 1866 and by Verchère in 1867. The correctness of this identification has been doubted by Waagen (Salt Range Foss., Pal. Indica, ser. XIII, Vol. IV, Geological Results, p. 105), who united the Himálayan species with *Spirigerella Derbyi*, Waag. This latter view has not, however, been confirmed by my subsequent examination of a considerable number of types, collected by Godwin-Austen, Verchère and Lydekker in the carboniferous strata of the Kashmir Valley. In none of these specimens is the beak of the ventral valve firmly appressed to the apex of the dorsal one, but always distinctly exhibits the moderately large foramen truncating its extremity. In the list of synonyms of *Spirigerella Derbyi*, given in my memoir on the permian Productus shales of Johár and Painkhánda (Pt. IV of the present volume), these Kashmir shells have been quoted on the authority of Waagen. This view I find, however, no more tenable.

My specimens agree almost perfectly with some of the best figures which have been published up to the present of the American types of *Athyris subtilita* by Geinitz and Meek, and of the European ones by L. de Koninck.

The ovoid shell is, as a rule, longer than wide, its greatest width being situated a little in advance of the middle. The ventral valve is somewhat tapering at the beak, which is prominent, distinctly incurved and always pierced by a moderately large foramen. In the majority of my Kashmir specimens the two valves are almost equally convex, but not very strongly inflated. As has been noticed by L. de Koninck in some Belgian types of *Athyris subtilita*, the inflation occasionally becomes so strong that they can barely at all be distinguished from *A. globularis*, Phill. Among my Himálayan specimens a similar shape never occurs. None of them is decidedly gibbous, not even in adult age. A mesial sinus is nearly always present, but indicated only by a low median impression, which originates in the visceral region of the shell and corresponds to an elevated convex curve in the front margin. A distinct tongue-shaped process, corresponding to this frontal wave, so conspicuous in *Spirigerella Derbyi*, Waagen, especially in the var. *acutiplicata*, is but rarely developed.

On the dorsal side of the ventral valve, laterally from the beak, a narrow false area extends to the end of the hinge-line. It is marked off from the remainder of the shell by indistinct ridges.

The dorsal valve is not quite equally curved in the transverse direction but is of a somewhat roof-shaped appearance, sloping from a broadly rounded median crest in moderately convex planes towards the lateral margins. The beak is strongly incurved under that of the opposite valve.

The surface is nearly smooth or covered with irregular, concentric striæ of growth only. In the majority of my specimens, however, the shell substance has been too strongly injured by weathering to allow anything of the minor details of its ornamentation being noticed.

The measurements of a fairly complete specimen (fig. 1) are as follow :—

Entire length of the shell	. . . .	28·5 mm.
Length of the dorsal valve	. . . .	29·4 „
Entire breadth of the shell	. . . .	34 „
Thickness of both valves .	. . . .	18 „
Apical angle of the ventral valve	. . . .	81°
„ „ „ dorsal „	. . . .	96°

Of the internal structure nothing is to be made out in any of my specimens.

Locality and Geological position ; number of specimens examined.—Coarse, grey limestone, north of Khummúu, near Pampur, Kashmir Valley ; Coll. Lydekker, 9 ; Coll. Godwin-Austen, 5 ; Zawoor, Kashmir Valley, Zewán beds, Verchère.

Remarks.—Athyris subtilita is a very common and widespread species, ranging throughout the entire carboniferous system into permo-carboniferous and perhaps even into still higher permian deposits. It has been quoted from the United States of North America, from Nova Scotia, Bolivia, Brazil, Western Europe and Barents Island by various authors.

Meek advocates the identity of *Athyris subtilita*, Hall, with a shell from the upper coal-measures of Illinois, which had been described as *Terebratula argentea* by Shepard in 1838. The figure given by that author in the American Journal of Science (Vol. XXXIV, fig. 8) does not however agree with adult forms of *A. subtilita*. Shepard's original type not having been found up to present, its claims of priority to the name of that species are as yet very uncertain. Meek's question, whether the present shell ought not to be called *A. Peruviana*, because A. d'Orbigny figured it in 1847 under the name of *Terebratula Peruviana*, must also be answered in the negative. The name *Terebratula Peruviana* has been given erroneously in the Plate of d'Orbigny's monograph, whilst in the text it is applied to a very different species from devonian rocks (Voyage dans l'Amérique Méridionale, T. III, Paléont. p. 30, Pl. II, figs. 22-25).

Spirigera protea var. subtilita, Abich (Geologische Forschungen in den Kaukasischen Laendern, I, Th. Eine Bergkalkfauna aus der Araxesenge bei Djulfa, p. 59, Taf. VIII, figs. 10, 11) has nothing to do with the true *Athyris subtilita*, as has been remarked by Val. von Moeller and by Rothpletz.

A species which is probably very closely allied to the present one has been described and figured as *Athyris Buddhista* by Dr. Verchère (l. c. p. 210, Pl. II,

figs. 2, 2a, 2b). This species, which has been collected in the Zewán beds of
Kashmir, seems to differ from *A. subtilita* by its strongly attenuated apical region
and by the presence of a sharp median fold in the dorsal valve. No specimen
resembling Dr. Verchère's illustration has come to my notice. Provided this
illustration be correct, *Athyris Buddhista* should be considered as an independent
species.

Family: SPIRIFERIDÆ, d'Orbigny.

Subfamily: SUESSIINÆ, Waagen.

Genus: SPIRIFERINA, d'Orbigny.

1. SPIRIFERINA CF. KENTUCKENSIS, Shumard.   Pl. V, figs. 11, 12.

1852. *Spirifer octoplicatus*, Hall, in Howard Stansbury's Report on the Exploration of the Valley of the
Great Salt Lake in Utah, p. 409, Pl. XI, fig. 4.

1858. *Spirifer Kentuckensis*, Shumard, Geological Survey of Missouri, p. 208.

1858. *Spirifer Kentuckensis*, Hall, Pacific Railroad Report, Vol. III, p. 108, Pl. II, figs. 10, 11.

1866. *Spirifer Leminensis*, Geinitz (non M'Coy), Carbonformation und Dyas in Nebraska, p. 45, Taf. II,
fig. 19.

1867. *Spirifera octoplicata var. transversa*, Verchère, Kashmir, the Western Himalaya and the Afghan
Mts., Journ. Asiat. Soc. of Bengal, Calcutta, Vol. XXXVI, Pl. 2, p. 210, Pl. I, figs. 2, 2a, 2b.

1872. *Spiriferina Kentuckensis*, Meek, Palaeontology of Eastern Nebraska in Final Rep. upon the U.S.
Geol. Surv. of Nebraska, p. 166, Pl. VI, fig. 3, Pl. VIII, fig. 11.

1877. *Sp. Kentuckensis*, White, in Wheeler's Report upon U. S. Geological Surveys W. of the one hundredth Meridian, Vol. IV. Palaeontology, p. 154, Pl. X, fig. 4.

Numerous casts and external impressions of dorsal valves of a strongly transverse *Spiriferina* have been collected in the Zewán beds of Kashmir both by Capt.
Godwin-Austen and by Dr. Verchère. Curiously enough, dorsal valves only are
represented among the materials available to me for examination. Some of them
which are tolerably well preserved, though none is complete, closely resemble
Spiriferina Kentuckensis from the North American coal-measures. I consequently
thought it advisable to refer them provisionally to that species, although I do not
venture on a direct identification, having regard to the absence of any ventral
valves among my materials.

Among four better preserved dorsal valves two are quite as large as the specimen from the coal-measures of Nebraska, figured by Meek on Pl. VI, fig. 3 of his
above-quoted memoir. They are very strongly transverse, of a sub-fusiform shape,
terminating in slender attenuated ears. The hinge-line corresponds to the greatest
width of the shell. On each side of the mesial fold from three to five simple prominent ribs are distributed. All the ribs are slightly rounded on their crests and
separated from each other by deep V shaped valleys, which are sharply rounded at
the bottom. The first plications on each side of the mesial fold are but slightly
inferior in size to the latter.

The surface is ornamented with very numerous prominent and closely-
crowded lines of growth. This concentric sculpture is so strong as to hide almost
completely the granulated structure of the shell substance.

Of the ventral valve the impression of the beak and area only are preserved in one of my specimens (Pl. V, fig. 12). In this specimen the greatest width of the dorsal valve is 24 mm., corresponding to a length of 11 mm., and to a length of the entire shell of 13 mm.

Locality and Geological position ; number of specimens examined.—Zewán beds, Kashmir Valley ; shales and micaceous limestones with numerous *Bryozoa*, *Spirifer Musakheylensis*, *Derbyia cf. senilis*, etc.; Coll. Godwin-Austen, 4 ; Coll. Verchère, 2.

Remarks.—Among all the hitherto described species of *Spiriferina* there is only the present one to which my specimens from Kashmir may be advantageously compared. The sub-fusiform shape and the strong concentric ornamentation clearly distinguish them from *Spiriferina cristata*, Schloth., and *Sp. octoplicata*, Sow. They likewise differ by their sub-fusiform shape and long hinge-line from *Sp. insculpta*, Phill., and *Sp. ornata*, Waag., with which they have the distinct concentric ornamentation in common. The only European species, which my Kashmir forms resemble more closely, is *Sp. peracuta*, de Koninck (Faune du calcaire carbonifère de la Belgique, Annales du Musée Royal d' hist nat., T. XIV, 6 ème ptie., p. 101, Pl. XXII, figs. 56-61) from the mountain-limestone of Belgium and Ireland. A remarkable difference consists, however, in the larger size of the dorsal median fold in *Sp. peracuta* which, according to L. de Koninck's description, is twice as large as the neighbouring lateral ribs, whereas it scarcely surpasses them in size in my Himálayan shell.

Spiriferina laminosa, M'Coy, which has been erroneously identified with *Sp. Kentuckensis*, by Geinitz, is easily distinguished from the latter by its much larger size and the broad area of the ventral valve.

The shells described and figured as *Spiriferina octoplicata* var. *transversa* by Verchère, will probably fall within this species. I am led to this conclusion, which might scarcely be drawn from Verchère's exceedingly bad figures, by the examination of two specimens collected by that author in the Zewán beds of the Kashmir Valley.

As one of the chief characteristics of *Spiriferina Kentuckensis* lies in the shape of the sinus of the ventral valve, the reference of my specimens, which are dorsal valves only, to this American species, must yet remain provisional.

Spiriferina Kentuckensis is a common species in the carboniferous and permo-carboniferous strata of Kentucky, Illinois, Missouri, Iowa, Nebraska, Kansas, Utah, New-Mexico, Arizona and Texas, but does not descend into beds of sub-carboniferous

Sub-family : DELTHYRINAE, Waagen.

Genus : SPIRIFER, Sowerby.

The genus *Spirifer* is rather richly represented in the anthracolithic system of Kashmir and Spiti. Not less than ten species are counted among the materials examined by Prof. Davidson and myself, although one-half of this number only are sufficiently complete to allow of a satisfactory diagnosis of their specific characters being given.

These species may be grouped most conveniently in the following manner :—

I. GROUP OF SPIRIFER FASCIGER, Keyserl.
1. *Spirifer Musakheylensis*, Davidson.
2. *Sp. sp. ind., aff. Musakheylensis.*
3. *Sp. Nitiensis*, Diener.

II. GROUP OF SPIRIFER TRIGONALIS, Mart.
4. *Spirifer cf. Triangularis*, Martin.

III. GROUP OF SPIRIFER VIRGOIS, Sowerby.
5. *Spirifer Fibiosus*, Davidson.

IV. GROUP OF SPIRIFER RAJAH, Salter.
6. *Spirifer Rajah*, Salter.
7. *Sp. sp. ind., aff. Rajah.*

V. GROUP OF SPIRIFER CLARKEI, de Kon.
8. *Spirifer Lydekkeri*, nov. sp.
9. *Sp. sp. ind. ea. aff. Lydekkeri.*

VI. GROUP OF SPIRIFER ALATUS, Schloth.
10. *Spirifer Kashmericensis*, Davidson.

Of all these species eight are entirely restricted to the anthracolithic rocks of the Himálayas. Among them *Spirifer Lydekkeri* and its allies are of a special interest, on account of their relationship to the Australian *Sp. Clarkei*, whilst *Spirifer Rajah* exhibits a close affinity to Arctic types from Spitzbergen. Two species only occur also in the carboniferous system of Europe; these are *Sp. Musakheylensis* and *Sp. triangularis*, while a third one, *Sp. vihianus*, is very closely allied to the European *Sp. pinguis*.

I. GROUP OF SPIRIFER FASCIGER, KEYSERLING.

1. SPIRIFER MUSAKHEYLENSIS, Davidson, Pl. V, figs. 3—7.

1862. *Spirifer Musakheylensis*, Davidson, Quart. Jour. Geol. Soc., London, Vol. XVIII, p. 29, Pl. II, fig. 2.

To the list of synonyms, given in Part IV of the present volume, the following ought to be added :—

1867. *Sp. Musakheylensis*, Verchère, Journ. Asiat. Soc. of Bengal, Vol. XXXVI, Pt. 2, p. 210, Pl. III, figs. 1, 1a.

1890. *Sp. Musakheylensis var. australis*, Foord, Notes on the Palaeontology of Western Australia, Geol. Magazine, New ser., Decade III, Vol. VII, p. 147, Pl. VII, fig. 3.

This species is very common, both in the Zewan beds of Kashmir and in the Kuling shales of Spiti, but no complete specimen has been noticed among the materials available to me for examination, ventral and dorsal valves being nearly always met with separately.

To the detailed description of the specimens from the Productus shales of Johár and Painkhánda I have but little to add.

In spite of the great variability of the shapes there is not a single one among my specimens from the anthracolithic rocks of Kashmir and Spiti which agrees with either *Spirifer fasciger*, Keyserling, or with *Sp. tegulatus*, Trautschold. The folds, corresponding to the fasciculi of ribs, are invariably rounded but never provided with acute edges, as in Gruenewaldt's and Tschernyschew's type-specimens of *Sp. fasciger*. The lamellose character of the striæ of growth is distinctly developed in the majority of my forms, but the peculiar sculpture of *Sp. tegulatus*, reminding one of a tiled roof, has not been noticed in any of my Indian representatives of this group. Some of my specimens from the Kuling shales of Spiti agree perfectly well with the type from the Productus shales of Kuling, figured on Pl. V, fig. 1 of Pt. IV of the present volume by the unusually flat convexity of their folds.

The specimens from the Kuling shales of Spiti attain very considerable dimensions, the largest specimen collected by Dr. Gerard measuring about 60 mm. in length and 130 mm. in breadth.

Of the internal characters of the ventral valve of this species some information has been gathered by an examination of the casts figured on Pl. V, Figs. 5 and 7 of the present memoir. In the apical region the shell-substance is so extremely thickened that the dental plates and the outer walls of the valve are united into one solid shelly mass, on which the entire area rests. The muscular impressions are distinctly marked. On each side of them the internal surface of the shell is covered with numerous irregular grooves, which on the cast are exhibited as rounded granulations. The ornamentation of the cast is very simple, consisting of a few broad and flat wavy folds only, while the ribbing of the external shell-surface is completely absent.

Locality and Geological position ; number of specimens examined.—Zewán beds, Kashmir Valley, Coll. Godwin-Austen, 6, Coll. Lydekker, 1 ; Kuling Shales, Spiti Valley, Coll. Gerard, 3, Coll. Stoliczka ; Khar, Coll. Griesbach, 1.

Remarks.—The specimen from the carboniferous rocks of West Australia, described and figured as *Spirifer Musakheylensis var. Australis*, by Foord, does certainly belong to the present species. Foord, stating the close resemblance of the Australian fossil to Davidson's types, notes the only difference between them to be that "the ornaments of the Australian species are perhaps a little coarser than those of the Indian one, i.e., the former has slightly larger and consequently fewer small ribs (comparing together individuals of the same size) than the latter and the imbricating lamellæ exhibit the same divergence of character. It seems however scarcely necessary to regard these slight differences as of more than varietal importance, especially if one takes into account the variations in any large assemblage of brachiopods, as Davidson himself has so often demonstrated in his plates".

That the Australian variety is included among the variations of the Himálayan *Sp. Musakheylensis*, is clearly evident from an examination of the specimen from the Productus shales of Kiunglung, figured on Pl. IV, Fig. 1, of my monograph of the fauna of the Productus shales (Pt. IV of the present volume).

This specimen exactly agrees with the Australian type of our species by its coarse ribs and lamellæ.[1]

2. SPIRIFER CF. NITIENSIS, Diener.   Pl. V, fig. 9.

1897. *Spirifer Nitiensis*, Diener. The Permian fauna of the Productus Shales of Johár and Painkhánda. Pal. Indica, ser. XV, Himálayan Fossils, Vol. I, Pt. 4, Pl. IV. figs. 4, 5.

The fragment of a ventral valve from the Kuling shales of Spiti agrees so perfectly well in its outline and ornamentation with this remarkable species that I do not hesitate to refer it to the latter, although its fragmentary condition may forbid a direct identification.

This fossil is almost of exactly the same size as my type specimen from Kiunglung (Pl. IV, fig. 5), though perhaps even a little more strongly transverse. It is moderately curved in the longitudinal, but quite flat in the transverse direction. Its tolerably broad, reclining, parallel-sided and vertically-striated area is overlooked by the little pointed beak. The ornamentation consists of numerous ribs of unequal strength, arranged into fasciculi, each of which is composed of a small number of ribs only. On the wings the fascicular arrangement of the ribs becomes gradually indistinct. All the ribs are flatly rounded at their crests.

Locality and Geological position; number of specimens examined.—Sandstone partings in the Kuling shales, with *Athyris Gerardi*, Po, Spiti Valley; Coll. Gerard; 1.

Remarks.—Spirifer *Nitiensis* is a very characteristic species of the permian Productus shales of Gurhwál. I know of no other species of the genus *Spirifer* to which the present fragment could be referred, its peculiar ornamentation combined with the strongly fusiform shape distinguishing it from all the rest of congeneric forms.

3. SPIRIFER SP. IND. AFF. MUSAKHEYLENSIS.   Pl. V, fig. 10.

This interesting species, which in the Geological Survey's Himálayan collection from Kashmir is represented unfortunately by a single ventral valve only, seems to hold an intermediate position between *Spirifer Musakheylensis*, Davidson, and *Sp. Johárensis*, Diener (Part IV of this volume, Pl. IV, Fig. 3) from the permian Productus shales of Johár, but is more closely allied to the former.

It chiefly differs from Davidson's species by its stronger folds which are highly prominent and composed of a small number of secondary ribs only. There are not more than three ribs present in each of the fasciculi in the vicinity of the sinus. Among them the median primary rib is a little stronger than the rest. The sinus not having been preserved in my fragment, I have not been able to state the pre-

---

[1] *P. S.*—After having sent this paper to Calcutta for the press I had an opportunity of studying the beautiful collections of the Comité géologique de la Russie at St. Petersburgh. Prof. Tscherayschew, to whom I am indebted for many valuable information regarding the carboniferous fauna of the Ural Mts., was kind enough to show me a large number of specimens of *Spirifer fasciger*. Keyserl. By their examination I have been convinced of the insufficiency of the characters on which I thought a specific distinction of *Sp. Musakheylensis* and *Sp. fasciger* ought to be based; consequently I no longer object to Prof. Tscherayschew's view as to the identity of the two species. For the more populated, *Tingtsehoid*, a varietal rank ought, however, to be retained.

sence of a mesial fold, as it is developed in *Sp. Johorensis*. The radial ornamentation is crossed by raised striæ of growth as in *Sp. Musakheylensis*.

The present specimen is too fragmentary to allow of any exact measurements being taken.

Locality and Geological position ; number of specimens examined.—Dark, shaly, micaceous limestones with *Spirifer Musakheylensis*, *Productus cf. undatus* and numerous Bryozoa, Barus, Kashmir Valley ; Coll. Godwin-Austen ; 1.

Remarks.—A species, which the present specimen most closely resembles, has been figured and described by Toula (Permo-Carbon-Fossilien von der Westkueste von Spitzbergen, Neues Jahrb. f. Mineralogie, 1875, p. 240, Taf. VII, Fig. 3) as a variety of *Sp. cameratus*, Morton. The Spitzbergen specimen from Axel Island for the loan of which I am indebted to Dr. Fuchs, Director of the Imperial Museum of Natural History in Vienna, is likewise distinguished from *Sp. cameratus* and from *Sp. Musakheylensis* by its strong folds, composed of a small number of secondary ribs only, as has been correctly remarked by Toula. Nevertheless I do not think that the Himálayan and Spitzbergen forms should be united in the same species, since the ribs composing the fasciculi are equally strong in the latter but different in size among one another in the former specimen.

The fragment, figured as *Spirifer sp. ind.* by Gruenewaldt (Beiträge zur Kenntniss der sedimentaeren Gebirgsformationen etc. Mém. Acad. impér. des sciences de St. Pétersbourg, VII, ser. T. II, 1860, p. 99, Pl. V, fig. 4) might perhaps fall within the relationship of this species or of *Spirifer Johorensis*.

II. Group of SPIRIFER TRIGONALIS, Mart.

4. Spirifer cf. triangularis, Martin. Pl. V, fig. 5.

1409. *Conchiliolithus Anomites triangularis*, Martin, Petref. Derb., p. 10, Pl. XXXVI, fig. 2.

1827. *Spirifer triangularis*, Sowerby, Min. Conch. Great Brit., Vol. VI, p. 120, Pl. 542, figs. 5, 6.

1836. *Spirifer triangularis*, Phillips, Geology of Yorkshire, Vol. II, p. 217, Pl. IX, fig. 13.

1840. *Sp. triangularis*, (la v. Buch), Mém. Soc. Géol. de France, T. IV, p. 162, Pl. VIII, fig. 6.

1843. *Sp. triangularis*, L. de Koninck, Descr. des animaux foss. de terrain carbonif. de la Belgique, p. 234, Pl. XV, fig. 1.

1844. *Spirifera attenuata*, M'Coy, Synopsis of the characters of the carbon. limestone foss. of Ireland, p. 133, Pl. XXI, fig. 3.

1844. *Sp. triangularis*, Semenow, Zeitschr. Deutsche Geol. Ges. Bd. VI, p. 330.

1854. *Spirifera attenuata*, M'Coy, Descr. of British Palæozoic, Foss. in the Cambridge Museum, p. 415, Pl. III, fig. 27.

1857. *Sp. triangularis*, Davidson, Monogr. British Carb. Brachiopoda, pp. 37 and 233, Pl. V, figs. 16-24, Pl. L, figs. 10-17.

1867. *Sp. triangularis*, L. de Koninck, Faune du Calc. carbonifère de la Belgique, Annales du Musée Royal, d'hist. nat. de Belgique, T. XIV, 6ème ptie., p. 124, Pl. XXIX, figs. 7-15.

1879. *Sp. triangularis*, Kiotaw, Mém. Com. Géol. de la Russie, St. Pétersbourg, Vol. VI, p. 415.

The existence of this as a Himálayan species depends upon a fragmentary dorsal valve, collected by Lydekker in a quartzitic sandstone near Bishmakam (Kashmir Valley). In shape it is very like some of the types figured by Davidson and L. de Koninck, but slightly surpasses them in size. It is about twice as broad as long, although exact measurements cannot be given, both the wings and the frontal portion of the mesial fold having been broken off. The most prominent character in the present valve is its strongly elevated, sharp, mesial fold, which "assumes the shape of a single, produced and acutely angular, cuneiform ridge or rib." Along the lateral slope of this central ridge two very low secondary folds are distinctly indicated, such as have been figured by L. de Koninck in his type specimen from Visé (Pl. 29, fig. 7), or in Martin's original form, reproduced on Pl. V, Fig. 16 of Davidson's monograph.

On either side of the mesial fold from six to seven single, obtusely rounded ribs ornament the lateral portions of the valve. They gradually diminish in strength towards the wings. Those situated near the extremities of the latter are but very faintly marked.

My specimen having been partly injured by weathering, the delicate concentric ornamentation, which is peculiar to well preserved forms of *Spirifer triangularis* is but indistinctly marked. Nevertheless in a few spots it is sufficiently well indicated to allow its presence to be stated with full certainty.

Locality and Geological position; number of specimens examined.—Quartzitic sandstone of a rusty brown colour, near Eishmakam, Kashmir Valley ; Coll. Lydekker ; 1.

Remarks.—The present specimen, although very incomplete, so closely agrees with *Spirifer triangularis* in its general shape and sculpture, especially in the presence of a prominent, cuneiform, mesial ridge, that I cannot refer it to any other species of the group of *Sp. trigonalis*. It will at any rate tend to show the existence of this remarkable group in the anthracolithic rocks of the Kashmir Valley.

Spirifer triangularis has been quoted from the mountain-limestone of Great Britain, Ireland, Boligum and Silesia by various authors, and from the upper carboniferous strata of Central Russia by V. von Moeller and by Krotow.

III. GROUP OF SPIRIFER PINGUIS, Sowerby.

5. SPIRIFER VIHIANUS, Davidson.

1866 *Spirifer Vihianus*, Davidson, Quart. Journ. Geol. Soc., Lond-a, Vol. XXII, p. 41, Pl. II, fig. 4.
1879. *Sp. Vihianus*, Waagen, Salt Range Fossils, Pal. Ind., ser. XIII, Vol. IV, Geological Results, p. 106.

This beautiful species is, unfortunately, not represented in the Himálayan collection entrusted to me for examination, but the figure and description given by Prof. Davidson are so excellent that from them a very good idea of its character can be formed.

Davidson himself compared his new species to *Spirifer pinguis*, Sow., from which he found it to differ by its constant, well marked, median rib in the sinus of the ventral valve. The resemblance of *Sp. Vihianus* to this common British form is indeed a very close one, especially to such shapes of *Spirifer pinguis* as are characterised by the presence of a strongly marked, longitudinal groove in the mesial fold of the dorsal valve.

ʰ 2

In the face of such features, Waagen's opinion "that *Spirifer Vihianus* apparently belongs to the same group of forms as *Spirifer duodecimcostatus*, M'Coy, and appears to differ from that form solely by its greater number of ribs," has hardly ever been admissible. It is certainly no more tenable, since M'Coy's Australian species has been proved to belong to the genus *Spiriferina* by Etheridge, jun. (Geology and Palæontology of Queensland and New Guinea, London, 1892, p. 234).

A species, which needs a closer comparison to *Sp. Vihianus*, has been described and figured as *Sp. Parryanus* by Toula (Permo-carbon-Fossilien von der Westkueste von Spitzbergen, Neues, Jahrb. f. Miner. 1875, p. 256, Taf. VII. fig. 8) from Hinlopen Straits in Spitzbergen. The name of this Spitzbergen shell must, however, be changed, since the priority of the denomination of *Sp. Parryanus* is claimed by a species from the devonian rocks of Iowa, described in 1858 by J. Hall (Report on Geological Survey of Iowa, Vol. I, Pt. 2, Palæontology, p. 509, Pl. IV, fig. 8). This species, for which I venture to propose the name of *Sp. Loreni*, seems to hold an intermediate position between *Sp. Vihianus* and the group of *Sp. Rajah*, Salter. With the former it agrees in the presence of a deep longitudinal groove in the mesial fold of the dorsal valve, with the latter in the tendency to develop secondary ribs, originating from the massive primary costæ. But in *Sp. Loreni* this tendency affects the two ribs bordering the mesial sinus and fold only, whereas in the two following costæ it is restricted to the immediate vicinity of the front margin. Whether the dorsal valve, marked erroneously as Fig. 7 instead of 8*d* on Pl. VII of Toula's memoir, actually belongs to the same species as the three ventral valves which by the kindness of Director Fuchs I have been able to examine, is yet doubtful, though highly probable. The reconstruction of the wings (not preserved) by the draughtsman is entirely misleading.

The specimens described by Davidson were obtained from the anthracolithic rocks of Barus in the Kashmir Valley.

IV. GROUP OF SPIRIFER RAJAH, Salter.

6. SPIRIFER RAJAH, Salter, Pl. IV., Figs. 1—7, Pl V., Fig. 1.

1863. *Spirifer Rajah*, Salter, Palæontology of Niti, etc., pp. 59 and 111.

1865. *Sp. Keilhavii*, Stoliczka, Geological Sections across the Himalaya Mts. from Wangtu bridge on the river Sutlej to Sungdo on the Indus, etc., Mem. Geol. Surv. of India, Vol. V, Pt. I, p. 27.

1866. *Sp. Rajah*, Davidson, Quart. Journ. Geol. Soc., London, Vol. XXII, p. 40, Pl. II, fig. 3.

1867. *Sp. Verchèrei*, de Verneuil, in Verchère: Kashmir, the Western Himálaya and the Afghan Mts., Journ Asiatic Soc. of Bengal, Calcutta, Vol. 36, p. 203, Pl. I, figs. 1, 1*a*.

1867. *Sp. Rajah*, Verchère, ibidem, p. 210.

This elegant species, one of the most remarkable among the brachiopoda of the anthracolithic system in the Himálayas, is very variable in shape, dimensions and relative proportions. The shell is, as a rule, longitudinally oval, or square-shaped, sometimes sub-circular, as wide as long, very rarely even broader than long. The hinge-line is always shorter than the greatest width of the shell. The cardinal angles, which are but exceptionally preserved, are acutely rounded.

The ventral valve is considerably deeper than the opposite one, and strongly vaulted in the either direction. The beak is strongly incurved and prominent, in full-grown specimens (Pl. IV, Fig. 5) approaching very nearly the apex of the dorsal valve. The area is moderately broad, distinct, concave and divided in the middle by a proportionately large, triangular fissure of equal height and width. It exhibits in well-preserved specimens an indistinct horizontal striation. A deep mesial sinus extends from the extremity of the beak to the front. It is in general broadly rounded, but in a very small number of specimens, however, acutely so, at its bottom, and is invariably ornamented by a narrow, median, thread-like rib.

The sculpture is rather variable. The surface of this valve is ornamented with from twelve to twenty broadly rounded ribs, which become gradually indistinct in the vicinity of the cardinal angles, whilst those bordering the mesial sinus are the largest and most prominent. Each of these flat broadly-vaulted primary ribs is ornamented by a variable number of lower, secondary costæ. In some forms, as in Davidson's type specimen or in the types figured on Pl. IV, Figs. 1 and 7 of this memoir, these secondary costæ are of equal strength and equidistant throughout their entire length. In the majority of my specimens, however, one or two of the secondary costæ are again sub-divided into smaller ones of irregular strength and distance. The specimen figured on Pl. IV, Fig. 4, is a good instance of this shape. The manifold transitions between these two extreme shapes prove their specific identity; an opinion, which I have founded on the minute examination of more than fifty individuals.

If the shelly substance has been partly injured by weathering, the secondary costæ gradually disappear, the flat primary ribs remaining solely. On such weathered specimens *Spirifer Verckerei* has been founded, which must consequently be erased from the number of independent species. Among the forms collected by Dr. Verchère and contained in the Geological Survey's Himálayan collections, there are several which agree pretty well with the illustrations of *Sp. Verckerei* given by that author, but are certainly nothing else but strongly-weathered individuals of *Sp. Rajah*.

The dorsal valve is less strongly convex than the ventral one. It is almost equally curved in the longitudinal direction on its lateral parts and along the median fold. A narrow but distinctly developed hinge-area is noticed below the tolerably prominent apex. The mesial fold is considerably elevated and shaped into a single, acutely rounded crest. At some distance from the apex smaller lateral ribs are produced on either side, which in larger specimens become again sub-divided before reaching the front line. The lateral parts are likewise ornamented by rounded folds, simple at their origin, but soon producing on either side a smaller lateral rib, which is either single or dichotomous but always inferior in strength to the main rib. The intercostal depressions are regularly rounded.

In weathered specimens, in which the details of the sculpture are lost, the ornamentation consists of simple coarse radiating ribs only, as in the type figured as *Sp. Verckerei* by Verchère (*l. c.* Fig. 1a).

In perfectly preserved individuals the surface is marked on both valves with very delicate longitudinal striæ. The radiating sculpture is crossed by indistinct marks of growth of greater or lesser strength at irregular distances.

The internal characters of the ventral valve are well exhibited in several of my specimens. The hinge-teeth are supported by two large diverging dental plates, forming the walls of the triangular fissure and extending into the interior of the valve for some distance. Between them a large portion of the free space at the bottom of the shell is occupied by the muscular impressions. On each side of the muscular impressions the interior of the shell is covered by numerous coarse granulations.

The measurements of a pretty large specimen (Pl. IV, fig. 5.) are as follow:—

Entire length of the shell	. . . .	65 mm.
Length of the dorsal valve	. . . .	51 „
Greatest breadth of the shell	. . . . . .	ca. 63 „
Length of the hinge-line	. . . .	40 „
Thickness of both valves	. . . .	42 „
Apical angle of the ventral valve	. . . . .	88°
„ „ „ „ dorsal „	. . . .	125°

Locality and Geological position ; number of specimens examined.—Kuling shales of Spiti, associated with *Marginifera Himalayensis*, Kuling, Coll. Gerard, 12, Coll. Stoliczka, 10, Coll. Griesbach, 3 ; Muth, Coll. Stoliczka, 2 ; Lilang, Coll. Stoliczka, 6 ; Spiti Valley (exact locality not known), Coll. Gerard, 9. All the forms from Kashmir, contained in the Geological Survey's Himálayan collection, are loose specimens. The majority, 21, have been obtained from Barus by Capt. Godwin-Austen. The exact locality, where the specimens collected by Dr. Verchère (6) and by Lydekker (9), were picked up, has not been given on the accompanying labels. It cannot therefore be made out with certainty whether the species actually occurs in the Zewán beds, or is restricted to the beds with *Productus Abichi* and *Marginifera Himalayensis*, corresponding to the Kuling shales of Spiti, although the section, published by Capt. Godwin-Austen on p. 33 of his memoir on the carboniferous rocks of the Kashmir Valley (Quart. Journ., Geol. Soc., London, Vol. XXII.) is strongly in favour of the former view.

Remarks.—*Spirifer Rajah* belongs to a very remarkable group of this genus, which is distinguished by the presence of coarse fasciculate ribs, and seems to exhibit a distant similarity to *Sp. integricosta*, Phill., among the mountain limestone forms of Western Europe. This group is represented by *Sp. Tasmaniensis*, Morris, in Australia, by *Sp. interplicatus*, Rothpletz, in Timor, by *Sp. Tibetanus, Sp. Rajah*, and a third yet very incompletely known species in the Himálayas, by *Sp. Keilhavii*, v. Buch., *Sp. Loveni*, Diener (= *Sp. Parryanus*, Toula) and *Sp. Wilczeki*, Toula, in Spitzbergen.

Among these species *Spirifer Keilhavii*, v. Buch (Ueber Spirifer Keilhavii über dessen Fundort und Verhæltnisssuähulichen Formen, Abhandlgn. Kœnigl. Akad. d. Wissensch. Berlin, 1876, p. 65) seems to be most closely related to the present one. In general shape and sculpture they are indeed very similar, although points of difference forbidding their identification are not absent. These characters chiefly consists in the absence of a median thread-like rib in the sinus of *Sp. Keilhavii*, and in the shape of the mesial fold of the dorsal valve. The latter is divided in the middle by a broad and shallow longitudinal depression or groove, whereas it is shaped into a sharply-rounded crest in *Sp. Rajah*.

Another species, which closely approaches the present one, is *Spirifer Tibetanus*, Diener (Pt. III, of this volume, Pl. VI, Figs. 1—7,) from the permo-carboniferous limestone crag of Chitichun No. I, in Hundés. It is, however, readily distinguished from *Sp. Rajah* by its short hinge-line and area, which is much more triangular and broader in proportion to its width. In its ornamentation *Sp. Tibetanus* chiefly resembles such shapes of *Sp. Rajah* as are characterised by the irregular strength of their secondary ribs (Pl. IV, Fig. 4), but the dichotomous character of the dorsal median fold in the Tibetan species has not been noticed in the present form.

Spirifer Loczi (= *Sp. Porryanus*, Toula, Neues. Jahrb. f, Min. 1875, p. 256, Taf. VII, fig. 5) seems to hold an intermediate position between *Sp. Rajah* and *Sp. Fihiavus*, as has been explained more fully in the description of the latter species.

The similarity to *Spirifer Wilczeki*, Toula (Kohlenkalk-Fossilien von der Suedspitze von Spitzbergen, Sitzgsber, Kais. Akad. d. Wiss, Wien., LXVIII. Bd. p. 271, Taf. I, fig. 3), *Sp. Tasmaniensis*, Morris (in Strzelecki's Physical Description of New South Wales and Vandiemensland, p. 280, Pl. XV, figs. 3, 4) and *Sp. interplicatus*, Rothplets (Die Perm-Trias-und Jura Form. auf Timor und Rotti, Palæontographica, 39, Bd. 1892, p. 78, Taf. IX, fig. 6) is a more distant one. Among these species *Spirifer Tasmaniensis*, as figured by Morris, L. von Buch and L. de Koninck, is certainly least closely related to *Sp. Rajah*, from which it differs by its strongly-transverse shape, the less prominent and differently sculptured dorsal median fold and by the longitudinal ornamentation of the sinus.

A species, which might also be compared to the group of forms as the prototype of which *Spirifer Rajah* ought to be considered, has been described and figured as *Sp. Waageni*, Tschernyschew (Mém. Comité Géol. 8. Pétersbourg, Vol. III, No. 4, p. 208, Taf. V, fig. 2), but the fasciculate arrangement of the secondary ribs seems to be less distinctly developed in this Artinskian form.

It is rather remarkable that no representative of this group has as yet been met with in the Productus limestone of the Salt Range, whereas both *Sp. Rajah* and *Sp. tibetanus* are among the most common and characteristic types of brachiopods in the anthracolithic rocks of the Himálayas.

No representative of this group has even been found in deposits of a lower carboniferous age.

7. SPIRIFER sp. ind. ex aff. SP. RAJAH, Pl. IV, Fig. 6.

A species which seems to be very closely allied to the preceding one, is represented among Dr. Gerard's collections from the Kuling shales of Spiti, although unfortunately, by an isolated dorsal valve only. This valve, which has been slightly deformed by crushing, is somewhat square-shaped, wider than long, and provided with slightly-produced, attenuated wings, terminating in acutely rounded angles. The mesial fold is not strongly elevated above the general convexity of the valve, but nevertheless considerably surpasses in strength all the rest of the radiating ribs. The latter are of the same character as in weathered specimens of *Sp. Rajah*. The majority of them are single, but occasionally traces of the original secondary bifurcating costæ are still to be noticed. Concentric imbricating marks of growth are strongly developed.

The chief difference between this species and *Sp. Rajah* consists in the shape of the median fold. It is composed of a single rib at its origin, and continues so for some distance, when it becomes dichotomous. The two ribs, into which the original fold is thus splitting up, gradually increase in size towards the front, and are separated by a deep longitudinal valley, which is much more deeply excavated than the corresponding depression in *Sp. Keilharii*, v. Buch.

The measurements of this valve are as follow :—

Entire length of the dorsal valve	. . . .	35 mm.
,, breadth ,, ,, ,,	. . . .	53 ,,
Apical angle ,, ,, ,,	. . . .	ca. 116°

Locality | and *Geological position; number of specimens examined.*—Kuling shales near Muth, Spiti ; Coll. Gerard ; 1.

Remarks.—If we set aside *Spirifer Rajah* and *Sp. Keilharii*, there is a species from the anthracolithic rocks of Spitzbergen described by Toula (Kohlenkalk-Fossilien von der Suedspize von Spitzbergen, Sitzgsber. Kais. Akad. d. Wissensch Wien. Bd. LXVIII, p. 273, Taf. II, figs. 1, 2), to which the present one might be more especially compared. This species, of which the partly weathered dorsal valve only is known, agrees with my Himálayan specimen in the presence of a strongly dichotomous mesial rib, the two branches of which are separated by a deeply excavated valley.

V. GROUP OF SPIRIFER CLARKEI, de Kon.

8. SPIRIFER LYDEKKERI nov. sp., Pl. III, Figs. 1-4.

1866. *Spirifer sp. ind.*, Davidson, Quart. Journ., Geol. Soc., Vol. XXII, p. 36, Pl. 1, fig. 6.
1883. *Sp. striatus.* Lydekker, Mem. Geol. Surv. of India, Vol. XXII, Pt. II, fig. 6.

No complete specimen of this interesting species has come to my knowledge, but besides a large number of well-preserved dorsal valves one internal cast and two decorticated external impressions of ventral valves have been discovered among the fossil materials collected by Lydekker. Thus a fairly clear idea of the characteristic features of the species may be formed from a comparison of the different specimens.

The shell is rather variable in its dimensions, though always strongly transverse, more than twice as wide as long, the hinge-line corresponding to the greatest breadth of the shell. The lateral margins of each valve rapidly converge towards the extremities of the hinge-line, being thus produced into attenuated little wings, terminating in acute cardinal angles. Both valves are only moderately and almost equally convex.

I shall begin with the description of the dorsal valve which is more completely known to me. This valve is but little curved in either direction. Transversely it is slightly depressed in the vicinity of the small, attenuated and pointed wings, and is divided by a prominent, large, median fold. This median fold is considerably elevated above the general convexity of the valve, broadly rounded above, and without any trace of secondary ribs. On each side of the median fold the lateral parts of the valve are ornamented by a variable number of single, straight, radiating ribs, which become gradually obsolete in the vicinity of the wings. In the specimen figured on Pl. III, Fig. 2, about fifteen distinct ribs are counted on each side of the mesial ridge. In some forms they occur to the number of twenty. All the ribs are very regularly rounded above and separated by narrow rounded furrows which are far inferior to them in width.

If the test is entirely preserved, the surface of the valve is covered all over by closely disposed, concentric, undulating laminæ. But in the majority of my specimens this beautiful sculpture has been greatly injured or completely destroyed by weathering.

In casts devoid of their test, as in the form figured on Pl. III, Fig. 6, or in Lydekker's type specimen (Pl. III, Fig. 7) faint concentric marks of growth only are occasionally noticed.

In none of my numerous specimens has the apex been preserved. It has been either broken off or has been destroyed by weathering, thus partly exhibiting some of the internal characters of the valve. The Figures 6, 7, 10 will give a better idea of them than any particular description. I need only remark their strong similarity to the internal features exhibited by *Spirifer Lonsdalei*, as illustrated by Prof. Davidson (in Davidson-Suess, Classification der Brachiopoden, Wien, 1856, Pl. III, fig. 4). The impressions of the adductor muscles are especially well preserved in the specimen, Fig. 6a.

The measurements of two specimens, the larger of which has been collected in the dark-blue carboniferous rocks near Eishmakam, are as follow :—

	I. (Fig. 2a.)	II. (Fig. 6a.)
Entire length of the dorsal valve	43 mm.	31 mm.
„ breadth „ „ „	38 „	78 „
„ thickness „ „ „	10 „	10 „

The ventral valve of this species is represented among Lydekker's collection by an internal cast from Eishmakam, and by two external impressions from the Quartz sandstone of the Ladakh valley.

This valve is provided with a moderately deep, mesial sinus, regularly rounded at the bottom and devoid of any longitudinal sculpture. Its lateral borders are sharply rounded. The ornamentation of the lateral parts is of exactly the same pattern as in the opposite valve. In the specimen from Eishmakam the test has been partly preserved in the vicinity of the front line. It exhibits the concentric

L

lamellose condition, peculiar to complete forms of this species. The plaster-casts of the two external impressions, which have been reproduced in the Figures 8 and 9 of Pl. III, represent strongly crushed and deformed individuals, but show pretty well both the radiating sculpture of the surface and the shape of the rounded sinus strongly enlarged frontwards. Neither the apex nor the front line has been preserved. The anterior outline, as given in Fig. 5, ought not to be mistaken for the actual front of that specimen.

The cast from Kishmakam exhibits the two strong diverging dental plates extending to a distance of more than one-third the entire length of the valve.

The measurements of this specimen are as follow :—

Entire length of the ventral valve	. . . . . .	27 mm.
„ breadth „ „ „	. . . . . .	. ca. 30 „
Width of the area	. . . . . .	12 „
Apical angle	. . . . . .	135°

Locality and Geological position ; number of specimens examined.—Dark-blue shales and limestones with numerous *Fenestellidæ*, N. of Kishmakam, Coll. Lydekker, 4 ; yellowish-grey quartz sandstone, reminding of the devonian Spiriferen-Sandstein of the Rhenish region in its lithological aspect, Ladakh Valley, Coll. Lydekker, 12.

The identity of the specimens from the Barus beds of Kishmakam and from the quartz sandstones of the Ladakh Valley does not seem to me in any way doubtful.

Remarks.—Among the carboniferous *Spiriferidæ* the present species seems to be most closely allied to *Spirifer Clarkei*, L. de Kon. (Recherches sur les fossiles Paléozoiques de la Nouvelle Galles du Sud, p. 236, Pl. XIII, fig. 3). The only essential difference between the two forms consists in the presence of small attenuated wings in *Sp. Lydekkeri*, whereas in *Sp. Clarkei* the lateral margins unite with the cardinal border in rounded angles without any proper wings being developed. If we set aside this difference, the two species agree very well in their general shape, in the presence of a wide, slightly raised, smooth mesial fold in the dorsal and of a moderately deep, smooth and rounded sinus in the ventral valve, in the character of their radiating sculpture, which is composed of numerous single rounded ribs, and in their beautiful laminose concentric ornamentation.

In the face of such features there can be but little doubt, I think, that our Himálayan species belongs to the same group of forms as the Australian *Spirifer Clarkei*.

There is quite a number of carboniferous and devonian species which show a more distant similarity to the present one, but there is none among them which might justly put in a claim for a closer comparison.

9. SPIRIFER SP. IND. EX AFF. SP. LYDEKKERI, PL. V, fig. 2.

This species is perhaps still more closely allied to *Spirifer Clarkei*, de Kon., than the preceding one, but my materials are too scanty to allow anything to be asserted more positively.

Among the Geological Survey's Himálayan collection the present species is represented by a single incomplete dorsal valve only, with parts of the area of the ventral valve adhering to the rocky matrix. It is less strongly transverse than *Sp. Lydekkeri*, considerably less than twice as wide as long, without any proper wings and, so far as can be made out from the indistinct outline of the shell, with rounded cardinal angles. Its sub-elliptical shape readily distinguishes this species from the preceding one, whereas in the ornamentation no difference can be noticed. The specimen being devoid of its test, concentric marks of growth are but very faintly indicated.

The measurements of this specimen are approximately as follow :—

Entire length of the valve	. . . .	28 mm.
breadth „ „	. . . .	ca. 40 „
thickness „ „	. . . .	9 „

Locality and Geological position ; number of specimens examined.—Yellowish grey quartz sandstone, associated with *Spirifer Lydekkeri*, Ladakh Valley, Kashmir ; Coll. Lydekker ; 1.

VI. GROUP OF SPIRIFER ALATUS, Schloth.

10. SPIRIFER KASHMIRENSIS, Davidson.

1866. *Spirifer Kashmeriensis*, Davidson, Quart. Journ. Geol. Soc. London, Vol. XXII, p. 41, Pl. II, fg. 5.
1883. *Spirifer Kashmeriensis*, Waagen, Salt Range Foss., Pal. Ind., ser. XIII, Vol. I, Prod. Lamell. Foss., p. 521.
1891. *Spirifer Kashmeriensis*, Waagen, ibidem, Vol. IV, Geological Results, p. 166.

The following is Davidson's description of this species :— " Shell transversely fusiform, hinge-line long and straight, the lateral margins becoming gradually attenuated. Ventral valve ornamented with about twenty simple ribs. The sinus deep and divided along the middle by a small, median, slightly projecting rib, which, commencing at a short distance from the beak, extends to the front. Length 7, width 18 lines."

Of this species a few ventral valves only from the Zewán beds of Barus and Khoonmoo were obtained by Captain Godwin-Austen and entrusted to Mr. Davidson for examination. My materials are still more incomplete, consisting of a single incomplete ventral valve, which I refer with some hesitation to *Sp. Kashmeriensis*. Thus our knowledge of the characters of this interesting form remains yet very limited.

Davidson compared *Spirifer Kashmeriensis* to the devonian *Sp. macropterus*, Goldf, on account of its very transverse spindle shape, but marked the presence of a median rib in the sinus as an easily recognisable point of difference. Waagen in 1883 believed the present species to be closely allied to *Sp. alatus*, but afterwards, rejecting his former opinion, placed it in the group of *Sp. vespertilio*, Sow., from which he assumed it to differ solely by its more numerous ribs.

There is a good deal of confusion about *Spirifer vespertilio*, Etheridge, jun., differing entirely from L. de Koninck and Waagen in the synonymy of that species. I believe, however, that Mr. Etheridge (Geology and Palæontology of Queensland, p. 23) is perfectly right in retaining Sowerby's original name for the strongly transverse form, figured by Morris on Pl. XVII., Figs. 1 and 2, of his memoir in Count Strzelecki's "Physical description of New South Wales" and in proposing a new denomination, *Sp. Stutchburii.*, for the species, figured by Morris on Pl. XVII., Fig. 3, and by L. de. Koninck on Pl. XIII., Figs. 4*b*, 4*c*, and Pl. XIV., Fig. 3, of his memoir on the palæozoic fossils of New South Wales. Waagen's shell from the *Conularia*-nodules of the Salt Range consequently ought to be identified with *Sp. Stutchburii*, not with *Sp. vespertilio*, if an identification may be based altogether on so fragmentary materials.

A third species of the group of *Spirifer vespertilio* is indicated by the form figured by L. de Koninck on Pl. XIII., Figs. 4, 4*a*, of his above-quoted memoir. This species approaches *Sp. Kashmericnsis* by the presence of a mesial fold in the sinus, but strongly differs from it by its sharp angular ribs. Neither in the true *Sp. vespertilio* nor in *Sp. Stutchburii* a mesial rib is developed in the bottom of the sinus. Nor does the shape of the sinus agree in any way with that exhibited in the Kashmir shell. To me, therefore, the affinity of the latter to the Australian group of *Sp. vespertilio* appears, to say the least, very uncertain.

The permian *Spirifer alatus*, Schloth., appears to me more closely allied to the present form than any of the above quoted species. In this respect I perfectly agree with the view expressed by Prof. Waagen in 1883, when he stated his specimens of *Sp. alatus* from the Amb beds of the Salt Range Productus limestone to differ from *Sp. Kashmericnsis* by some minor details only.

Subgenus : SYRINGOTHYRIS, Winchell.

SYRINGOTHYRIS CUSPIDATA, Martin, Pl. IV., Figs. 9, 10.

1796. *Anomites cuspidatus*, Martin, Transactions. Linnean Soc., Vol. IV., p. 44, Pl. III., figs. 1—4, 5, 6.
1809. *Conchyliolithus Anomites cuspidatus*, Martin, Petrifacata Derbiensia, Vol. I., p. 10, Pl. XLVI. figs. 3, 4, Pl. XLVII, fig. 5.
1818. *Spirifer cuspidatus*, Sowerby, Mineral Conchology, Vol. II., p. 42, Pl. 120, figs. 1-3.
1825. *Sp. cuspidatus*, Sowerby, ibidem, Vol. V., p. 90, Pl. 461, fig. 2.
1836. *Spirifera cuspidata*, Phillips, Geology of Yorkshire, Vol. II., p. 216, Pl. IX., figs. 1—4
1840. *Spirifer cuspidatus*, L. v. Buch, Mém. Soc. Géol. de France, T. 1. V., p. 187, Pl. IX., fig. 18.
1843. *Sp. cuspidatus*, L. de Koninck, Déscr. des animaux fossiles du terrain carbon. de Belgique, p. 248. Pl. XIV., fig. 1.
1847. *Spirifera cuspidata*, Davidson, Monograph British Carb. Brachiopoda, pp. 44 and 234, Pl. VIII., Figs. 9—34, Pl. IX., figs. 1, 2.
1877. *Spirifer cuspidatus ?* Meek, Report U. S. Exploration of the fortieth parallel, Vol. IV., Palæontology, p. 87, Pl. III., figs. 11, 11*a*.
1884. *Syringothyris cuspidata*, Walcott, Palæontology of the Eureka District, Monogr. U. S. Geol. Surv., Washington, Vol. VIII., p. 219.

A ventral and a dorsal valve, which, however, do not belong to the same individual, from Kuling, in Spiti, are referable to this common mountain limestone species. They agree perfectly with typical specimens of *Syringothyris cuspidata* from Tournay, which I have been able to procure for comparison.

The ventral valve, which is not complete and slightly distorted, exhibits the characteristic, transversely pyramidal shape of a medium-sized form of *S. cuspidata*. Its straight hinge-line is terminating in acutely rounded off cardinal angles. The arched triangular area is almost vertical, and curved slightly forward in the vicinity of the apex. The apical angle is obtuse. The narrow triangular fissure is nearly twice as high as broad. Of the flat, concave sinus the umbonal portion only has been preserved in my specimen, since the frontal portion of the valve, situated anteriorly to the hinge-line, has been broken off. The sculpture consists of single straight radiating ribs, of which from eighteen to twenty are counted on each side of the sinus.

The measurements of this specimen are as follow :—

Length of the hinge line	. . . .	56 mm.
Width of the area	. . . .	18 „
Width } of the triangular fissure	. . . . . .	15 „
Height }		9.5 „
Apical angle of the ventral valve	. . . . . .	110°

The specimen represented by the dorsal valve, figured on Pl. IV., Fig. 10, has been considerably larger than the preceding one. This valve is strongly transverse, moderately and rather regularly convex, with nearly acute cardinal angles and a barely prominent apex. The large smooth median fold is but little elevated above the general convexity of the shell. It is divided along its middle by an indistinct longitudinal depression, extending from the extremity of the apex to about one-half the length of the fold.

There are about twenty-two simple radiating ribs present on each side of the mesial fold.

The measurements of this specimen are as follow :—

Entire length of the dorsal valve	. . . . . .	37.5 mm.
„ breadth „ „	. . . . . .	72 „
„ thickness „ „	. . . . . .	19 „
Apical angle „ „	. . . . . .	ca. 125°

Locality and Geological position ; number of specimens examined.—The two specimens were obtained near Kuling, in Spiti, by Dr. Stoliczka from a black crinoidal limestone, entirely different from any of the rocks which form part of the typical Kuling shales of Spiti or of the Productus shales in Johár and Painkhánda. There is some probability of this rock-specimen having been derived from the crinoid limestone horizon, which Griesbach has demonstrated to underlie the white quartzite of Spiti and which he correlates with the mountain limestone of Europe. If this probability could be proved, the presence of *Syringothyris cuspidata* would be strongly in favour of Griesbach's correlation, as neither in Europe nor in America this species has hitherto been ever met with in beds of an upper carboniferous age.

Sub-family: MARTINIINAE, Waagen.

Genus: MARTINIOPSIS, Waagen.

1863. *Martiniopsis.* Waagen, Salt Range Fossils, Pal. Ind., ser. xiii, Vol. I., Prod. Liment. Foss., p. 624.

MARTINIOPSIS (?) sp. ind. AFF. SUBRADIATA, Sowerby. Pl. VI., Fig. 9.

[Compare: *Spirifera subradiata*, Sowerby, Darwin's Geological Observations on the volcanic islands visited by H. M. ship Beagle, etc., 1844, p. 159.]
Spirifer subradiatus, Morris, in Strzeleck's Physical description of New South Wales and Vandiemensland, 1845, p. 281, Pl. XV., fig. 5, Pl. XVI., figs. 1—4.
Spirifer glaber, Dana, Geology, Wilkes' U. S. Exploring Expedition, etc., 1849, Vol. X., p. 683, Pl. I, fig. 6.
Spirifer glaber, L. de Koninck, Recherches sur les fossiles paléozoiques de la Nouvelle Galles du Sud, 1877, Pl. III, p. 227, Pl. XI., figs. 8, 9, Pl. XII., figs. 1a—c.
Spirifer Darwini, L. de Koninck (non Morris), ibidem, p. 230, Pl. X., fig. 11a, b, Pl. XI., fig. 10, 10a, Pl. XII., fig. 1.
Martiniopsis subradiata, Etheridge jun., in Jack and Etheridge, Geology and Palæontology of Queensland, and New Guinea, 1892, p. 298, Pl. XI., fig. 14.

It is with some hesitation that I introduce this name for a very fragmentary and somewhat crushed shell from the Zewán beds of the Kashmir Valley, but among the forms with which the present specimen could be compared, *Martiniopsis subradiata*, Sowerby, seemed to be the most similar one.

I consider this specimen to be the fragment of a ventral valve, with the internal cast partially exposed in its apical region. The sharp left border of the cast seems to correspond to the inner wall of a dental plate. The linguatiform impressions of the cardinal muscles are divided by a large double median septum. The median septum is more strongly developed than in any of the types of *Martiniopsis* figured by L. de Koninck. It must, however, be borne in mind that the present form if complete, would probably exceed in size even the type specimen of L. de Koninck's figure 10c on Pl. XI. of his memoirs on the palæozoic fossils of New South Wales.

My fragment is too incomplete to allow any exact reconstruction of its original outlines, but what can be made out of its general shape is not contradictory to an identification as a species of *Martiniopsis*, allied probably to *M. subradiata*. With this identification the sculpture of the shell surface agrees very well. It consists of a comparatively small number of strong concentric laminæ of growth which are crossed by a very faint radial ornamentation. A few obtuse radiating costæ are, however, more strongly marked. The shell structure exhibited in the vicinity of the front margin only is moderately punctate and fibrous.

Locality and Geological position; number of specimens examined.—Greenish shales of the Zewan beds, associated with *Protoretepora ampla*, Lonsd., Kashmir Valley; Coll. Vorchère; 1.

Remarks.—It is to be regretted that the present specimen is too poorly preserved to allow a definite identification. Could I have satisfactorily proved its appertenance to the group of *M. subradiata*, it might very justifiably have been quoted among the species indicating an affinity between the faunæ of the Himâlayan Zewan beds and of the carboniferous rocks of Australia.

Sub-order: ANCISTROPEGMATA, Zittel.

Family: *RHYNCHONELLIDÆ*, Gray.

Sub-family: RHYNCHONELLINÆ, Waagen.

Genus: RHYNCHELLA Fisch. v. Waldh.

Three species of this genus have been described by Davidson from the collections of Captain Godwin-Austen made in the Zewan beds of Kashmir. Only one of them was identified with a European form by that learned author, whilst the two others were considered to be new. These three species are:—

1. *Rhynchonella pleurodon* var. *Derxenxiana*, de Kon (*Rh. triplex*, M'Coy).
2. " *Barusicasis*, Davidson (l. c. p. 441, Pl. II, fig. 8. 1.)
3. " *Kashmericnsis*, Davidson (l. c. p. 42, Pl. II, fig. 91.)

The genus seems to be extremely rare in the anthracolithic rocks of Kashmir. Only very few specimens were available to Prof. Davidson, who himself stated the study of a larger number of specimens to be necessary for arriving at a satisfactory conclusion regarding the affinities of his new species to the rest of congeneric forms. Unfortunately, however, I am not in a position to add anything to elucidate this point. Both in the quartz sandstone with *Conularia tenuistriata* and *Spirifer Lydekkeri* from the Ladakh Valley, collected by Lydekker, and in a slab of rock from the Zewan beds of Barus collected by Major Collet, a small number of *Rhynchonellidæ* have been discovered, but they are so badly preserved that I could not even decide whether they actually belong to either *Rhynchonella* or *Camarophoria*. In a sericitic slate from the Ladakh Valley very badly preserved specimens of *Rhynchonella* have likewise been noticed.

Sub-family: CAMAROPHORIINÆ, Waagen.

Genus: CAMAROPHORIA, King.

CAMAROPHORIA CF. PURDONI, Davidson. Pl. VII., Fig. 4.

1863. *Camarophoria Purdoni*, Davidson, Quart. Journ. Geological Soc., London, Vol. XVIII., p. 30, Pl. II, fig. 4.

For a complete list of synonyms I refer to my memoir on the fauna of the perm-–carboniferous limestone crag of Chitichun No. I (Pt. 3 of this volume).

Two specimens from the anthracolithic rocks of Kashmir are referable to this species, with which I should have united them without the slightest hesitation had not their insufficient state of preservation prevented me from doing so.

In its general shape and outline the more complete specimen of the two agrees almost perfectly with moderately inflated forms of *Camarophoria Purdoni* from Chitichun No. I. This specimen, which has served as type for the illustration on Pl. VII. of this memoir, having been strongly injured by weathering, the exact number of ribs cannot be made out. One of the ribs in the median fold of the dorsal valve can be traced to the proximity of the apex. The two lateral portions of the shell are distinctly asymmetrical.

The measurements of this specimen are as follow :—

Entire length of the shell	31 mm.
Length of the dorsal valve	28 „
Entire breadth of the shell	32 „
Breadth of the median fold	16 „
Thickness of both valves	16 „
Apical angle of the ventral valve	110°
„ „ „ dorsal „	145°

The specimens under consideration are probably identical with *C. Purdoni*, but there are several species of this genus which so closely approach the latter form, especially *C. alpina*, Schellwien, from the upper carboniferous Fusulina limestone of Carinthia, that a direct identification is only possible if completely preserved forms were available.

Locality and Geological position ; number of specimens examined.—Black, micaceous shells, Kashmir, exact locality not known ; Coll. Lydekker ; 2.

From the character of the matrix adhering to the loose shells it appears doubtful whether they have been obtained from the Zewan beds or from the geologically younger shales with *Productus Abichi* and *Marginifera himalayensis* corresponding in age probably to the Kuling shales of Spiti.

Sub-order : ANCYLOPEGMATA, Zittel.

Family : *TEREBRATULIDÆ*, King.

Sub-family : TEREBRATULINÆ, Waagen.

Sub-genus : DIELASMA, King.

DIELASMA HASTATUM, Sowerby, Pl. VII., Fig. 6.

1824. *Terebratula hastata*, Sowerby, Mineral Conch. of Great Britain, Vol. V., p. 66, pl. 446, fig. 2, media (fig. 2, dextra et sinistra excisae).

1857. *T. hastata*, Davidson, Monograph British carb. Brachiopoda, p. 11, Pl. I., fig. 1 (cetera exclusa).

1870. *T. hastata*, Roemer, Lethaea Palaeozoica I., Th. atlas, tab. XLIII., fig. 1.

1880. *T. sacculus* var. *hastata*, Davidson, Monogr. British Fossil Brachiopoda, supplement, Vol. IV., p. 260, Pl. XXX., fig. 17.

1887. *Dielasma hastatum*, L. de Koninck, Faune du Calcaire carbonifère de la Belgique, p. 9, Pl. III., figs. 1—96, Pl. IV., figs 9—22, var. figs. 23—25.

Among the fossil material collected by Lydekker in the anthracolithic rocks of the Kashmir valley, there is one single but perfectly complete specimen of *Dielasma*, which, I think, may be safely identified with *Dielasma hastatum*, even if this latter species is accepted in the narrow circumscription proposed by L. de Koninck. In the genus *Dielasma* an interpretation of species has been introduced by that learned author, as has scarcely been adopted in any other group of brachiopoda, their distinction being based on very subordinate details to which in other genera of this class barely a varietal importance would have been accorded.[1] Of some of those species L. de Koninck himself candidly admits it to be solely a matter of taste, whether the difference be considered or not as sufficient for the distinction of separate species. If, notwithstanding this fact, I venture on a direct identification of the present specimen with *Dielasma hastatum* in the narrow circumscription proposed by L. de Koninck, it is on the ground of its agreeing entirely with some of the type specimens of that author from the mountain limestone of Belgium.

The shell is of a somewhat pyriform shape, elongated, truncated in the frontal region, and provided with a slightly arched frontal line. Its largest transverse diameter is situated a little anteriorly to the middle of its entire length.

The ventral valve is strongly inflated, especially so in the vicinity of the beak, gradually tapering towards the front line where a broadly excavated sinus is formed. This sinus originates in the middle of the entire length of the valve. The beak is thick, regularly curved, and slightly prominent beyond the apex of the dorsal valve. It is pierced by a large, longitudinally oval, oblique foramen. Its lateral portions are somewhat flattened, indistinct ridges extending down the beak on both sides of the foramen and limiting off a very ill-defined false area.

The dorsal valve is less strongly inflated than the opposite one, regularly vaulted in either direction, and provided with a shallow broad sinus, which is considerably less deep than the corresponding sinus in the ventral valve. The lateral margins of the two valves meet in a very flatly curved line, the convexity of which is turned towards the ventral valve. The front line is slightly raised in the opposite direction.

The surface of the shell is nearly smooth, ornamented by a few irregularly disposed, concentric striæ of growth only, which are restricted to the proximity of the front.

The measurements of this specimen are as follow :—

Entire length of the shell	.	.	.	.	.	.	.	.	.	.	29	mm.
„ „ „ smaller valve	.	.	.	.	.	.	.	.	.	.	25	„
„ breadth of the shell	.	.	.	.	.	.	.	.	.	.	21	„
Thickness of the shell	.	.	.	.	.	.	.	.	.	.	16	„
Apical angle of the ventral valve	.	.	.	.	.	.	.	.	ca.	107°	
„ „ „ dorsal valve	.	.	.	.	.	.	.	.	„	90°	

Locality and Geological position; number of specimens examined.—N. of Lishmakam, Kashmir Valley, obtained probably from the Zewan beds; Coll. Lydekker; 1.

[1] In strict contrast to this narrow interpretation of species stands L. de Koninck's identification of a *Dielasma* from Loping, described by Kayser (Oberearbonische Fauna von Loping, Richthofen, China, IV. Bd., p. 176, Taf. XXIII, fig. 9) with *D. normale*. L. de Kon. (l. u. p 2?, Pl. VI., figs. 49—63). If these two forms are to be united, we may as well accept species of so monstrous an extension as *D. sacculus* in the interpretation of Davidson.

M

Remarks.—L. de Koninck insists on this species being restricted to the mountani limestone of Belgium and Ireland only, and does not admit its identity with permian forms described as varieties of *Dielasma hastatum* by Davidson and by Kirkby. It is, however, impossible to decide whether among the forms from upper carboniferous beds, quoted under that name by European and American authors, the true *D. hastatum* may be represented or not, as a very different definition of this species has been adopted by the majority of palæontologists and by L. de Koninck.

Order: INARTICULATA, Huxley.

Family: *DISCINIDÆ*, Gray.

Genus: DISCINA, Lamarck.

DISCINA KASHMIRIENSIS, Davidson.

1866. *Discina Kashmiriensis,* Davidson, Quart. Journ. Geol. Soc., London, Vol. XXII., p. 43, Pl. II., fig. 19.

Two specimens of this interesting species were obtained by Captain Godwin-Austen in the Barus beds of Khoonmoo and described by Davidson. Waagen considers them to be very closely allied to *Discinisca Worthi* (Salt Range Fossils, Palæont. Indica, ser. XIII, Vol. IV., Geological Results, p. 134, Pl. V., figs. 12-15), from the *Conularia* nodules of the lower speckled sandstone in the Eastern Salt Range and to differ from the latter species solely by their larger dimensions and by their more strongly inflated upper valves.

No specimens of any *Discinidæ* have been discovered among the fossil material from Kashmir and Spiti entrusted to me for examination by the Director of the Geological Survey of India.

The genus *Discina* is of so scarce an occurrence in Asiatic deposits of an anthracolithic age that the presence of this Himálayan species in the carboniferous rocks of Kashmir is of some importance.

Class: BRYOZOA.

Although this class of fossils is very richly represented in the anthracolithic rocks of Kashmir, their examination is rendered exceedingly difficult by their unsatisfactory state of preservation. Not only they are, as a rule, found in the condition of impressions only, but even in the few cases, if the polyzoarium itself has been preserved, I have, with one single exception, either not succeeded in clearing the poriferous side from the matrix, or have found it so strongly injured from weathering that no definite information concerning the nature of the cells could be obtained. In the absence of this character I deemed it preferable to abstain from a specific identification of my specimens and to be rather content with comparing them with such species as I thought to be probably their nearest allies. In one single case only did I dare to make an exception from this treatment of my materials. This species is *Protoretepora ampla*, Lonsdale. The examination of some excellently preserved impressions from the Zewán beds of Kashmir revealed to me the nature and arrangement of the cells, agreeing perfectly with those exhibited in the Australian types of this remarkable form.

Thus the present monograph does not give an adequate idea of the rich fauna of *Bryozoa* contained in the anthracolithic system of the North-Western Himálayas. No doubt a systematic search of the localities mentioned will bring to light a much larger number of species than those here described.

Order : GYMNOLÆMATA, Allen.

Sub-order : CYCLOSTOMATA, Busk.

Family : *FENESTELLIDÆ*, King.

Sub-family : FENESTELLINÆ, Waagen.

Genus : PENESTELLA, Lonsdale.

1. PENESTELLA SP. IND. AFF. F. FOSSULA, Lonsdale. Pl. VII., Fig. 8 ; Pl. VIII., Fig. 4.

[Compare : *Fenestella fossula*, Lonsdale, in Darwin's Geol. Observations on Volcanic islands, 1844, p. 166, and in Count Strzelecki's Physical Description of New South Wales, etc., 1845, p. 269, T. IX., fig. 1.] *Fenestella fossula*, Etheridge, Quart. Journ., Geol. Soc., London, Vol. XXVIII, 1872, p. 333, Pl. 16, fig. 1. *Fenestella fossula*, Etheridge jun. in Jack and Etheridge, Geology and Palæontology of Queensland, etc., [p. 317, Pl. IX., figs. 4, 5].

A number of specimens of a *Fenestella* from the Zewán beds of the Kashmir valley appear to be very closely allied, if not actually identical with the present species. But as the poriferous side is not sufficiently well preserved to allow the character and arrangement of the cells to be studied, I dare not venture on a direct identification. Since a very narrow circumscription of the single species of this genus has been adopted by Stuckenberg in his monograph of the Russian carboniferous corals and Bryozoa, a specific determination of *Fenestellæ* is barely any more possible if one has not to deal with perfectly preserved specimens. Being not in this happy position, I must be satisfied with referring my forms to *Fenestella fossula* as their probably nearest ally.

All my specimens are characterised by their very regular structure and by their densely retiform appearance, being composed of very delicate and slender dissepiments and interstices. The branches are of equal thickness throughout their entire extent and are not frequently divided dichotomously. They are not swollen at the point of division. They are considerably thinner and bifurcate less frequently than in *F. plebeia*, M'Coy. The interstices form together with the dissepiments rectangular fenestrules, bordered by nearly straight bars, and provided with ovally rounded off corners. Within the space of 5 mm. there can be counted generally 10 meshes or fenestrules in the direction of the extension of the branches, and 14 in the transverse direction.

M 2

On the non-poriferous side the branches are flatly vaulted. Whether they have been ornamented by a longitudinal striation or not I am not able to decide, as in none of my numerous forms the surface has been perfectly well preserved. Of the poriferous side very little is known to me. Traces of a median keel are occasionally noticed in the hollow spaces left in the cast by the impression of the branches.

The species attains considerably large dimensions, one of my specimens, though incomplete, reaching a length of 100 mm. and a width of 90 mm.

In the specimen, figured on Pl. VII, Fig. 8, the natural colouring of the species, a dark Indian red, seems to have been preserved.

Locality and Geological position ; number of specimens examined.—Numerous forms of this species have been collected in the Zewán or Barus beds of the Kashmir Valley by Major Collet, of the Ladakh Valley and at Eishmakam by Lydekker, near Barus by Captain Godwin-Austen.

Remarks.—The dense arrangement of the very thin, but rarely dichotomising branches is a good character of this species which very closely approaches *Fenestella fossula,* Lonsd., from the carboniferous rocks of New South Wales, Tasmania and Queensland. In its general shape it agrees especially well with the specimen from Gympie figured by Etheridge, sen., in Volume XXVIII of the Quarterly Journal.

Waagen united together *Fenestella fossula* with *F. reverís,* Fisch., and *F. indiensis,* Waag., in a special group of forms, distinguished from the rest of congeneric species by a different aspect of the two faces of the colony, " on the poriferous side the fenestrules appearing more or less rectangular, while on the other side they appear oval or nearly circular." This peculiarity is however not mentioned by Etheridge, jun., in his minute description of Lonsdale's species in the " Palæontology of Queensland and New Guinea " (p. 217).

2. FENESTELLA SP. AFF. F. INTERNATA, *Lonsdale.*

Pl. VII., Fig. 9 ; Pl. VIII, Fig. 3.

[Compare *Fenestella internata,* Lonsdale, in Darwin's Geological Observ. on Volcanic Islands, etc., 1844, p. 166.]

F. internata. Lonsdale, in Strzelecki's Physical Description of New South Wales, etc., 1845, p. 280, Pl. IX, fig. 2.

F. internata, Dana, in Wilkes, U. S. Exploring Expedition, Geology, 1849, p. 710, Pl. X, fig. 12.

F. internata, Etheridge, jun., in Jack and Etheridge, Geology and Palæontology of Queensland, etc., 1892, p. 218, Pl. IX, figs. 5-7.]

Although among my materials this species is more numerously represented than the preceding one, I have seen it only in the condition of impressions in which the casts of the fenestrules and the hollow spaces left by the removed dissepiments and branches have been preserved. Thus the determination of the species remains yet more uncertain, being necessarily based on a general resemblance of my specimens to *Fenestella internata* only.

The present species exhibits the same densely retiform appearance, the rarely dichotomising branches, the regular arrangement of the straight interstices and disepiments enclosing rectangularly oval fenestrules as has been noticed in *F fossula*. It chiefly differs from the latter in being of a larger habit. Within the space of 5 mm. there can generally be counted 5 to 6 fenestrules in the direction of the extension of the branches, and 7 to 8 in the transverse direction.

Locality and Geological position ; number of specimens examined.—I have examined about twelve specimens of this species which were obtained at different localities from the Zewán beds of the Kashmir and Ladakh Valley, by Captain Godwin-Austen, Major Collet and Lydekker.

Subfamily: POLYPORINÆ, Waagen.

Genus : PROTORETEPORA, de Koninck.

PROTORETEPORA AMPLA, Lonsdale, Pl. VII, Fig. 10 ; Pl. VIII, Figs. 1, 2.

1844. *Fenestella ampla*, Lonsdale, in Darwin's Geolog. Observ. on Volcanic Islands, etc., p. 162.
1845. *F. ampla*, Lonsdale, in Strzelecki's Physical Descr. of New South Wales, etc., p. 268, Pl. IX, figs. 2.
1849. *F. ampla*, Dana in Wilkes, U. S. Exploring Expedition, Geology, p. 10, Pl. II, figs. 1, 1a.
1876. *Polypora ampla*, Etheridge, jun., Transactions Royal Soc. Victoria, Vol. XII, p. 66, fig. 1.
1883. *Protoretepora ampla*, Lydekker, Geology of the Kashmir and Chamba territories, etc., Mem. Geol. Survey of India, Vol. XXII, Pl. 11, fig. 1.
1892. *P. ampla*, Etheridge, jun., in Jack and Etheridge, Geology and Palæontology of Queensland, etc., p. 231.

This is the only species of *Bryozoa* contained in the Geological Survey's collections from Kashmir which is represented by sufficiently well preserved forms to warrant a certain identification. Mr. Etheridge, jun., has fully discussed the value of the generic name *Protoretepora* and redefined it. I have also relied on his arguments in adopting the present species in a narrower circumscription than had been introduced by L. de Koninck.

My specimens agree perfectly well with the Australian *Protoretepora ampla*, if the latter is accepted in the interpretation proposed by Etheridge, jun., excluding the *var. Konincki* and *var. Woodsi*. They form very large funnel or cup-shaped colonies with often strongly contorted or crumpled expansions. The largest among them seems to have attained at least 160 mm. in length and 180 mm in width. The cell-bearing face of the polyzoarium is internal. The interstices are tolerably straight and bifurcating at moderately long intervals. They are broad, flatly arched and expanding previous to bifurcation. The dissepiments are only one-half to one-third the length of the interstices, from which they are, however, not very distinctly defined. The numerous elongate oval fenestrules are arranged radially in regular rows, some of which can be traced from the very root to the margins of the colony. There are generally four fenestrules within the space of 10 mm. along the longitudinal direction.

On the poriferous face of the colony the interstitial interspaces between the fenestrules are occupied by from five to eight rows of circular cell-apertures. The dissepiments as well as the interstices are celluliferous

The outer surface has been too much injured in my specimens by weathering to allow its ornamentation to be studied in detail.

Locality and Geological position; number of specimens examined.—This remarkable species is not at all rare in the Zewán beds of the Kashmir Valley. The chief localities where it has been collected by Verchère, Captain Godwin-Austen and Lydekker are the following: Mandakpál, N. W. of Wasterwan, Leitipur, S. E. of Srinagar, Pailgam, Barus, Marhai Pam (associated with *Lyttonia sp. ind.*).

Remarks.—The identity of the Kashmir specimens with the typical *Protoretepora ampla* from the anthracolithic rocks of Tasmania and Queensland has been advocated by Lydekker on the authority of Dr. Feistmantel and fully confirmed by my examination of his materials. It is of no small geological importance, pointing as it does, to slight Australian affinities in the carboniferous fauna of Kashmir in a more direct way than the rest of similar indications.

Family : *THAMNISCIDÆ*, King.

Genus : ACANTHOCLADIA, King.

ACANTHOCLADIA sp. ind. Pl. VII, Fig. 7.

It is with considerable hesitation that I refer a single, incomplete and badly preserved specimen from Loodoo to this genus.

The arborescent colony consists of a primary branch, from which a few secondary branches take their origin, bifurcating at long intervals, but being never connected by dissepiments. All the branches are situated in one plane. The majority of them are ornamented either on both sides or on one side only with numerous little branchlets, which are narrower than the principal stems, very short, pointed and nearly parallel to each other.

Although the poriferous face is exposed in my specimen, the character and arrangement of the cellules cannot be made out with certainty on account of its unsatisfactory state of preservation.

Locality and Geological position; number of specimens examined.—Loodoo Vihi Valley, Kashmir, Zewán beds; Coll. Godwin-Austen ; 1.

FAUNISTIC RESULTS.

Among the faunæ of the anthracolithic system in the Himálayas of Kashmir and Spiti described in the present memoir, the fauna of the Kuling shales of Spiti is the geologically youngest in age. It comprises the following forms, being rather poor in species, though rich in individuals :—

1. *Marginifera himalayensis*, Diener.
2. *Chonetes cf. Lissarensis*, Diener.
3. *Athyris Gerardi*, Diener.
4. *Spirifer Rajah*, Salter.
5.   ,,   *sp. ind. aff. Rajah*.
6.   ,,   *Musakheylensis*, Davids.
7.   ,,   *cf. Nitiensis*, Diener.

So far as numbers go, *Marginifera himalayensis*, Diener—a species very clearly allied to *M. typica*, Waagen—and *Spirifer Rajah*, Salt., play the principal part. The class of Lamellibranchiata is represented by a few very poorly preserved fragments only, which do not allow a specific determination. The state of preservation of the brachiopoda is also, as a rule, rather indifferent. There is barely one single fairly complete form among the numerous specimens collected by Dr. Gerard, Stoliczka and Griesbach.

Griesbach correlated the Kuling shales of Spiti (*sensu stricto*) with the permian Productus shales of Johár and Painkhánda. The palæontological evidence afforded by the examination of the fossil materials handed over to me is certainly not adverse to this correlation, which was chiefly based on stratigraphical and lithological characters. Out of seven species of brachiopoda composing the fauna of the Kuling shales of Spiti, four are probably identical with forms from the permian Productus shales of the Central Himálaya. These four species are :—

> *Chonetes cf. Lissarensis*, Dien.
> *Athyris Gerardi*, Dien.
> *Spirifer Musakheylensis*, Davids.
> ,,   *cf. Nitiensis*, Dien.

Three of these species are restricted exclusively to the Productus shales of the Himálaya and do not occur in any other deposit. Among the three remaining forms, forming part of the brachiopod-fauna of the Kuling shales of Spiti also, not a single one occurs outside the sedimentary belt of the Himálaya. Thus the affinity of the present fauna to that of the permian Productus shales is undoubtedly more strongly marked than to the faunæ of any non-Indian strata of anthracolithic age. To this evidence it may be added that *Athyris Gerardi* and *Spirifer cf. Nitiensis* have been obtained from sandstone partings intercalated in the black micaceous Kuling shales, which lithologically agree so perfectly well with similar intercalations of sandstone partings in the Productus shales of Kiunglung that the specimens in the collection could not be separated without the labels attached to them.

An astonishing fact which must not be overlooked is the total absence of the two chief leading fossils of the Kuling shales of Spiti, *Marginifera himalayensis* and *Spirifer Rajah* in the Productus shales of Johár and Painkhánda. This fact, it is true, strongly diminishes in its importance if we bear in mind that the leading fossils of the Productus shales themselves are very unequally distributed at different localities. *Chonetes Lissarensis*, e. g., which is most abundant in the Productus shales of Johár, whole rock-specimens being made up of its shells only, is entirely absent in Painkhánda. On the other hand, *Productus cancriniformis*, which is very common in the Niti district, has not been found in the Productus shales of Johár.

This rather unequal distribution of species throughout the Himálayan Productus shales does not however exclude the possibility of another explanation of the absence of *Marginifera himalayensis* and *Spirifer Rajah* in the eastern portion of the Central Himálaya. To this explanation a passage in Griesbach's description of the sequence of anthracolithic rocks in the Niti area (Geology of the Central Himálayas, Mem. Geol. Surv. of India, Vol. XXIII, p. 120) gives a clue. In his description Griesbach makes reference to the development of a calcareous thick-bedded dark-grey sandstone, intermediate between the carboniferous white quartzite and the Productus shales of a ravine at the foot of the Niti pass. This dark-grey sandstone he found full of brachiopods, which he identified with *Productus semireticulatus*. "I have compared"—he further adds—"both the specimens and the matrix with specimens contained in the Geological Survey Museum, which had been collected at Kuling by Dr. Stoliczka; both are so close in form and lithological character that they might easily have come from the same locality."

Among the materials handed over to me for examination no fossils from this rock group are unfortunately represented. Could the presence of *Marginifera himalayensis*—corresponding to Griesbach's *Productus semireticulatus*—in the calcareous sandstones from the foot of the Niti Pass satisfactorily be proved, this evidence might point to the probability of a correlation of these sandstones with part of the Kuling shales of Spiti characterised by the presence of *M. himalayensis* and *Spirifer Rajah*. Nor is the possibility excluded that these beds are slightly lower in their position within the Kuling shales than the main mass of the latter from which the species identical with forms peculiar to the Himálayan Productus shales have been derived. I have hinted at this possibility to draw the attention of future observers to this question, as it may easily be decided by collecting more extensive fossil materials, according to single geological horizons. To say more on the poor evidence of a few vague palæontological indications only, which are not supported by a thorough knowledge of the actual stratigraphical sequence, would far transgress the limits of sound geological reasoning.

My recent examination of the anthracolithic fossils of Kashmir collected by Captain Godwin-Austen, Verchère, Major Cullet and Lydekker, has led me to recognize two fairly well differentiated faunæ among them. The geologically younger fauna is represented by a small set of brachiopods, collected by R. Lydekker in a dark micaceous shale with occasional intercalations of sandstone on the summit of a ridge north-east of Prongam Tríl. The species composing this faunula are the following :—

> *Productus Abichi*, Waagen.
> *Marginifera himalayensis*, Dien.
> *Chonetes grandicosta*, Waagen.
> *Streptomena analoga*, Phill.

So far as a correlation of horizon may be based on so small a number of fossils, all the evidence goes to prove that the rocks from Prongam Tríl correspond in age to the Kuling shales of Spiti. The few slabs of rock from this locality, contained in the Geological Survey's collections, are full of *Marginifera himalayensis* and exactly agree in their lithological character with specimens from the Kuling shales, collected by Stoliczka and Griesbach. Each of the three remaining species is represented in Lydekker's collection by one single individual only. Among them

Strophomena analoga is an ubiquitous form of a tolerably wide geographical and geological distribution, ranging throughout the entire carboniferous system into permian strata. *Productus Abichi* and *Chonetes grandicosta* decidedly point to a permian age. *Productus Abichi* is among the most characteristic permian species of the Salt Range, of Armenia and Timor. *Chonetes grandicosta* has as yet not been discovered outside the upper Productus limestone of the Salt Range.

In considering these facts it appeared to me that from a *palæontological* point of view a correlation of the shales with *Marginifera himalayensis* from Prongam Trál with the Kuling shales of Spiti will best express our present state of knowledge regarding their true stratigraphical position. This view is corroborated by the fact that none of the few fossils from this locality is identical with a species from the Zewán beds of Barus, Khoonmoo or Eishmakam. Stratigraphical evidence to ascertain this correlation of the Prongam Trál beds with the Kuling shales of Spiti is, however, unfortunately wanting. In Lydekker's report on the geology of the Kashmir district no reference is made to the position of those beds, which he probably failed altogether to distinguish from the rest of anthracolithic rocks.

Provided the correlation of the Prongam Trál beds in Kashmir and of the Kuling shales in Spiti with the Productus shales of Johár and Painkhánda be correct—a view which, I suppose, will better express the facts hitherto known than any other—the permian system appears to play an important part in the anthracolithic series of the Himálayas and to represent a distinct horizon of great geographical distribution.

The richest of all the anthracolithic faunas described in the present memoir is contained in the Zewán or Barus beds of the Kashmir Valley. This fauna, to which attention has first been drawn by the valuable memoirs of Godwin-Austen and Davidson, is composed of the following species:—

TRILOBITA.

1. *Phillipsia sp. ind. aff. seminifera*, Phill.

LAMELLIBRANCHIATA.

2. *Modiola sp. ind.* (?)
3. *Aviculopecten nov. sp. ind.*
4. *Pecten nov. sp. ind.*

BRACHIOPODA.

5. *Productus Cora*, d'Orb.
6.   „   *nodosus*, Defr.
7.   „   *semireticulatus*, Mart.
8.   „   *cf. longispinus*, Sow.
9.   „   *cf. scabriculus*, Mart.
10.   „   *cf. spinulosus*, Sow.
11.   „   *pustulosus*, Phill.
12.   „   *punctatus*, Mart.
13.   „   *aculeatus*, Mart.
14.   „   *mongolicus*, Diener.

15. *Streptelasma sp. ind., aff. costata*, Waag.
16.   „   *cf. (?) leonupina*, Waag.
17. *Chonetes lavis*, Davids.
18.   „   *Hardreana var. Kashmeriensis*, Dav.
19.   „   *Austeniana*, Davids.
20.   „   *Baruviensis*, Davids.
21. *Lyttonia sp. ind.*
22. *Derbyia cf. senilis*, Phill.
23. *Eumetria cf. grandicosta*, Davids. (*an compressa*, Maes ?)
24. *Athyris subtilita*, Hall.
25.   „   *Baddhista*, Verchère.
26.   „   *cf. expansa*, Phill.
27. *Spiriferina cf. Kentuckensis*, Shum.
28. *Spirifer Moosakheylensis*, Davids.
29.   „   *sp. ind., aff. Moosakheylensis.*
30.   „   *Rajah*, Salter.
31.   „   *cf. irruegularis*, Mart.
32. *Spirifer Lydekkeri*, Diener.
33.   „   *Kashmeriensis*, Davids.
34.   „   *Vihianus*, Davids.
35. *Martiniopsis (?) sp. ind., aff. subradiata*, Sow.
36. *Rhynchonella triplex*, M'Coy.
37.   „   *Barunensis*, Davids.
38.   „   *Kashmeriensis*, Davids.
(?) 39. *Camarophoria cf. Purdoni*, Davids.
40. *Dielasma hastatum*, Sow.
41. *Discina Kashmeriensis*, Davids.

BRYOZOA.

42. *Fenestella sp., aff. fossula*, Lonsdale.
43.   „   *sp., aff. internata*, Lonsd.
44. *Protoretepora ampla*, Lonsd.
45. *Acanthocladia sp. ind.*

I have not been able to ascertain if *Camarophoria cf. Purdoni* ought not to have been included rather in the list of fossils from Prongam Trál than in the present one.

The fossils of the Zewan beds are contained in variously coloured shales, sandstones and limestones, but are as a rule rather indifferently preserved. This fact will explain the large number of species marked in the preceding list as "*sp. ind.*" or as "*cf.*" only.

In this list are contained 45 species altogether, of which, however, barely more than 30 could be identified with tolerable certainty. Among them brachiopods by far predominate, both in number of species and individuals, composing with 37 species five-sixths of the entire fauna. Although this proportion may be partly due to the circumstance that among the Lamellibranchiata and Bryozoa available for examination very few specimens only were found worthy of a specific description, the predominance of brachiopoda over the other classes of organic remains may be considered a well-established fact.

Judging by its general zoological character, the fauna of the Zewán or Barus beds can only be looked upon as of upper carboniferous age. Leaving out such forms as are either restricted to the Zewán beds only or specifically undeterminable, the remaining species may be divided into two groups. The first group is represented by species which occur both in the mountain-limestone and in deposits of an upper carboniferous age. The second group, which from a palæontological point of view is much more important, is composed of such species as are restricted to younger carboniferous strata but are absent in the mountain-limestone of lower carboniferous age. These species are :—

> *Productus mongolicus*, Diener.
> *Strophalosia cf.* (?) *Lewisiana*, Wang.
> *Chonetes Barusiensis*, Dav.
> *Lyttonia sp. ind.*
> *Eumetria cf. grandicosta*, Davids.
> *Spiriferina cf. Kentuckensis*, Shum.
> *Spirifer Musakheylensis*, Davids.
> „   *Rajah*, Salt.
> *Camarophoria cf. Purdoni*, Davids.
> *Protoretepora ampla*, Lonsd.

The frequent occurrence of *Strophalosia*, the presence of the strange genus *Lyttonia*, of the group of *Chonetes grandicostatæ*, of *Spirifer Musakheylensis* and of a group of *Spirifer*, distinguished by coarse fasciculate ribs (*Sp. Rajah*) are characters of such high importance that in the face of them a correlation of the Zewán beds with the upper carboniferous series of other countries can hardly be questioned. To this may yet be added another fact, pointing in the same direction, and this is the absence of any species in the whole list which has hitherto only been met with in strata of a lower carboniferous age.

Though the question as to the age of the Zewán beds may thus be settled in a general way, it is hardly possible to decide to which particular horizon of the upper carboniferous series in the standard stratigraphic scale these beds may correspond. In elucidating this point it will be necessary to deal first with the relations which exist between the fauna of the Zewán beds and the faunas of anthracolithic deposits of other countries.

A tolerably large percentage of species, at least 16 out of the 45 species, quoted in the preceding list, are peculiar to the fauna of the Zewán beds. The majority among them are, however, closely allied to carboniferous forms, as has been indicated in the special descriptions. Nevertheless one-third of the entire Zewán fauna appears to be made up of species which have not been found hitherto outside the Kashmir territory. But their importance from a stratigraphic point of view is considerably lessened by the fact that one of them only, *Spirifer*

Lydekkeri, is among the chief leading fossils of the Zewán beds. All the rest of leading fossils are well known anthracolithic species, *viz.* :—

> *Productus semireticulatus*, Mart.
> *Derbyia cf. senilis*, Phill.
> *Athyris subtilita*, Hall.
> *Spiriferina cf. Kentuckensis*, Shum.
> *Spirifer Musakheylensis*, Davids.
>   „   *Rajah*, Salt.
> *Protoretepora ampla*, Lonsd.

One of the most striking features of the Zewán fauna is their comparatively slight affinity to any of the faunæ of the Salt Range Productus limestone. Five species only are identical, three of which have not been quite safely determined. These species are the following :—

> *Productus Cora*, d'Orb.
>   „   *semireticulatus*, Mart.
>   „   *cf. longispinus*, Sow.
> *Spirifer Musakheylensis*, Davids.
> *Camarophoria cf. Purdoni*, Davids.

An affinity to the fauna of the Productus limestone of the Salt Range is further indicated by a small number of forms, which are very closely allied to Salt Range species. Such species are :—

> *Strophalosia sp. ind., aff. costata*, Waag.
>   „   *cf. (?) lamispina*, Waag.
> *Lyttonia sp. ind.*
> *Eumetria cf. (?) grandicosta*, Davids.
> *Discina Koshmerenensis*, Davids.

These are rather slight affinities only. Much more close are the relations to the carboniferous deposits of Europe. Chiefly the list of *Producti* from the Zewán beds contains a large number of European carboniferous forms. Ten species are directly identical, but the number of species probably identical is increased to 17, if such forms are included as are marked in the above quoted list as "*cf.*" only. The species pointing to European affinities are the following :—

> *Productus Cora*, d'Orb.
>   „   *costatus*, Defrance.
>   „   *semireticulatus*, Mart.
>   „   *cf. longispinus*, Sow.
>   „   *cf. scabriculus*, Mart.
>   „   *cf. spinulosus*, Sow.
>   „   *pustulosus*, Phill.
>   „   *punctatus*, Mart.
>   „   *aculeatus*, Mart.
> *Derbyia cf. senilis*, Phill.
> *Athyris subtilita*, Hall.
>   „   *cf. expansa*, Phill.

Spirifer cf. triangularis, Mart.
„ *Musakheylensis*, Davids.
Rhynchonella triplex, M'Coy.
Camarophoria cf. Purdoni, Davids.
Dielasma hastatum, Sow.

This predominance of European carboniferous types is a very remarkable fact. It is the more astonishing if we take into consideration that very close relations exist between the faunæ of Chitichun No. I and of the Salt Range Productus limestone, more than one-half of the entire brachiopod fauna of Chitichun No. I being composed of identical species.

There are similar relations between the faunæ of the Zewán beds of Kashmir and of Loping in China, as between the former and the faunæ of European carboniferous deposits. Seven or eight species are probably identical, and the number of very closely allied forms is still larger. The fauna of the Zewán beds certainly bears a greater similarity to that of Loping than to any of the Salt Range faunæ, though the latter region is geographically less distant.

A third element in the faunæ of the Zewán beds is constituted by a few forms which point to an affinity with the carboniferous fauna of Australia. These Australian affinities are indicated by the following species :—

Spirifer Lydekkeri, Dien.
Martiniopsis (?) *sp. ind., aff. subradiata*, Sow.
Protoretepora ampla, Lonsd.
Fenestella sp. ind., aff. foveola, Lonsd.
„ *sp. ind., aff. internata*, Lonsd.

Among these five species *Protoretepora ampla* only is actually identical with an Australian carboniferous type. *Spirifer Lydekkeri* is very closely allied to *Sp. Clarkei*, de Kon. The determination of the specimen, figured on Pl. VI, Fig. 9, of the present memoir, as *Martiniopsis*, is not beyond all doubt. The two species of *Fenestella* may probably have their nearest allies in the carboniferous rocks of Australia, but their exact identification is impossible, owing to the unsatisfactory state of preservation of my materials available for examination.

Another Himálayan species identical with an Australian form is *Strophalosia Gerardi*, King, but the geological age of the beds in which the Himálayan type specimen was collected by Dr. Gerard is unfortunately unknown.

The faunistic affinities between the Zewán beds of Kashmir and the carboniferous deposits of Australia, as indicated by the above-mentioned fossils, appear therefore to be very slight only. They are certainly less strongly marked than the Australian affinities of the fauna contained in the *Conularia* nodules and in the *Eurydesma* sandstones of the Salt Range. Nevertheless I consider them to be of no small geological importance, because similar affinities are entirely absent in the Productus limestone.

A correlation of the Zewán beds with the lower speckled sandstone of the Salt Range has been advocated by Waagen. Although Davidson's description of the brachiopods, collected by Godwin-Austen, was only available to him as a base for correlation, he justly noticed the predominance of European carboniferous types in the Kashmir fauna, and correctly inferred that the Zewán beds should be placed rather high in the carboniferous series, "that they should either be placed on a level with the lower speckled sandstone of the Salt Range, or else they should be considered as intermediate in age between the latter beds and the lower Productus limestone (Amb beds)."[1]

The remarkable palæontological separation of the faunæ of the Zewán and Amb beds is indeed no evidence in favour of a correlation of these two rock-groups. So far as the testimony of fossils in the correlation of the faunæ of two distant regions can be relied on to the extent and with the precision which our ability to interpret them will permit, the predominance of European and the slight admixture of Australian types point to a geological horizon slightly lower in age than the Amb beds. A correlation with the lower speckled sandstone is nevertheless yet far from being established safely. Only by obtaining new and abundant materials and giving them an exhaustive study could the problem of the relations existing between the Kashmir and Salt Range anthracolithic faunæ be practically solved.

The question whether the Zewán beds should be placed on a level with the Moscovian or Gshelian stages of the carboniferous system in Europe, is likewise an open one. Those who try to establish a natural classification of Himálayan rocks will probably come to the conclusion that the minor divisions of the carboniferous system, which are thoroughly adapted to the stratigraphic order in Eastern Europe, cannot be recognized in the carboniferous series of extra-peninsular India. I seriously doubt that the sub-divisions of the upper carboniferous strata, which are locally distinguishable in the Ural or in Central Russia, are satisfactory for purposes of correlation in the Indian province. All attempts to establish a correlation of the Zewán beds with any of these sub-divisions on biological evidence will prove forced and artificial.

There is yet one horizon among the anthracolithic series of the North-Western Himálayas, to which, upon the scanty palæontological data available, an upper carboniferous age must probably be attributed. This horizon is represented by a quartz-sandstone, collected in the Ladakh Valley by Lydekker, and containing the following fossils :—

> *Productus* sp. ind.
> *Spirifer Lydekkeri*, Dien.
> *Sp. ind. aff. Lydekkeri*.
> *Conularia tenuistriata*, M'Coy.

Neither the stratigraphic position of this sandstone nor the locality is mentioned in Lydekker's memoir. The assemblage of species strongly exhibits an affinity to Australian carboniferous types. Especially the presence of *Conularia tenuistriata*, occurring also in the boulder group of the Salt Range, points in this direction.

A horizon lower in age than all those hitherto mentioned is perhaps indicated by the specimen of a crinoidal limestone collected by Stoliczka near Kuling in Spiti.

---

[1] W. Waagen, Salt Range Fossils, l. c. Vol. IV, Geological Surveke. p. 104.

This rock specimen yielded two forms of *Syringothyris cuspidata*, Mart., a very common mountain limestone form in Western Europe. There is some probability of the specimen having been derived from the crinoid limestone horizon, which Griesbach has proved to form the base of the carboniferous series in the Central Himálayas, and which he correlated with the lower carboniferous beds of the European standard.

This is all I am able to say with regard to the faunistic features of the anthracolithic series in Kashmir and Spiti. My indications are rather vague, I regret to say, and the results of my studies less certain than I could have wished. But it must be borne in mind that in many instances the exact stratigraphic position of the fossils entrusted to me for examination was not known, and that it would require a personal study of the anthracolithic deposits *in situ* and the collection of extensive materials, exactly to single geological horizons, to obtain safer results. I can only express my earnest hope that the interesting problems connected with the stratigraphy of the anthracolithic system in the North-Western Himálayas may be solved in time by a detailed survey of the Kashmir Valley. If the details of both stratigraphy and palæontology in this district are worked out with sufficient minuteness, they may not only permit fuller correlations with the anthracolithic system in other parts of the world, but may probably lead to the solution of one of the most important problems in the natural history of the anthracolithic epoch, *i.e.*, of the relations between the carboniferous deposits of Europe and of Australia.

PLATE I.

Figs. 1, 2. PHILLIPSIA, sp. ind. aff., SEMINIFERA, Phill.

    Two pygidia from the Zewán beds, N. of Eishmakam, Kashmir Valley ; coll. Lydekker.

    1a, 2a natural size ; 1b, 2b twice enlarged.

Fig. 3. AVICULOPECTEN, sp. ind.

    Right valve from the Zewán beds of Eishmakam, Kashmir Valley ; coll. Lydekker.

Fig. 4. PECTEN, sp. ind.

    Left valve from the same locality ; coll. Lydekker.

Fig. 5. MODIOLA (?) sp. ind.

    Right valve from the same locality ; coll. Lydekker.

Figs. 6, 7. PRODUCTUS ACULEATUS, Martin.

    Zewán beds, Kashmir Valley ; coll. Verchère.

    6a, 7 ventral view 6b lateral view.

Fig. 8. PRODUCTUS ABICHI, Waagen.

    Summit of a ridge N. E. Prongsan Trkl, Kashmir ; coll. Lydekker.

    8a ventral view, 8b lateral view.

Figs. 9, 10. PRODUCTUS UNDATUS, Defrance.

    Two casts of dorsal valves from the Zewán beds, N. of Eishmakam, Kashmir Valley ; coll. Lydekker.

Fig. 11. PRODUCTUS cf. LONGISPINUS, Sow.

    Incomplete ventral valve from the Zewán beds of Barus, Kashmir Valley ; coll. Lydekker.

    11a ventral view, 11b front view, 11c lateral view.

Fig. 12. PRODUCTUS COSTA, d'Orb.

    Ventral valve, Zewán beds, Barus, Kashmir Valley ; coll. Lydekker.

    12a ventral view, 12b apical view, 12c lateral view.

Fig. 13. PRODUCTUS PUSTULOSUS, Phill.

    Cast of a dorsal valve from the Zewán beds of Barus, Kashmir Valley ; coll. Godwin-Austen.

    13a dorsal view, 13b lateral view.

Fig. 14. PRODUCTUS SEMIRETICULATUS, Martin.

    Cast of a dorsal valve from the Zewán beds, N. of Khoonmo, near Pampar ; coll. Lydekker.

    14a dorsal view, 14b lateral view.

Figs. 15, 16. STROPHALOSIA, sp. ind. aff., S. COSTATA, Waagen.

    Two ventral valves from the Zewán beds, N. of Eishmakam, Kashmir Valley ; coll. Lydekker.

    15a, 16 ventral view, 15b lateral view.

Fig. 17. STROPHALOSIA cf. (?) TENUISPINA, Waagen.

    Ventral valve from the Zewán beds of Barus, Kashmir Valley ; coll. Godwin-Austen.

    17a ventral view, 17b lateral view.

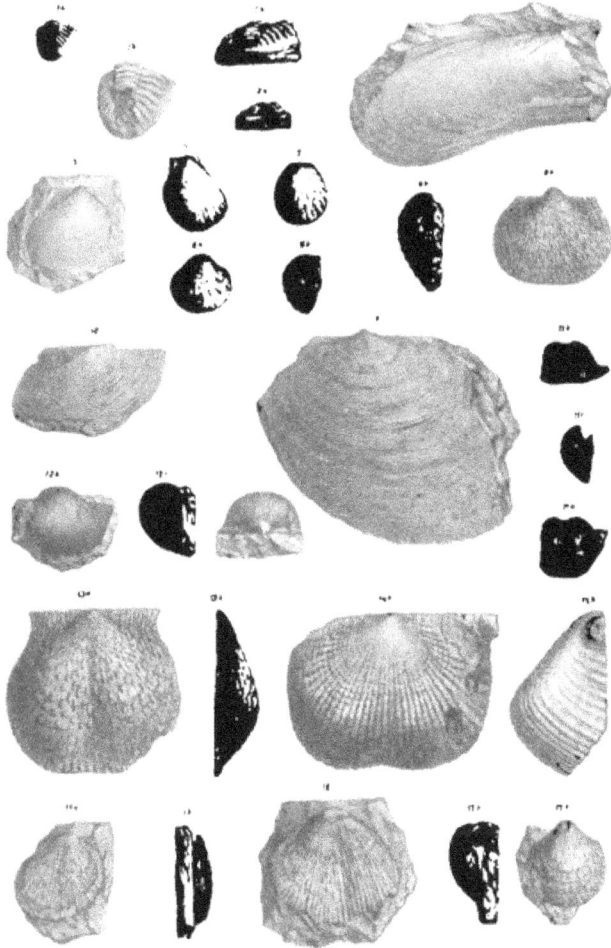

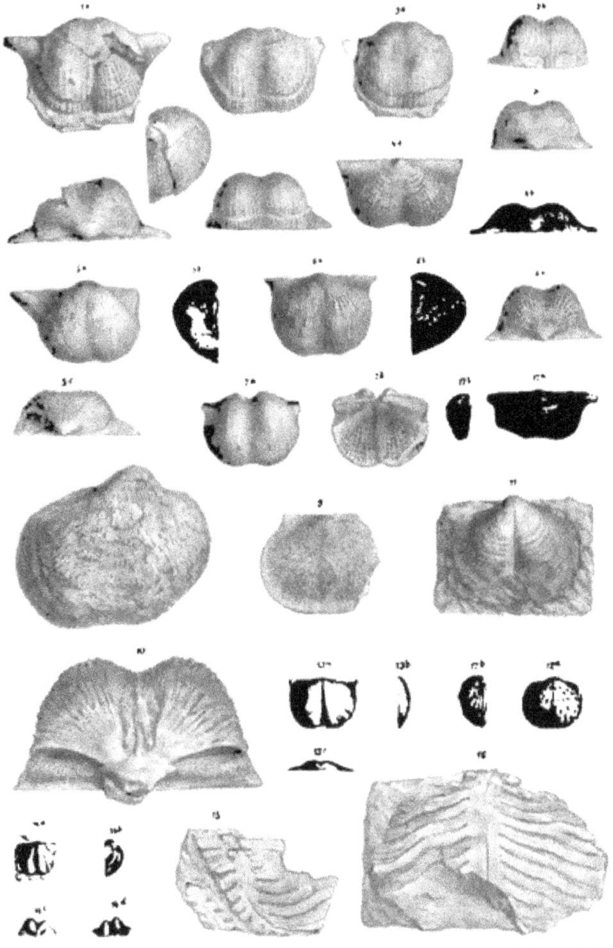

A. Swoboda del et lith.

Th. Bannwarth print.

PLATE III.

Figs. 1—4. Spirifer Lydekkeri, Diener.
From the Zewán beds, N. of Kishmakam, Kashmir Valley ; coll. Lydekker.

Figs. 5—11. From a quartz-sandstone of the Ladakh Valley (exact locality unknown) ; coll. Lydekker.

1, 2a, 3 dorsal valves.
2b cardinal region, slightly enlarged.

4, internal cast of a ventral valve, with impressions of cardinal teeth.
4a ventral view, 4b apical view.

5, 9, plaster-casts of external impressions of two ventral valves. Frontal region not preserved.

6, dorsal valve.
6a dorsal view, 6b lateral view.

7, dorsal valve, with cardinal region of ventral valve adhering, Lydekker's type-specimen (Mem. Geol. Surv. of India, Vol. XXII, Pl. II, Fig. 4).

8, 11, plaster-casts of external impressions of two dorsal valves.

10, dorsal valve with muscular impressions preserved.

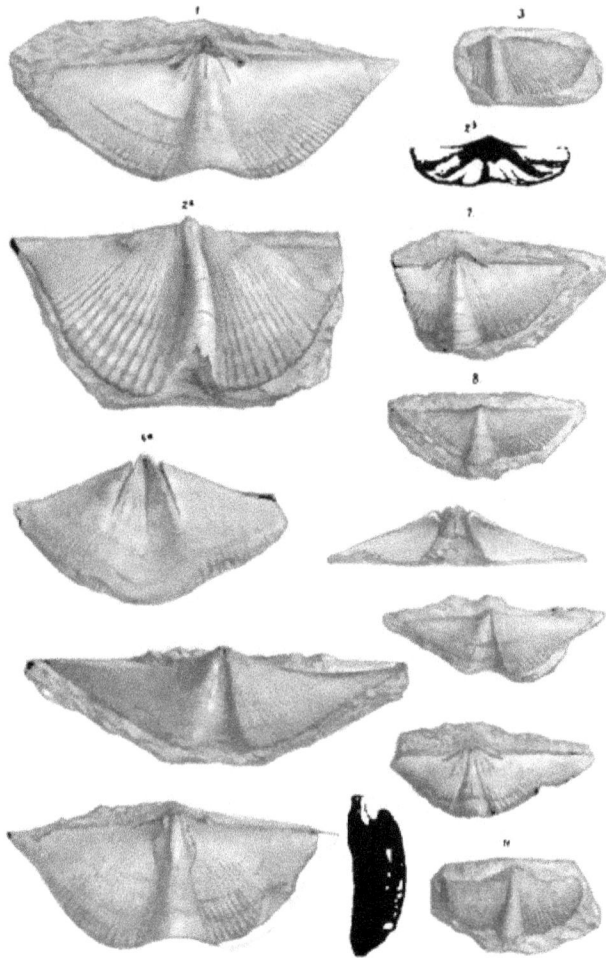

PLATE IV.

Figs. 1—7. SPIRIFER BALAR, Salter.

    1, ventral valve, from Baras, Kashmir Valley ; coll. Godwin-Austen.

    2, dorsal view of a specimen from the Spiti Valley ; coll. Gerard.

    3, internal cast of a ventral valve, from the Spiti Valley ; coll. Gerard.

    4, ventral valve, from Kuling, Spiti ; coll. Griesbach.

    5, largest specimen, known to me, from Kuling, Spiti ; coll. Griesbach.
    5a dorsal view, 5b lateral view.

    6, internal side of a ventral valve, from the Spiti Valley ; coll. Stoliczka.

    7, ventral valve, from the Spiti Valley ; coll. Gerard.

Fig. 8.   SPIRIFER sp. ind., or aff. Sp. BALAR.

    Dorsal valve from Muth, Spiti ; coll. Gerard.

Figs. 9, 10.   STRIROGMITHYRIS CUSPIDATA, Martin.

    Two specimens from a black, crinoidal limestone, Kuling, Spiti ; coll. Stoliczka.

    9, ventral valve.
    9a ventral view, 9b lateral view, 9c apical view.

    10, dorsal valve.
    10a dorsal view, 10b lateral view.

PLATE V.

Fig. 1.   Spirifer Rajah, Salter.
Dorsal view of a specimen from the Spiti Valley ; coll. Gerard.

Fig. 2.   Spirifer sp. ind., ex aff. Lydekkeri, Diener.
Dorsal valve from a quartz-sandstone of the Ladakh Valley, Kashmir ; coll.
Lydekker.

Figs. 3—7.   Spirifer Moraanettensis, Davids.
3, dorsal valve from Muth, Spiti ; coll. Stoliczka.
4, ventral valve from the Spiti Valley ; coll. Gerard.
5, ventral valve from Kiar, Spiti ; coll. Griesbach.
6, ventral valve from the Zewán beds, Kashmir Valley ; coll. Godwin-Austen.
7, internal cast of a ventral valve from the Spiti Valley ; coll. Godwin-Austen.

Fig. 8.   Spirifer cf. trigonularis, Martin.
Fragment of a dorsal valve, from the Zewán beds of Eishmaham, Kashmir ; coll.
Lydekker.

Fig. 9.   Spirifer cf. Nitiensis, Diener.
Fragment of a ventral valve from Po, Spiti ; coll. Gerard.

Fig. 10.   Spirifer sp. ind., aff. Moraanettensis.
Fragment of a ventral valve from the Zewán beds of Kashmir ; coll. Godwin-
Austen.

Figs. 11, 12.   Spiriferina cf. Kentuckensis, Shumard.
Two external impressions of dorsal valve, from the Zewán beds of Kashmir ; coll.
Verchère.
12b. Reproduction of a plaster-cast taken from the specimen, fig. 12a.

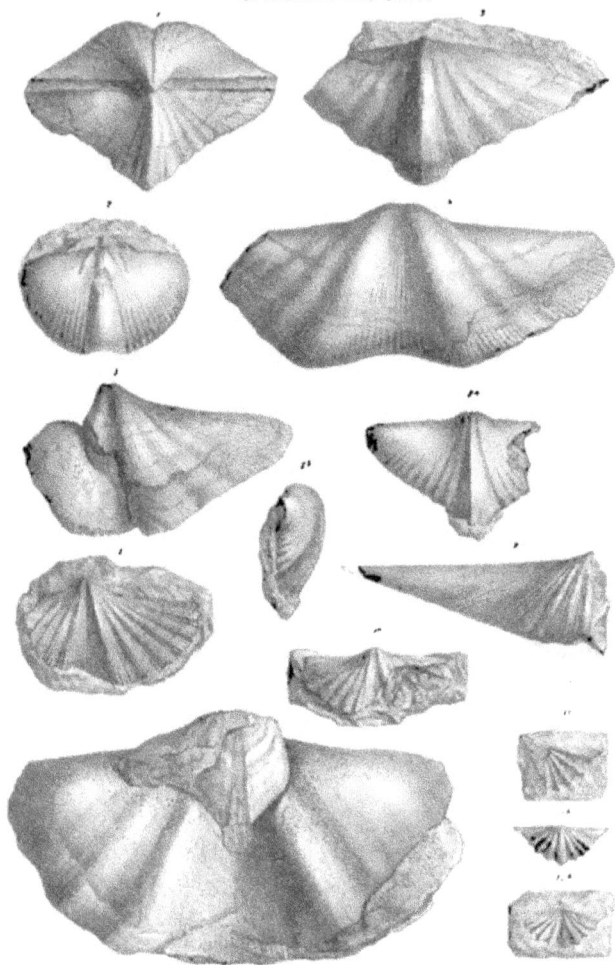

PLATE VI.

Geol. Surv. of India   ANTHRACOLITHIC-FOSSILS   Pl. VI

OF KASHMIR AND SPITI.

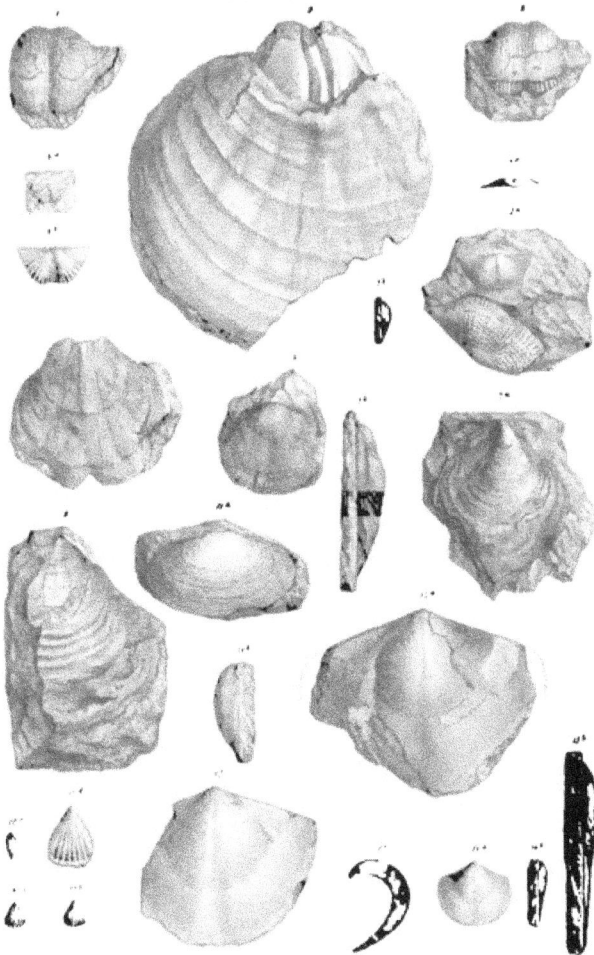

PLATE VII.

Figs. 1—3. ATHYRIS SUBTILITA, Hall.
Three specimens from the Zewán beds of the Kashmir Valley; coll. Godwin-Austen.
1a, 2a, 3a dorsal view, 1b ventral view, 1c, 2b, 3b lateral view, 1d frontal view.

Fig. 4. CAMAROPHORIA cf. PURDONI, Davids.
From Kashmir (horizon unknown); coll. Lydekker.
4a dorsal view, 4b ventral view, 4c lateral view.

Fig. 5. DIELASMA HASTATUM, Sowerby.
Complete specimen from the Zewán beds of Kishmakam; coll. Lydekker.
5a dorsal view, 5b ventral view, 5c lateral view, 5d frontal view.

Fig. 6. CONULARIA TENUISTRIATA, M'Coy.
Fragment from a quartz-sandstone of the Ladakh Valley, Kashmir; coll. Lydekker.
6a frontal view, 6b transverse section, 6c lateral view; all of natural size; 6d and 6, parts of the surface, enlarged.

Fig. 7. ACANTHOCLADIA sp. ind.
Zewán beds, Loodoo, Vihi Valley, Kashmir; coll. Godwin-Austen.

Fig. 8. FENESTELLA sp. aff. ROBUSTA, Lonsd.
Colony from the Zewán beds of Kashmir, coll. Collet, showing the non-poriferous face.
8b twice enlarged.

Fig. 9. FENESTELLA sp. aff. INTERNATA, Lonsd.
Colony from Mandakpal, N. W. of Wastarwan; coll. Lydekker.
Impression of the poriferous face.

Fig. 10. PROTORETEPORA AMPLA, Lonsdale.
Impression of the poriferous face of a colony from the Zewán beds of Mandakpal, N. W. of Wastarwan; coll. Lydekker.

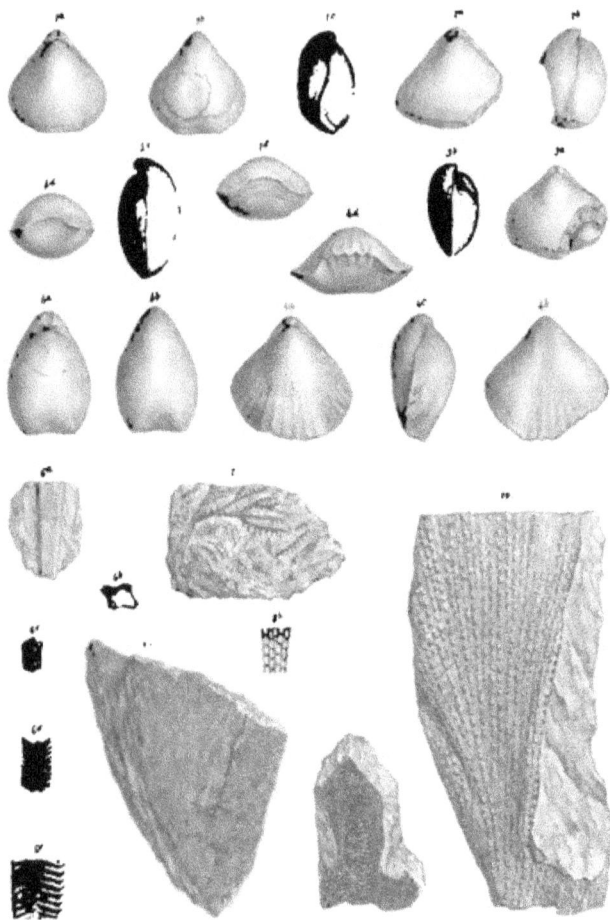

PLATE VIII.

Fig. 1.   Protoretapora ampla, Lonsdale.
   A very large colony from the Zewán beds of Mandakpal, N. W. of Wastiawan, Kashmir ; coll. Lydekker.
   1b transverse section of the polyzoarium.

Fig. 2.   Protoretapora ampla, Lonsd.
   Part of the surface of the specimen, figured on Pl. VII, Fig. 10, enlarged.

Fig. 3.   Fenestella sp. aff. internata, Lonsd.
   Impression of the poriferous face of a colony from the Zewán beds of the Kashmir Valley ; coll. Collet.

Fig. 4.   Fenestella sp. ind., aff. foretla, Lonsd.
   External impression of the poriferous face of a colony from the Zewán beds of the Kashmir Valley ; coll. Collet.

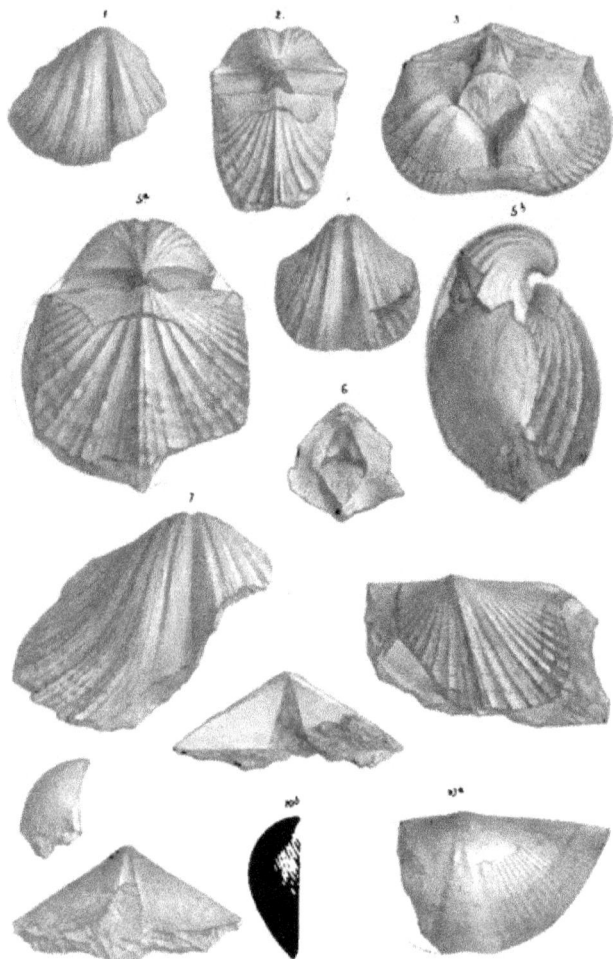

Pl. VIII

MEMOIRS

OF

THE GEOLOGICAL SURVEY OF INDIA.

Palaeontologia Indica,

BEING

FIGURES AND DESCRIPTIONS OF THE ORGANIC REMAINS PROCURED DURING THE
PROGRESS OF THE GEOLOGICAL SURVEY OF INDIA.

PUBLISHED BY ORDER OF HIS EXCELLENCY THE GOVERNOR GENERAL OF INDIA IN COUNCIL.

Ser. XV.

HIMÁLAYAN FOSSILS.

Vol. I, Part 3.

THE PERMOCARBONIFEROUS FAUNA OF CHITICHUN No. I.

By CARL DIENER, Ph.D.,

Professor of Geology at the University of Vienna.

Plates I to XIII

CALCUTTA:

SOLD AT THE

GEOLOGICAL SURVEY OFFICE.

LONDON: KEGAN PAUL, TRENCH, TRÜBNER & CO.

MDCCCXCVII.

PRINTED BY THE SUPERINTENDENT OF GOVERNMENT PRINTING, INDIA, 8, HASTINGS STREET, CALCUTTA.

MEMOIRS OF THE GEOLOGICAL SURVEY OF INDIA.

The price fixed for these publications is 1 Re. (10s.) each volume.

HIMALAYAN FOSSILS.

Vol. I, Part 3.

THE PERMOCARBONIFEROUS FAUNA OF
CHITICHUN, No. I.

THE PERMOCARBONIFEROUS FAUNA OF CHITICHUN, No. I.

BY

CARL DIENER, Ph.D.

PROFESSOR OF GEOLOGY AT THE UNIVERSITY OF VIENNA.

INTRODUCTION.

In the introduction to the chapter " Cephalopoda of the triassic limestone crags of Chitichun " in Vol. II, Part 2 of the present work, a short notice of the limestone crag of Chitichun No. I (17,710 feet) in Hundés is given. This crag was discovered in 1892 by the expedition to Johár, Painkhánda and the adjoining district of Hundés, in which C. L. Griesbach, C. S. Middlemiss, and myself took part. Even then it was highly probable, that its rich fossil contents were representatives of the middle Productus limestone fauna in the Salt Range, although a more thorough examination had to be waited for, to form the base of an exact determination of their age. Both the large number of forms composing this fauna, which is undoubtedly the richest palæozoic fauna hitherto discovered in the Himálayas, and the peculiar conditions under which it occurs, have led to a special part of this volume being devoted to its description.

A preliminary note on the geological features of this country, one of the most interesting in the Central Himálayas, has been published by C. L. Griesbach in the Records of the Geological Survey of India for the year 1893 (Vol. XXVI, Part I, page 19). In the Denkschriften der kais. Akademie der Wissenschaften in Wien for the year 1895 (Math. Nat. Classe, LXII, p. 588) I have given a more detailed description of the Chitichun region. It is, however, in the present memoir only, that all the geological conclusions will be collected that may be drawn from a minute study of the faunas of the different crags.

Leaving a discussion of the geological facts to the last chapter of this memoir, I shall restrict myself to the following introductory remarks.

The fauna, which I am going to describe, is contained in a white, partly semicrystalline limestone, alternating with layers of red, arenaceous or earthy limestones, and lenticular intercalations of a red crinoid limestone. This white limestone forms the main mass of the peak Chitichun No. I (17,740 feet) in the Tibetan area

between the Laptal ranges and the head of the Dharma valley. It is from 300 to
450 feet in thickness, and apparently rests on Spiti shales or is associated with
intrusive rocks (diabase prophyrite) which penetrate both the limestone crag and the
Spiti shales. The Spiti shales are very rich in fossils, especially in ammonites of the
genera *Hoplites* and *Olcostephanus*, which, according to information from Dr. Uhlig
who examined them, mark this division of the Spiti shales as a representative of the
Berrias stage of South-western Europe.

Throughout the limestone crag fossils are common, especially brachiopods,
bryozoa and corals. The two trilobites and the only ammonite, which we have been
able to secure, came from the upper portion of the block. Although during our
expedition in 1892 we could only stay about a week at Lochambelkichak encamping
ground, the starting point for an ascent of Chitichun No. I, we visited the crag four
times altogether and succeeded in obtaining finally a very large suite of fossils, to
Mr. Middlemiss belonging the honour of having found the first trilobite (*Phillipsia
Middlemissi*).

The fossils are, as a rule, well preserved, at least such as are contained in the
white limestone, whereas those which have been collected in the red arenaceous
layers, are frequently crushed and considerably deformed. A disadvantage in the
condition of the majority of fossils is their fragility, owing to the numerous planes
of cleavage, which intersect them in almost every direction. Nor is the internal
structure of the brachiopods, the chief constituents of this fauna, accessible to
observation, with the exception of a very small number of specimens, the rest
being casts, either hollow or filled by crystals of calcite.

Before entering into the specific descriptions, I feel obliged to express my most
sincere gratitude to Professor Waagen, to whom I am greatly indebted for his most
valuable assistance and advice.

DESCRIPTION OF FOSSILS.

Class: CRUSTACEA.

Section: ENTOMOSTRACA.

Order: TRILOBITÆ.

Family: *PROETIDÆ*, Barrande.

Genus: PHILLIPSIA, Portlock.

1. PHILLIPSIA MIDDLEMISSI, nov. sp. Pl. I, fig. 3.

Head and thorax unknown. The single pygidium, by which this species is represented in our collections from Chitichun No. I, was found by Mr. C. S. Middlemiss, in whose honour its specific denomination is proposed.

The pygidium is considerably broader than long, and strongly vaulted. At its anterior border the axial lobe occupies a little more than one third of the entire breadth of the pygidium. Its posterior portion has been partly broken off, but from the outlines preserved we may judge, that its extremity was regularly rounded. Its profile is of a semicircular shape.

The axial furrows are distinctly marked in the vicinity of the proximal border only. Near the posterior extremity of the axial lobe they gradually disappear. The number of axial segments can only be made out approximately. About ten coalesced rings, separated by deep furrows, seem to be present. The lateral lobes consist of seven or eight pleuræ, which terminate within the margin. The lateral lobes are regularly arched, the pleuræ are simple and united at the margin by a smooth rim, which occupies a little more than a third of the breadth of the lobe.

Neither axis nor pleuræ have any ornamentation upon them.

The measurements of this pygidium are as follow:—

Length of the entire pygidium	12·5 mm.
Breadth „ „ „	16·6 „
Length of the axial lobe	10·5 „
Breadth „ „ „ at its anterior margin	6 „
Breadth of the smooth border	3 „

Number of specimens examined. — 1.

Remarks. — Among British carboniferous trilobites of the genus *Phillipsia*, *Ph. Cliffordi* Woodw.[1] or *Ph. articulosa* Woodw.[2] might be compared to the present species as regards the general outlines and the absence of any ornamentation. But

[1] H. Woodward, Monograph of the British Carboniferous trilobites : Pal. Soc., 1883-84, p. 69, Pl. X, fig. 8—12.
[2] H. Woodward, loc. cit., p. 70, Pl. X, figs. 6, 13.

Ph. Cliffordi is of a broader shape, its axial lobe is bluntly rounded at the posterior extremity, and the lateral pleuræ bifurcate as they approach the margins. In *Ph. articulosa* the shape of the pygidium and the characters of the axial lobe and pleuræ are very similar, but it differs in possessing a greater number of coalesced segments (17) and lateral pleuræ (13), a feature, which seems sufficient to justify the specific separation of the Tibetan specimen.

From the caudal shields of *Griffithides globiceps*, Phill., and *Gr. obsoletus*, as figured by Prof. Woodward, this pygidium differs principally by its simple lateral pleuræ, these being double in the two species mentioned above. *Griffithides globiceps* is, moreover, distinguished by the corrugated character of its axial lobe.

Among the Russian representatives of the genus *Phillipsia*, which have been described and figured by V. von Möller *Ph. Eichwaldi* Fisch. (Ueber die Trilobiten der Steinkohlenformation des Ural, Bull. Soc. imp. des naturalistes de Moscou, 1867, No. I. Pl. II, fig. 3) somewhat recalls our Tibetan specimen in its outlines and in the development of a broad, smooth marginal rim. It differs, however, by the granulated character of its surface, and by the presence of flat transverse furrows on each of the lateral pleuræ.

Among the trilobites of the permian Fusulina limestone from Sosio in Sicily as figured and described by Gemmellaro (Mem. Soc. Ital. Sci., Napoli, VIII, ser. 3a, No. I) there is no species, which might be advantageously compared with *Ph. Middlemissi*.

Family *INDETERMINABLE*.

Genus: CHEIROPYGE, nov. gen.

Until 1884 no trilobite, in which the pleural ridges of the pygidium extend beyond the border, was known from carboniferous or permian rocks. The only slight exception from the types possessing a pygidium with definite, even outlines, was *Philippsia lodiensis* from the carboniferous deposits of Ohio, as described by Meek (Rep. Geol. Surv. Ohio, Vol. II, Pt. II, Palæontology, p. 324, Pl. 18, fig. 3). According to Meek, this species is distinguished by the "fimbriated character of the posterior and lateral margins of the pygidium." Prof. Meek says of this feature: "the segments are continued down and across the sloping border, at the edge of which they terminate in little, pointed projections, so as to present a fimbriated appearance around the posterior and lateral margins." This character is, however, so slight, that it has not been represented in the figure of Meek's type specimen. Nor has Claypole been able to observe the crenate character of the margin of the pygidium in other specimens of *Ph. lodiensis*.

In 1884, however, Prof. E. W. Claypole described and figured[1] the pygidium of a trilobite from the Cuyahoga shales, the uppermost member of the lower

---

[1] E. W. Claypole.—On the occurrence of the genus *Dalmanites* in the Lower Carboniferous rocks of Ohio; Geol. Mag. Decade iii, I, 1884, p. 303; woodcut fig. A).

carboniferous system, of Akron in Ohio. This pygidium is characterised by the absence of any marginal tract, and by its segments being "produced for the most part about half their length beyond the marginal line formed by their union, and ending in points, the third, seventh and ninth produced to double the distance and having the appearance of spines." This species has been assigned provisionally to the genus *Dalmanites* Barr. (*Dalmania* Emmr.) by its author under the denomination of *Dalmanites* (?) *Cuyahoga*.

This reference to the genus *Dalmania* is rather astonishing. In all the typical species of *Dalmania*, which have been described by Barrande, the pygidium is surrounded by a broad, smooth border, across which the pleuræ never extend. In my opinion, this pygidium from Akron more probably belongs to the subgenus *Phaëton*, Barrande, a section of the genus *Proctus*.

A second pygidium of a trilobite, provided with pleuræ extending beyond the border, has been collected by myself in the permocarboniferous limestone of Chitichun No. I. I dare not class the species, which I propose as the designation of this pygidium, in the same genus with *Dalmanites* (?) *Cuyahoga*, Claypole. While the American form may after all be only a representative of *Phaëton*, my Tibetan species must certainly belong to a new genus, although we must await the discovery of other parts of the carapace for a complete diagnosis of the latter.

The reasons, which prevent me from classing this form with any of the hitherto known genera of trilobites, will be discussed after a detailed description of its specific features has been given.

1. CHEIROPYGE HIMALAYENSIS, nov. sp. Pl. I., fig. 2 a, b, c.

Head and thorax unknown. Pygidium nearly as long as broad, distinctly trilobate, and moderately vaulted. The axis is distinctly marked off from the lateral lobes, and is composed of 15 coalesced segments. It is strongly conical, with a nearly semicircular profile, elevated considerably above the lateral lobes and equalling a little more than one third of the entire breadth of the pygidium at its anterior end.

The axial segments are rounded on their tops, and separated by deep furrows, which are as broad as their elevated portions. The latter are regularly covered with granulations, whilst the furrows between them remain perfectly smooth. The lateral lobes are more flatly curved than the axis, but are bent down rather abruptly towards their margins. No marginal space or limbus is formed. Each of the lateral lobes consists of six well defined pleuræ, which are interrupted by deep, narrowly rounded depressions. These pleuræ originate as sharp ridges in the furrows, limiting the axial lobe, broaden considerably, as they approach the margin, where they bend suddenly downwards. The posterior pleuræ are broader than the anterior ones. They are not distinctly carinate, although their profile in general takes the shape of a slightly pointed arch. The posterior termination of the axial lobe is surrounded by an axial lappet, which is exactly like the lateral pleuræ, but slightly exceeds them in width. Both the lateral pleuræ and the posterior axial lappet are ornamented with numerous rounded granulations of equal size. The interpleural depressions are smooth.

The strong projection of each pleura imparts to the margin a crenated appearance, recalling the pygidium in *Cromus*, Barr., of the family of *Encrinuridæ*.

The measurements of the present pygidium are, as follow :—

Length of the entire pygidium	. . . .	11 mm.
Breadth	. . . .	13·5 ,,
Length of the axial lobe	. . . .	8 ,,
Breadth ,, ,, ,, at its anterior margin	. . . . .	4·5 ,,

Number of specimens examined.—1.

Remarks.—Among the pygidia of hitherto described trilobites those of the subgenus *Phaëton* (*Phaëtonides*), Barr., may be first compared.

This subgeneric section has been introduced by Barrande for such species of the genus *Proëtus*, which are distinguished by a crenated margin of their pygidia.[1] Of the three species attributed by Barrande to this subgenus, *P. Archiaci* and *P. striatus* are of upper silurian age, whereas *P. planicauda* occurs in deposits of lowest devonian age. This last species has been considered by Novák[2] as the prototype of a proper subgenus, *Phaëtonellus*. No axial lappet or pleura is developed on the prolongation of the axial lobe, either in the Bohemian species of *Phaëton* or *Phaëtonellus*, or in American *Proëtus Verneuili*, which, judging from the figure given by Prof. Hall in his Illustrations of Devonian Fossils (Pl. XV, fig. 18) may also belong to *Phaëton*. On the contrary an emargination of the border always corresponds to the posterior termination of the axial lobe. This is the first important difference between *Phaëton* and *Cheiropyge*. A second difference is in the character of the lateral pleurae. In *Phaëton* they are flat and divided by a median furrow, reaching from the axial lobe to the pointed extremity of each segment. So remarkable are these two points of difference, that they not only make a reference of the present species to *Phaëton* impossible, but even induce me to believe, that *Cheiropyge himalayensis* belongs to an altogether different family from the *Proëtidæ*.

A certain resemblance seems to exist between this Tibetan trilobite and a few silurian forms, which have been included by Barrande in his genus *Cromus*. *Cromus transiens* Barrande (Système Silurien, p. 828, Pl. 43, figs. 18, 19) is a form which at a first glance recalls *Cheiropyge*. This similarity is, however, more apparent than real. There is no axial lappet or pleura, but the posterior element of the axial lobe is produced towards the margin and bordered on both sides by lateral segments. In the external character of the pleurae there are likewise differences, which make a distinction easy. The resemblance of the present specimen to pygidia of *Cheirurus* is still more distant. I do not know of any other genera of trilobites, which might advantageously be compared with *Cheiropyge*. Nor am I

---

[1] J. Barrande. Notes préliminaires sur le Système Silurien et les trilobites de Bohême, Leipzig, 1846, p. 62 ; Système Silurien du centre de la Bohême, Vol. I, 1852, p. 452.

[2] O. Novák, Vergleichende Studien an einigen Trilobiten aus dem Herzynn von Bicken, Wildungen, Greifenstein und Böhmen : Palaeontologische Abhandlungen, neue Folge, I., 1890, Heft 3, p. 16.

able to give any clue as to its systematic position among this order. The character of its pygidium is however so peculiar, that I believe to be justified in establishing a new genus, although without the knowledge of the cephalothorax a satisfactory diagnosis of the latter cannot be given.

MOLLUSCA.

Class: CEPHALOPODA.

Order: AMMONOIDEA.

Suborder: AMMONEA LEIOSTRACA.

Family: *CYCLOLOBIDÆ*, Zittel.

Genus: POPANOCERAS, Hyatt.

Subgenus: STACHEOCERAS, Gemmellaro (WAAGENINA, Krotow).

The genus *Popanoceras* has been introduced by Hyatt in his Genera of Fossil Cephalopods (Proc. Boston Soc. Nat. Hist., XXII, April 1883, p. 337). In this genus he included three Russian species, which E. de Verneuil had formerly classed among *Goniatites*, viz., *G. subolewskyanus*, *G. kingianus* and *G. koninckianus*, and, as "extreme form" *Arcestes antiquus*, Waagen, from the cephalopoda beds of the upper Productus limestone.

E. von Mojsisovics[1] in his memoir on the triassic faunae of the Arctic region accepted Hyatt's genus and gave a more appropriate diagnosis, referring to it five Spitzbergen forms and a species from the permian or triassic rocks of Timor, which had been described by Beyrich (Monatsber. k. Akad. Wiss. Berlin, 1867, p. 66) as *Ammonites megaphyllus* and considered as a type of his group of *Megaphylli*.

In the meantime Krotow[2] united three species of *Popanoceras*, from the Artinskian stage of the Ural Mountains, in a new genus, *Waagenia*, a name which he changed afterwards in *Waagenina*[3] as the priority of the former denomination had been claimed by Neumayr for a group (*hybonoti*) of *Aspidoceras*.

In 1887 Gemmellaro[4] described and figured a large number of species of *Popanoceras* from the permian Fusulina limestone of Sosio in Sicily. His examination of a very rich material led him to a generic distinction of two groups of forms. *Popanoceras sobolewskyanum* furnishes the type of the first group, for which the

[1] E. v. Mojsisovics, Arktische Triasfaunen : Mém. Acad. Imp. Sci., St. Pétersbourg, sér. vii, XXXIII, No. 6, 1886, p. 65.

[2] P. Krotow, Artinskian stage, Kama, 1885, (in Russian) p. 204.

[3] Mém. Com. Géol. Russ., St. Pétersbourg, VI, 1888, p. 474.

[4] G. Gemmellaro, La fauna dei calcari con fusulina della valle del Fiume Sosio, nella provincia di Palermo, 1887, Fasc. I, p. 25

name of *Popanoceras* is retained, while the second group, consisting of such species, as are more nearly allied to *P. antiquum*, Waagen, is elevated to the rank of a separate genus, *Stacheoceras*. To the latter genus likewise belong the forms formerly united in *Waagenia* (*Waagenina*) by Krotow.

The new genus *Stacheoceras* has, however, not been accepted without opposition. It was rejected by E. von Mojsisovics,[1] who considered the characters of difference between *Stacheoceras* and *Popanoceras*, as enumerated by Gemmellaro, to be of a very subordinate importance and not sufficient for a generic or even subgeneric distinction. Karpinsky[2] came to a similar conclusion in his beautiful memoir on the ammonites of the Artinskian stage, and maintained his view against the arguments of Gemmellaro, who actively defended the introduction of a separate genus for the group of *Popanoceras antiquum* in a special appendix[3] to his above quoted monograph. A different view of this question was however taken by K. A. von Zittel, who in his recently published "Elements of Palæontology"[4] accepted *Stacheoceras* as a proper genus, although he cannot be considered an advocate of too narrowly circumscribed genera.

In face of the results of the somewhat contradictory observations of E. von Mojsisovics, Gemmellaro and Karpinsky, it can scarcely be denied, that among the points of difference between *Popanoceras* and *Stacheoceras*, as enumerated by Gemmellaro, some are either problematical or of only very small importance. This remark applies both to the length of the body chamber and to the character of the sutures. Nevertheless a distinction can be based on the following characters.

Stacheoceras is of a more globose shape, turning discoidal only in later stages of growth. It is provided with a more delicate sculpture, its ornamentation consisting of thin radial striæ only, but often distinguished by the presence of varices. The importance of these characters has been compared by E. von Mojsisovics to that of the differences between two groups of the genus *Arcestes*, viz., the sections of *intralabiati* and of *subumbilicati*. Since the latter groups have meanwhile been elevated to the rank of subgenera by E. von Mojsisovics[5] himself, I do not think that this learned author will object any more to a similar treatment of *Popanoceras* and *Stacheoceras*.

A subgeneric distinction between *Popanoceras* and *Stacheoceras* will certainly meet the approval of palæontologists, who agree with Haug as to the advisability of introducing a special subgeneric designation for the Spitzbergen and Siberian species of *Popanoceras*. For these species, which are distinguished by a more complicate sutural line, the name *Parapopanoceras* has been proposed.[6]

[1] E. v. Mojsisovics, Über einige arktische Tris—ammoniten des nördlichen Sibirien: Mém. Acad. Imp. Sci. St. Pétersbourg, sér. vii, XXXVI, No. 5. 1888, p. 10.

[2] A. Karpinsky, Ueber die Ammoneen der Artinsk-Stufe und einige mit denselben verwandte carbonische Formen: Mém. Acad. Imp. Sci., St. Pétersbourg, sér. vii, XXXVII, No. 2 1887, p. 67, 84.

[3] G. Gemmellaro, loc. cit., Appendix. Palermo, 1888, p. 10.

[4] K. A. von Zittel, Grundzüge der Palæontologie (Palæozoologie). 1895, p. 168.

[5] E. v. Mojsisovics, Das Cephalopoden der Hallstatter Kalke: Abhandlgn. K. K. Geol. Reichsanstalt VI, ii Theil, p. 745.

[6] E. Haug, Les Ammonites du Permien et du Trias: Bull. Soc. géol. France, sér. iii, XXII, 1894, p. 396.

Another subgeneric denomination will be found necessary for a species from the triassic rocks of British Columbia, which has been described by Whiteaves as *Popanoceras McConnelli*.[1] It is readily distinguished from the Spitzbergen species of *Paropopanoceras* by its very distinctly angulated periphery. The lobes are minutely incised at their base and margins, but the saddles are not distinctly mega-phyllic, as is the case in the true *Popanoceratidae*.

Up to now neither *Popanoceras* nor *Stacheoceras* has been found in deposits of an older than permocarboniferous age. In permocarboniferous and permian strata of a pelagic facies they are the most common ammonites.

The occurrence of a species of *Stacheoceras* in the limestone of Chitichun No. I is therefore of no small geological importance, as, in correspondence with other evidence, it clearly proves these limestones to be younger than upper carboniferous in age.

1. POPANOCERAS (STACHEOCERAS) TIBETICUM, nov. sp. Pl. I, fig. 1 a—f.

Of this beautiful species the chambered portion of the inner whorls of a specimen attaining a diameter of 55 mm. and a few chambered fragments of the outer volutions are available for examination. It belongs to a group of forms, which are most nearly allied to *Stacheoceras mediterraneum*, Gemmellaro (La fauna dei calcari con Fusulina, p. 20, Pl. IV, fig. 2-6, VII, fig. 11-13) from the permian rocks of Sicily.

In its general shape and outlines my type specimen is very similar to *Stacheoceras Tietzei*, Gemm. (Pl. V, fig. 1-8). It is thickly lenticular, with a very narrow umbilicus and a broadly rounded siphonal part. The involution of the slowly increasing whorls is very considerable. The projection of the spiral of the penultimate whorl meets the last volution above the middle of the height of the latter.

The transverse section is broadly oval. The lateral parts are but slightly vaulted, and slope with regularly increasing convexity towards the umbilicus. No umbilical edge is developed. In the direction of the siphonal area the lateral parts show a very equal curve, and unite with the regularly arched external part without any proper demarcation.

The greatest transverse diameter of the volutions is situated a short distance below the middle of their height.

My specimen is an internal cast without any trace of a sculpture. I am not able to decide whether the shelly substance was smooth or covered with a delicate ornamentation, the fragments of the shelly layer, which have been preserved, being too incomplete and, moreover, deteriorated by weathering. But if any sculpture was ever present, it must have been but very faint.

No varices have been noticed.

[1] Whiteaves. The fossils of the triassic rocks of British Columbia : Contributions to Canadian Palaeontology, Vol. I, Part ii, Montreal, 1889, p. 136, Pl. XVIII, fig. 2, 3.

The following measurements have been taken from the specimen, corresponding to a diameter of 52 mm.:—

Diameter of the shell	. . . .	53 mm.
" " " umbilicus	. . . .	25 "
Height of the last volution { from the umbilical suture	. . .	25.5 "
" preceding whorl	. . .	11 "
Thickness of the last volution	. . . .	29 "

Body chamber.—Unknown.

Sutures.—The projection of the spiral of the penultimate whorl touches, in the last volution, the third lateral lobe at its inner margin. *Stacheoceras Trimurti* must therefore be classed among the species with three lateral lobes, whereas, as a rule, in this subgenus the normal number of lateral lobes (two) prevails.

The sutural line exhibits the semicircular arrangement, which, according to Gemmellaro, is peculiar to the group of *Stacheoceras mediterraneum*. It is only in the vicinity of the anterior margin of the last volution, that I have succeeded in tracing it out entirely. Corresponding to a height of the volution of 27 mm., the sutural line is composed of the following elements.

A very deep and broad siphonal lobe is divided by a median prominence, which reaches only half the height of the siphonal saddle. This median prominence is slightly enlarged at its top, and interrupted by a broad funnel for the entrance of the siphonele. The two branches of the siphonal lobe, which are very strongly individualised, are longer than any of the following lobes. Each of these two branches is provided with a secondary indentation. The principal lateral lobe is bipartite. All the following lobes are tripartite. The two branches of the principal lateral lobe are quite symmetrically arranged, of equal length, and terminate in sharp points. In the following lobes the central indentation is the largest, but the two lateral dentations on each side are as a rule, not symmetrical.

There are seven auxiliary lobes present.

The saddles are club shaped, and evenly rounded above. They regularly diminish in size from the siphonal saddle in the direction towards the umbilical suture. This is a type of sutures, to which the term "serial" as proposed by Blake,[1] might be advantageously applied.

In later stages of growth a further complication of the sutural line is indicated by the development of an accessory indentation in each of the two branches of the principal lateral lobe. The trifid termination of the following lobes remains however unchanged. This character of the sutural line is clearly developed in the fragment of an outer whorl, the height of which can scarcely have been less than 45 mm. As this fragment is entirely chambered, the species seems to have attained a rather remarkable size.

Number of specimens examined.—1.

Remarks.—Among the Sicilian species of the subgenus *Stacheoceras* the present one is most nearly allied to *St. Tietzei*, Gemmellaro, from which it chiefly differs by

[1] J. F. Blake, on the bases of the classification of Ammonites . Proc. Geol. Ass., XIII, pt. ii, 1893, p 8.

the absence of varices, and by the shape of the lobes, which in Gemmellaro's species exhibit a bipartite termination. Regarding the character of the sutural line, my Tibetan specimen approaches more nearly *St. mediterraneum*, Gem. In both species the siphonal lobe stands deepest and is of an exactly similar shape, the principal lateral lobe is bipartite, while the remaining lobes show a trifid termination. The only difference consists in the presence of a small accessorial dentation in the outer branch of the principal lateral lobe in *St. mediterraneum*. In their external characters however the two species differ more strongly, the Sicilian form being provided with a comparatively large and funnel shaped umbilicus.

Regarding the development of the sutural line there is no species more nearly allied to the present one, than *Stacheoceras Krasnopolskyi*, Karp.[1], (= *Waagenina subinterrupta*, Krotow[2]) from the Artinskian sandstones of Russia. In *St. Krasnopolskyi* the principal lateral lobe is bipartite and the second lateral lobe tripartite, corresponding to a height of the volution of 4 mm. Corresponding to a height of the volution from 7 to 10 mm., accessory dentations are developed in the branches of the principal lateral lobe. The number of lateral and auxiliary lobes together is six. From a comparison of the development of the sutural line in both species it clearly results, that the siphonal and principal lateral lobes show a bipartite arrangement, each of them gradually developing accessory dentations, whereas in all the following lobes the tendency of a tripartite differentiation prevails.

A similar character of the sutural line has been noticed in *Stacheoceras Parkeri*, Heilprin (Proc. Acad. Nat. Sci. Philadelphia, I, 1884, p. 53, fig. 1-2) from the permocarboniferous rocks of Texas, and in *Stacheoceras sp. ind. aff. Parkeri*, Karpinsky (Mém. Acad. Imp. Sci., St. Petersbourg, p. 75, Pl. V, fig. 5) from the Artinskian deposits of [Russia. It is a pity, that only incomplete fragments of all these species have hitherto been discovered, as the largest specimen of *St. Krasnopolskyi* reaches a diameter of scarcely more than 20 mm. A comparison of outlines with *Stacheoceras Trimurti* is therefore of very little use, as the globose shape of the Russian forms may be peculiar to young individuals only.

In the Indian zoogeographical region the subgenus *Stacheoceras* is represented by three species, the present one from the permocarboniferous rocks of Chitichun No. I, *St. antiquum*, Waagen, from the Jabi beds of the upper Productus limestone of the Salt Range, and *St. tridens*, Rothpletz[3] from the permian rocks of Timor. From the two later species *St. Trimurti* is easily distinguished. *St. antiquum*, Waagen (Salt Range fossils, I, p. 28, Pl. I, fig. 10) is more strongly compressed, and its lateral and auxiliary lobes are all provided with a tripartite termination. In *St. tridens* not only are the auxiliary and lateral lobes tripartite but even the siphonal lobe exhibits a trifid termination in each of its two branches.

[1] A. Karpinsky, Ueber die Ammoneen der Artinisk-Stufe : Mém. Acad. Imp. Sci., St. Pétersbourg, sér. vii, XXXVII, No. 2, 1887, p. 73, Pl. V, fig. 10.

[2] A. Karpinsky, Zur Ammoneen-Fauna der Artinsk-Stufe : Mélange géol. et pal. tirés du Bull. Acad. Imp. Sci., St. Pétersbourg, I, 20 Nov. 1890, p. 76.

[3] A. Rothpletz, Die Perm-Trias-und Jura-formation auf Timor und Rotti: Palaeontographica, XXXIX, 1892, Pl. IX, fig. 4, p. 87.

Class: LAMELLIBRANCHIATA.

Order: ANISOMYARIA, Neumayr.

Family: *PECTINIDÆ*, Lam.

Genus: AVICULOPECTEN, M'Coy.

1. AVICULOPECTEN AFF. JABIENSI, Waagen.   Pl. I, fig. 4.

Only the cast of a left valve is available to me for description. My determination as *Aviculopecten aff. jabiensi* is based on the fact, that this shell appears to be more nearly allied to this Salt Range species,[1] than to any other form of the present genus. An identification is, however, impossible, Waagen's species itself having been founded on a fragmentary specimen, with only the inside of its left valve preserved.

The general outline of the slightly inflated shell is nearly circular, with continuous margins, and of equal height and breadth. The apex is about median in its position, shifted but very slightly towards the anterior portion of the valve, and but slightly prominent. It is limited on both sides by very unequal wings, the anterior of which is distinctly marked off from the remainder of the shell. The anterior wing is comparatively small, although its exact size cannot be made out, its most prominent portion having been broken off. The posterior wing is very large, nearly flat, and limited on its upper side by a long and straight hinge line. It is broadly emarginated along its posterior margin and terminates in a sharp point, which projects somewhat further than the most prominent point of the posterior shell margin.

The sculpture consists of numerous (20 to 30) radiating ribs, which are broader than the intercostal depressions, separating them. Neither dichotomous, nor intercalated ribs have been observed. They increase considerably in breadth, as they approach the ventral margin. The majority of them seems to be slightly carinate. The two wings are perfectly smooth. Delicate striæ of growth, extending parallel to the margin, are occasionally visible as impressions on the cast.

My specimen is of very small dimensions. Its measurements are as follows:—

Entire length of the shell	.	11·5 mm.
„ height „ „	.	10 „
Thickness of the left valve	.	app. 1 „
Length of the hinge line, probably	.	10·5 „
Apical angle without the wings	.	about 106° „

Apart from its smaller dimensions, my specimen is very similar to *A. jabiensis*. The two forms agree perfectly well in their most characteristic features, viz., in the outlines, in the general character of the sculpture, and in the development of a very large and pointed posterior wing.

[1] *W. Waagen, Salt Range fossils: Pal. Indica, ser. xiii. 1, Productus Limestone fossils, p. 309, Pl. XXIII, fig. 2.*

Slight differences are observed in the details of the sculpture. In particular the apparently carinate shape of the broadly vaulted ribs seems to be absent in the Salt Range species. The left valve of the true *Aviculopecten jabiensis*, moreover, seems to be still less inflated, so far as one may judge from Waagen's figure. In spite of these differences I do not think the specimen under consideration worthy of a proper name, although its identity with *Aviculopecten jabiensis* is improbable. But the discovery of better materials, both of the Himálayan shell and of the Salt Range species, must be awaited for, before anything positive can be stated in this matter.

Number of specimens examined.—1.

Remarks.—Waagen considers *Aviculopecten indianensis*, Meek and Worthen, from the Keokuk group of North America as the probable ancestor of *A. jabiensis*. The single fragmentary specimen, on which the latter species has been founded, was collected by Waagen in the Cephalopoda beds of the upper Productus limestone.

Regarding the generic position of *Aviculopecten* I refer to the description of *A. hiemalis*, Salter, in part 4 of the present volume.

This is the only specimen of lamellibranch in the collections from the limestone crag of Chitichun No. I, which is worth noticing. All the rest are so badly preserved, that no specific description of them can be given.

MOLLUSCOIDEA.

Class: BRACHIOPODA.

Order: TESTICARDINES, Bronn.

Suborder: APHANEROPEGMATA, Waagen.

Family: *PRODUCTIDÆ*, Gray.

Subfamily: PRODUCTINÆ, Waagen.

Genus: PRODUCTUS, Sow.

The genus *Productus* takes the most important part among the fossils of the permocarboniferous limestone crag of Chitichun No. I, in the number of species, although, as regards the number of individuals, some species of *Athyris (Spirigera)* are the most frequent.

There are altogether nine different species, which must be attributed to this genus. Among the subdivisions of *Productus*, established by L. de Koninck, and partly emended by Waagen, five are represented among the fauna of Chitichun No. I, viz. the groups of *lineati*, *semireticulati*, *spinosi*, *fimbriati* and *irregulares*. Although these groups are quite artificial, having been based almost exclusively on the external sculpture, I am obliged to accept them in the following classification, as I do not know of any other, based on more sound principles.

A classified list of the *Producti*, which have been collected by our expedition in the permocarboniferous limestone crag of Chitichun No. I, is drawn up in the following scheme:—

I. Section. LINEATI.

a. Group of Productus Neffedievi, Vern.
1. *P. lineatus*, Waagen.

b. Group of Productus corrugatus, M'Coy.
2. *P. Cora*, Orb.

II. Section. SEMIRETICULATI.

c. Group of Productus semireticulatus, Mart.
3. *P. semireticulatus*, Mart.
4. *P. boliviensis*, d'Orb. var., *chitichunensis*, nov. var.

d. Group of Productus costatus, Sow.
5. *P. cf. subcostatus*, Waagen.

e. Group of Productus fortilocriatus, Norw. and Pratt.
6. *P. gratiosus*, Waagen.

III. Section. SPINOSI.

f. Group of Productus Cancrini, Vern.
7. *P. cancriniformis*, Tschern.

IV. Section. FIMBRIATI.

g. Group of Productus Humboldti, Orb.
8. *P. Abichi*, Waag.

V. Section. IRREGULARES.

h. Group of Productus striatus, Fisch.
9. *P. mongolicus*, nov. sp.

The majority of species, contained in this list, are identical with such, as occur in the Productus limestone of the Salt Range, especially in its middle division.

I. Section. LINEATI.

a. GROUP OF PRODUCTUS NEFFEDIEVI, Vern.

1. PRODUCTUS LINEATUS, Waagen. Pl. IV, fig. 2 a-c, 3 a-d, 4 a-d, 5 a-d.

1843 *Productus Cora* (d'Orbigny), Davidson, Quart. Journ. Geol. Soc. XVIII, p. 31
1876 *P. cora*, (Orb), Trautschold, Die Kalkbrüche von Mistschkowa, Pt. V, fig. 1, p. 63 (syn. exclus).
1884 *P. lineatus*, Waagen, Pal. Indica, ser. xiii, Salt Range Fossils, I, Productus Limestone Fossils, Pl. LXVI, fig. 1, 2, Pl. LXVII, fig. 3, p. 673.
1889 *P. lineatus*, Tschernyschew, Mém. Com. Géol. Russ., St. Pétersbourg, III, No. 4, p. 372, Pl. VII, fig. 96, 97.

1890. *P. lineatus*, Nikitin. Mém. Com. Géol. Russ., St. Pétersbourg, V, No. 5, p. 156.

1893. *P. lineatus*, Schellwien, Die fauna des Karnischen Fusulinenkalks, Palæontographica, XXXIX, p. 21, Pl. I, fig. 16-18, III, fig. 1.

This species, which is not rare in the permocarboniferous limestone crag of Chiti-chun No. I, agrees well with the figures and descriptions given by Waagen. As has been pointed out by this learned author, the shells, united formerly under the denomination of *Productus Cora*, may be conveniently divided into two series of forms, one with a median sinus in the visceral region, and another without any median sinus. Our Indian species belongs to the first group, of which *P. Neffedievi* Verneuil (Geologie de la Russie d'Europe, Vol. II, Paleontologie, p. 259, Pl. XVIII, fig. 11) is considered as a prototype.

My specimens vary very considerably in size and outlines. They never attain as large dimensions as Waagen's types from the Salt Range. The largest (Pl. IV, fig. 4) scarcely exceeds in this respect Schellwien's smaller types from the carnian Fusulina limestone. Specimens with an elongately oval outline are quite an exception. As a rule, the length and breadth of the ventral valve are nearly equal. Some of my specimens even exhibit a strongly transverse outline, recalling in this respect the North American *Productus multistriatus*, Meek (Rep. geol. expl. fortieth parallel, Washington, 1877, Pt. I, Palæontology, p. 76, Pl. VIII, fig. 3) from which they differ by their more shallow sinus.

The ventral valve is always strongly inflated. In one of my specimens the lateral parts expand considerably, below the level of the hinge line, as has been described by Trautschold and Waagen in specimens from Russia and the Salt Range.

The median sinus of the ventral valve is never strongly marked, although in the majority of my specimens it is distinctly developed. But there are certainly transitional forms between this species and *P. Cora*, whose only difference seems to be the entire absence of any median sinus. While there are such transitional forms between *Productus lineatus* and *P. Cora* among my specimens, no similar ones seem to exist between the present species and *P. Weyprechti*, Toula (Kohlenkalk-Fossilien von der Südspitze von Spitzbergen, Sitzgsber. Kais. Akad. Wiss. Wien, math. nat. Cl. LXVIII, November 1873, p. 13, Pl. V, fig. 2, 3, and Kohlenkalk-und Zechstein-Fossilien aus dem Hornsund an der Südwestküste von Spitzbergen, ibid, LXX, 1874, p. 6, fig. 4) or *P. impressus*, Toula (Permocarbon-Fossilien von der Westküste von Spitzbergen, Neues Jahrbuch, 1875, p. 236, Pl. V, fig. 1), in which the median sinus is always considerably deeper, than in any other species of the group of *Productus Neffedievi*.

The sculpture consists of numerous, delicate, radiating, straightly descending striæ, which are occasionally interrupted by irregularly distributed tubercles. It is the presence of these tubercles which serves for a distinction of *P. lineatus* from the very nearly allied *P. Neffedievi*. I am, however, bound to observe, in agreement with Nikitin, that transitional forms between the two species seem to exist, as in some of my specimens the number of tubercles is but very small. Nor is the

concentric sculpture of the wings a constant character, the folds being of a very unequal strength in different specimens and becoming even quite indistinct.

The dorsal valve is not known to me. Nor are the internal characters of the
shell accessible in any of my specimens.

The measurements from the largest specimen (fig. 4), are approximately, as
follows :—

Length of the shell	54 mm.
Breadth ...	38 „
Length of the hinge line	34 „
Thickness of the ventral valve	19 „

It has been suggested by Nikitin, that the internal characters of the Muscovian
types of *P. lineatus* might perhaps differ from those of the Salt Range form. But
as Nikitin himself does not give a detailed description of the internal structure of
his specimens and, notwithstanding his own statement, accepts *P. lineatus* as interpreted by Waagen, there is, for the present at least, no sufficient reason for a specific
separation of the Russian and Indian types of the species.

Number of specimens examined.— 8.

Remarks.— *Productus lineatus* is a species of considerable vertical distribution.
It has been mentioned by Waagen from the lower Productus limestone, where it is
however very rare, from the Virgal and Kalabagh beds of the middle Productus
limestone, and from the upper Productus limestone. In Russia the species occurs
both in the Muscovian and Gshelian stage of the central coal basin, in the
Fusulina limestone of the Ural of upper carboniferous age, and in the permocarboniferous Artinskian marls. By Schellwien the species is quoted from the
Carnian Fusulina limestone of the Krone in Carinthia of upper carboniferous age
(Gshelian stage).

b. Group of PRODUCTUS CORRUGATUS, M'Coy.

2. PRODUCTUS CORA, d'Orbigny (P. PRATTENIANUS, Norw. and Prat.).   Pl. IV,
fig. 1.

1842. *Productus Cora*, d'Orbigny, Voyage dans l'Amérique Méridionale, III, pt. iv. Paléontologie, p 55.
PL V, figs 8, 9.
1854. *P. Prattenianus*, Norwood and Pratten, Journ. Acad. Nat. Sci., Philadelphia, ser. ii, III, p. 17.
Pl. I, fig. 10.
For list of further synonyms, see Waagen, Pal. Indica, ser. xiii, Salt Range Fossils, I, Productus Limestone
Fossils, p. 677, to which the following must be added :—
1842. *Productus Cora*, Möller, Journal des Mines, St. Pétersbourg, p 178, Pl 9, fig 3.
1866. *P. sparsus*, Trautschold, Bull. Soc. Imp. Nat. Moscou, XL, Pl. V, fig. 1.
1859. *P. Cora*, Tschernyschew, Mém. Com. Géol. Russe, St. Pétersbourg, III, No. 4, p 283.
1843. *P. Cora*, Schellwien, Die Fauna des Karnischen Fusulinenkalkes, Palæontographica, XXXIX, p. 81,
Pl. III, fig. 8.

With the present species I identify two ventral valves, which do not exhibit any
trace of a median sinus, and by this character differ remarkably from *Productus
lineatus*. The presence of a few irregularly scattered spines on the surface marks

my specimens as specifically distinct from *P. corrugatus*, M'Coy (Synopsis of the characters of the Carboniferous Limestone Fossils of Ireland, Pl. XX, fig. 13, p. 107), which otherwise shows the same configuration and the same inflation of the ventral valve. Traces of indistinct spines may also be observed along the hinge line.

My specimens are of nearly equal length and breadth, so far as this can be made out in their incomplete state of preservation. They are provided with tolerably spread out wings. Their hinge line corresponds to the greatest breadth of the shell.

Concentric folds or wrinkles are but quite indistinctly developed. The character of the radiating striæ, which cover the surface of the valve, is exactly the same as in *Productus lineatus*.

The specimens are smaller than the largest types of *P. lineatus* from Chitichun No. I. The figured specimen is about 18 mm. long, 20 mm. broad, and 12 mm. thick, but it is barely possible to give exact measurements, as it has not been preserved entire. Neither the dorsal valve nor the internal characters are accessible to observation.

Number of specimens examined.—2.

Remarks.—Waagen refers the American *Productus prattenianus*, Norw. and Pratt., to the true *P. Cora*, but considers the Russian *Prod. Cora*, Möller, as specifically distinct from d'Orbigny's species. Tornquist (Das fossilführende Untercarbon am östlichen Rossbergmassiv in den süd-Vogesen, Abhandl. zur geologischen Special-Karte von Elsass-Lothringen, V. Heft 4, Strassburg 1895, p. 52) likewise asserts that the true *P. Cora* is probably altogether wanting in the carboniferous strata of Europe.

The Russian authors, who have been working out recently the description of the younger palæozoic faunas of their country, do not, however, follow this view. Tschernyschew positively asserts that the Russian examples cannot be distinguished from *P. Cora*, and that they will consequently require to be added to the synonyms of d'Orbigny's species.

Productus Cora, if taken with the definition attributed to this form by the Russian geologists, is a species of a rather wide geographical and geological distribution. In Russia, it has been found in the Moscovian stage, in the upper Fusulina limestone of the Ural where it is the leading fossil of a special horizon, and in the Artinskian stage of permocarboniferous age; in America in the coal measures of the Mississippi valley through the whole thickness of which it ranges[1], and in the carboniferous and permocarboniferous strata of Missouri, Nebraska, Yarbichambi (Bolivia), and Itaituba (Brazil); in the Salt Range it occurs off and on through the whole vertical extent of the Productus limestone but most numerously at the very base of the fossiliferous lower Productus limestone. It has been also obtained by Schellwien from the Carnian Fusulina limestone of upper carboniferous age.

[1] Meek in Cl. King's Exploration of the 40th parallel, Vol. IV, p. 73.

B

II. Section. SEMIRETICULATI.

c. GROUP OF PRODUCTUS SEMIRETICULATUS, Mart.

8. PRODUCTUS SEMIRETICULATUS, Martin. Pl. II, fig. 1, 3, 5, Pl. III, 1, 2.

1809. *Anomites semireticulatus*, Martin. Petrificata Derbyensia, Pl. XXXII, fig. 1, 2, Pl. XXIII, fig. 4.
1845. *Productus semireticulatus*, de Verneuil, Géologie de la Russie d'Europe, Vol. II, Paléontologie, Pl. XVI, fig. 1, non Pl. XVIII, fig. 9, 10, p. 262.
1847. *Productus semireticulatus*, L. de Koninck, Monographie des genres Productus et Chonetes, Pl. VIII, fig. 1, Pl. IX, fig. 1, Pl. X, fig. 1, p. 83.
1863. *Productus semireticulatus*, Davidson, Monograph of British Carboniferous Brachiopoda, p. 140, Pl. XLIII, fig. 1—11, Pl. XLIV, fig. 1—6.
1874. *Productus semireticulatus*, Meek and Hayden, Final Report of the U. S. Geol. Survey of Nebraska, p. 160, Pl. V, fig. 7.
1874. *P. semireticulatus*, Derby, Carboniferous Brachiopoda of Itaituba, Bull. Cornell University, Ithaca, I, No. 2, p. 47, Pl. IV, fig. 5, Pl. VI, fig. 18, Pl. VII, fig. 5, 6, 7, 16, 18.
1875. *P. semireticulatus*, Toula, Permocarbon-Fossilien von der Westküste von Spitzbergen, Neues Jahrbuch, p. 234, Pl. VI, fig. 1 a, b, c (non d).
1876. *P. semireticulatus*, Trautschold, Die Kalkbrüche von Mjatschkowa, Pl. V, fig. 3, p. 56.
1877. *P. semireticulatus*, Meek, Clarence King, Report of the Geological Exploration of the fortieth parallel, Vol. IV, Palaeontology, p. 62, Pl. VII, fig. 6.
1883. *P. semireticulatus*, Kayser, Obercarbonische Fauna von Loping, Richthofen's China, IV, p. 181, Pl. XXV, fig. 1—4.
1884. *P. semireticulatus*, Waagen, Pal. Indica, ser. xiii, Salt Range Fossils I, Productus Limestone fossils, p. 670, fig. 23 a, b, c.
1886. *P. semireticulatus*, Tschernyschew, Mém. Com. Geol. Russ. St. Pétersbourg, III, No. 4, p. 278.
1890. *P. semireticulatus*, Walther, Ueber eine Kohlenkalk-Fauna aus der egyptisch-arabischen Wüste, Zeitschr. Deutsche Geol. Gesellsch. XLII, p. 439, Pl. XXVI, fig. 9-11.
1892. *P. semireticulatus*, Schellwien, Die Fauna des Karnischen Fusulinenkalks, Palaeontographica, XXXIX, p. 23, Pl. II, fig. 1-3.
1892. *P. semireticulatus*, Rothpletz, Die Perm- Trias- und Juraformation auf Timor und Rotti, Palaeontographica, XXXIX, p. 77.
1894. *P. semireticulatus*, Schellwien, Ueber eine angebliche Kohlenkalkfauna aus der egyptisch-arabischen Wüste, Zeitschr. Deutsche Geol. Gesellsch. XLVI, p. 71.
1895. *P. semireticulatus*, Tornquist, das fossilführende Unterkarbon am östlichen Rothberg-Massiv in den süd-Vogesen. Abhandlgn. zur Geol. Spezial-Karte von Elsass-Lothringen, V, Heft 4, p. 60, Pl. XIV, fig. 10, 12.

This well known, extremely characteristic and far spread species of the genus *Productus* is very numerously represented in the permocarboniferous limestone of the Chitichun crag. The preceding list of synonyms might easily have been enlarged considerably, but the quotations given in respect to the present species, are sufficient to prove that it is the true *Productus semireticulatus*, with which my specimens, I believe, must be identified.

Notwithstanding the variability of the shells belonging to this species, their principal features are always sufficiently characteristic, to make their distinction comparatively easy. The majority of the specimens belong to the group of strongly sinuated varieties of *P. semireticulatus*, which in strata of upper carboniferous and permocarboniferous age considerably predominate over the types with a shallow median sinus. Nevertheless the latter shape, which closely resembles the typical form of the species, is not altogether absent. The specimen, figured Pl. III, fig. 2, is a representative of this group, being provided with a very shallow, indistinct depression along the median portion of its ventr valve. This depression extends only as far as the visceral part, but lower down disappears almost entirely, as in Davidson's type specimen (Pl. XLIII, fig. 1).[1]

The shape, which is most frequently met with among my materials of this species from Chitichun No. I, is the specimen figured Pl. II, fig. 1.

In its general outlines it resembles most closely the types of *P. semireticulatus*, collected by F. von Richthofen near Loping and figured by Kayser, or Trautschold's specimen from Miatschkowa, whose dimensions, however, it considerably exceeds. It is transversely oval and is provided with a hinge line, which is shorter than the greatest breadth of the shell. The ventral valve is moderately vaulted. Its auriculate expansions are distinctly developed, but are not produced beyond the lateral margins of the valve. The beak is strongly incurved but overhangs the hinge line but slightly. The apical region is reticulate. The reticulate sculpture extends for a distance of 55 mm. from the apex, measured along the curve, or across two-thirds of the length of the valve, excluding the trail. The radiating ribs, which cover the frontal and lateral parts of the shell, are rather regularly disposed and of nearly equal thickness. There are from 0 to 7 ribs within the space of 10 mm. The larger number of ribs are simple. Bifurcating or intercalated ribs are quite an exception. The number of spines is very limited. All my specimens agree pretty well in this respect. If any spines do occur, it is upon the wings, in the vicinity of the hinge line.

The dorsal valve is moderately concave. A flat median fold corresponds to the sinus in the ventral valve. Its ornamentation is exactly of the same character, as in the specimen figured by Walther from the upper carboniferous limestone of the Egyptian desert (Pl. XXV, fig. 11b).

The majority of the examples, more or less, closely resemble the specimen under consideration. Between this typical form of *P. semireticulatus* from the Chitichun limestone and the rest of the specimens differences are perceptible, with regard to both shape and sculpture. Some of them exhibit a very strong geniculation in the lower portions of the two valves, but in the majority the ventral valve, at least, is simply semicircularly curved. Their outlines are rather variable although not a single one is elongately oval. In one the greatest breadth of the shell is nearly twice its length, whereas in another the difference between these two dimensions is only insignificant.

Another character of variability is the depth of the sinus. The presence of forms with a shallow sinus, recalling the typical form from the carboniferous limestone of Belgium and England, has been mentioned in the preceding description of the species. Between this form (Pl. III, fig. 2), and the specimen, figured Pl. III, fig. 1, every degree of intermediate shapes may be observed. The latter example is characterised by its extraordinarily deep sinus. In this respect it may be compared to *P. semireticulatus* var. *bathykolpos*, Schellwien (Palæontographica, XXXIX, 1892, Pl. II, fig. 4-10, p. 22), from the upper carboniferous Fusulina

---

[1] Similar specimens have been mentioned by Kirkby from the coalmeasures of Fife as *Productus semireticulatus* var. *Martini*. (On the occurrence of marine fossils in the coalmeasures of Fife: Quart. Journ. Geol. Soc., London, XLIV, 1888, p. 780).

limestone of the Carnian Alps. But the smaller size, which Schellwien considers to be a constant character of the Carnian variety, is a point of difference between the latter and my Tibetan specimen.

The ornamentation varies with regard to both the width of the ribs and the interspaces between them. In an average sized specimen about 50 ribs may be counted on the frontal and lateral parts of the ventral valve, 5 to 8 occupying the space of 10 mm.

The dimensions of the species are also extremely variable. The largest specimen is 55 mm. in length, and 55 mm. in breadth. The average size of the specimens from Chitichun No. I is represented in Pl. II, fig. 1. The measurements of this specimen are as follows:—

Length of the shell in a straight line	47 mm.
„ „ „ along the curve	50 „
Breadth of the shell	51 „
Length of the hinge line	app. 35 „
Thickness of the ventral valve	26 „

In this specimen the trail has not been preserved.

Number of specimens examined.—30.

Remarks.—I scarcely need dwell upon the importance of *Productus semireticulatus* as a fossil of uncommonly wide geographical distribution. Unfortunately its geological range is scarcely less extended, as it ranges from the subcarboniferous rocks of North America and from the mountain limestone of Western Europe through the carboniferous system into permocarboniferous and probably even permian strata (Timor). From the predominance of forms, provided with a deeply impressed median sinus, in upper carboniferous and permocarboniferous strata the myth has originated, that the form with a shallow sinus, which is most frequently met with in the mountain limestone, is replaced by the sinuated variety in younger horizons. But the examination of my materials from Chitichun No. I gives evidence to the contrary, forms with a shallow sinus and strongly sinuated types being indiscriminately mixed together in this permocarboniferous fauna.[1]

4. Productus boliviensis, d'Orbigny, nov. var., chitichunensis.

Pl. II, fig. 2, 4.

1892 (?) *Productus sp. ind.*, Rothplets, Die Perm- Trias- and Jura-formation auf Timor and Rotti, Palæontographica, XXXIX, p. 77, Pl. X, figs. 17, 18.

Much difference of opinion has been expressed as to the specific value of *Productus boliviensis*, Orb. (Voyage dans l'Amérique Meridionale, III, pt. 4, Palæontologie, p. 52, Pl. IV, fig. 5-9). Some have considered it as only a variety of *P. semireticulatus*, whereas L. de Koninok, Tschernyschew and Nikitin maintained it as a separate species. As characters, which may serve for a distinction of the two closely allied forms, the strong curvature in the profile of the spirally inrolled

---

[1] As has been mentioned before, specimens of *P. semireticulatus*, which are devoid of any sinus, have been described by Kirkby from the coalmeasures of Fife, of upper carboniferous age.

ventral valve, the deep sinus, the transversely elongated shape, and the presence of the distinctly defined, strongly dilated ears have been enumerated by L. de Koninck (Mém. Soc. Roy. Sci., Liège, IV, p. 177, Pl. VIII, fig. 2, and Monographie des genres Productus et Chonetes, p. 77, Pl. VIII, fig. 2 a, b, c), who based his diagnosis on d'Orbigny's type specimen from Yarbichambi (near lake Titicaca).

The last mentioned character seems to be the most important one. Schellwien (Palæontographica, XXXIX, p. 23) lays a special stress on this remarkable feature, although he quotes *P. boliviensis* only among the varieties of *P. semireticulatus*. Nikitin (Mém. Com. Géol. Russ., St. Pétersbourg, 1890, V, No. 5, p. 57) likewise considers the "very large wings," combined with the involute shape and the deep sinus, a character of specific value, which permits the separation of *P. boliviensis* from *P. semireticulatus*. Grünewaldt (Beiträge zur Kenntnis der sedimentären Gebirgsformationen etc. Mém. Acad. Imp. Sci., St. Pétersbourg, 1860, ser. vii, II, No. 6, p. 110) in his description of a *Productus* from Saraninsk, which he identifies with *P. semireticulatus*, expressly remarks, that this form is distinguished from Martin's species by distinctly defined, strongly expanding ears, a character which has never been detected in the true *P. semireticulatus*. He further states that he would undoubtedly have compared this Russian form to *P. boliviensis*, had not the presence of transitional types between the former and the true *P. semireticulatus* prevented him from doing so.

Thus palæontologists are far from unanimous with reference to the specific or varietal rank of *P. boliviensis*. I am, however, inclined to believe that both L. de Koninck and Nikitin are substantially correct in separating this species from *P. semireticulatus*, on account of its remarkably well developed auriculate expansions, whereas the other characters of distinction, enumerated by the above-mentioned authors, appear to me of only somewhat slight importance.

Among the material from Chitichun No. I a considerable number of specimens are I think, most nearly allied to d'Orbigny's *P. boliviensis*. They differ from *P. semireticulatus*, which is, however, the more common species in the Chitichun crag, by their smaller size, the strongly transverse shape, a more delicate sculpture, but especially by the presence of unusually expanded ears. The ears are more or less distinctly marked off from the remainder of the shell and recall the wings of *P. giganteus*, Mart. They are inflated, slightly emarginated and obtusely rounded at their extremities.

Both the curvature of the profile and the depth of the sinus are as variable as in *P. semireticulatus*. The ventral valve, figured Pl. 11, fig. 4, resembles closely Nikitin's type specimen from Gshel (loc. cit. Pl. I, fig. 4). It is strongly inrolled, with a slightly flattened space not far from the apex, which extends almost as far as the reticulate portion of the valve. It is provided with a deep median sinus, whose shape seems to have been somewhat altered by pressure, and with a distinctly prominent beak. In the specimen, figured Pl. 11, fig. 2, the profile of the ventral valve is less strongly curved, and the sinus but shallow. The unusually expanded ears recall the figures of *P. giganteus*, given by L. de Koninck on Pl. 11 of his monograph of the genera *Productus* and *Chonetes*.

A third specimen is much more strongly inflated than the two figured types. Its ventral valve is regularly vaulted, and of a simple semiglobose shape, like d'Orbigny's type specimen of *P. boliviensis*, or the Russian examples from Saraninsk, described and figured by Grünewaldt (loc. cit. Pl. III, fig. 1).

In the Tibetan specimens the sculpture is much more delicate than in types of *P. semireticulatus* of the same size, and also more delicate than in any of the figures, which have hitherto been given, of *P. boliviensis*.

The longitudinal ribs are less thick than in Nikitin's type specimen from Gebel. They occur to the number of 12 to 15 to a space of 10 mm. at least in the vicinity of the front margin, where they augment considerably, mostly by bifurcation. In the reticulate portion of the valve about ten longitudinal striæ occupy a breadth of 10 mm. They are but slightly surpassed in width by the concentric wrinkles. The longitudinal ribs are occasionally flexuous. The spines are irregularly scattered on the surface of the shell and of unequal size.

The large number of longitudinal striæ, which, according to the figures given by L. de Koninck and Nikitin, do not seem to agree with the true average characters of d'Orbigny's species, may be found a sufficient reason for distinguishing my Tibetan examples by a proper varietal denomination. I think, the name *P. boliviensis* var. *chitichunensis* might be advantageously retained for them, in order to mark their sufficiently well defined differences from d'Orbigny's typical form.

The dorsal valve is strongly concave. A shallow median fold is not always present. Both in its shape and sculpture it is very similar to the dorsal valve in P. *semireticulatus*.

The measurements from the specimen Pl. II, fig. 4, which are however approximate only, on account of its incomplete state, are as follows :—

Length of the shell in a straight line	.	.	.	.	.	.	app.	25	mm.
" " along the curve	.	.	.	.	.	.	.	54	"
Greatest breadth of the shell	.	.	.	.	.	.	.	56	"
Thickness of the ventral valve	.	.	.	.	.	.	.	19	"

The internal characters of this species are not known to me, nor can I find any notice of them in the descriptions of the abovementioned authors.

Number of specimens examined.—8,

Remarks.—In his monograph of the permian and mesozoic fossils from Timor, Rothpletz described and figured two specimens of a *Productus*, which certainly belong to the section of *semireticulati*, but differ from the true *P. semireticulatus* by their much more delicate ornamentation, 12 to 16 radial striæ occupying a space of 10 mm, *i.e.*, nearly twice as many as in Martin's species. Rothpletz, although considering this *Productus* to be a proper species, quotes it only as *Productus sp. ind.* on account of the incomplete state of his specimens. I am inclined to believe that this species may prove identical with my Tibetan variety of *P. boliviensis*, but the fragments from Timor are too imperfectly preserved to establish their identity with full certainty.

The geographical and geological distribution of the true *P. boliviensis* is but imperfectly known. The species has been mentioned from the coalmeasures of

Yarbichambi in Bolivia and of Missouri (Norwood and Pratten), from the upper carboniferous limestone of Central Russia and of the Ural Mountains (Tschernyschew), and from the Artinskian horizon of permocarboniferous age.

d. GROUP OF PRODUCTUS COSTATUS, Sowerby.

5. PRODUCTUS cf. SUBCOSTATUS, Waagen. Pl. II, fig. 6 a, b, c.

1884. *Productus subcostatus*. Waagen, Pal. Indica, ser. xiii, Salt Range Fossils, I, Productus Limestone Fossils, p. 685, Pl. LXVII, figs. 4, 5, Pl. LXVIII, figs. 1, 2. Pl. LXIX, fig. 4.

The rather badly preserved cast of a dorsal valve with a portion of its trail which is alone available for description, agrees pretty well with the figures of the present species, as given by Waagen. An identification is however not possible, on account of the insufficiency of the material.

As it is the cast of the valve which is represented in the figure, its shape and sculpture are just the reverse of what they would be if the shell were visible from the outer side. If we describe it as from the outer side of the shell, it exhibits a transversely elongated, regularly concave shape, with a blunt geniculation marking off the trail from the remainder of the valve. A median fold is distinctly developed in the frontal region of the visceral part and extends over the trail, apparently corresponding to a deep and rather narrow sinus in the ventral valve.

The wings are flat and distinctly separated from the visceral part. They are covered with concentric wrinkles, which contrast sharply with the strong radial plications of the trail. There are 28 radial ribs present. The visceral part of the valve is strongly reticulate. The figure 6c gives a tolerably clear idea of this sort of sculpture. The deep grooves along the inner margin of the wings, which Waagen considers to be the most singular feature in the dorsal valve of *P. subcostatus*, are distinctly indicated.

As the present shell agrees in every respect with *P. subcostatus*, I think myself justified in provisionally attributing it to Waagen's species, without venturing however to pronounce it identical with the latter.

Remarks.—*P. subcostatus* has been collected by Waagen in the Virgal beds of the middle Productus limestone and in the Khund ghat beds of the upper Productus limestone of the Salt Range. It is a rare species and has not as yet been discovered outside the Punjab.

e. GROUP OF PRODUCTUS PORTLOCKIANUS, Norw. and Pratt.

6. PRODUCTUS GRATIOSUS, Waagen. Pl. III, figs. 3—7.

1865. *Productus semireticulatus* (Mart.). Beyrich, Ueber eine Kohlenkalkfauna von Timor, Abhandlgn. K. Akad. Wiss. Berlin, 1864, Pl. 11, fig. 2.

1884. *Prod. gratiosus*. Waagen, Pal. Indica, ser. xiii, Salt Range Fossils, I, Productus Limestone Fossils, p. 693, Pl. LXXII, figs. 3—7.

1891. *Prod. gratiosus*, Rothpletz, Die Perm- Trias- and Jura-formation auf Timor und Rotti, Palaeontographica, XXXIX, p. 76, Pl. X. fig. 15.

1892. *Prod. gratiosus* var *occidentalis*, Schellwien, Die Fauna des Karnischen Fusulinenkalks, Palaeontographica, XXXIX p. 27, Pl. III, fig. 6-9, Pl. VIII, fig. 24.

This elegant species is one of the most common types of the genus *Productus* in the permocarboniferous limestone of Chitichun No. I. The figures given on Pl. III of this memoir show on the one hand the absolute identity of the Tibetan and Salt Range specimens, and on the other hand the great variability of the species.

The largest specimen (fig. 5) is exactly of the same size as the largest type of this species which Waagen has met with from the Productus limestone of the Punjab. But in the Chitichun crag, as in the Salt Range or in the permian rocks of Timor, specimens of so large dimensions are quite the exception.

Regarding the general characters of *P. gratiosus*, which can scarcely be confounded with any other species of the genus, I have but very little to add to Waagen's description.

The ventral valve is always strongly inflated, slightly geniculated, and provided with a deep but comparatively narrow sinus, which originates in the immediate vicinity of the apex. In one of my specimens (fig. 7) the sinus is extraordinarily deep, considerably deeper than in any of the types from the Punjab. The ears are small, but distinctly defined, if they have been preserved at all, which is however rarely the case. The delicate reticulation is confined to the apical region. The strongly profiled radial ribs converge towards the mesial sinus and bifurcate rather frequently. The specimen fig. 6 is a good example of this sort of sculpture. The spines are very thin and numerous, but can only be observed in perfectly preserved specimens (fig. 4). On the casts the place of insertion of the spines is marked by small grooves, resembling the impressions of a sharp needle (fig. 3).

The dorsal valve is concave and provided with a prominent median fold. The reticulation extends considerably nearer to the front than in the ventral valve. The fan shaped character of the longitudinal ribs is very well exhibited in the figured specimen (fig. 3).

The measurements of my largest specimen are as follows :—

Length of the shell in a straight line	. . . .	27 mm.
" " " along the curve	. . . .	58 "
Breadth of the shell	. . . .	33 "
Thickness of the ventral valve	. . . .	17 "

Notwithstanding the great variability of my specimens, there is not a single one among them, agreeing with *P. gratiosus* var. *occidentalis*, Schellw., from the Carnian Alps. The prominent ridges, which in the Carnian types mark off the wings from the remainder of the ventral valve, have not been noticed in my Tibetan shells. Thus the Alpine variety seems to be distinguished from the typical form of the species by a constant, though rather subordinate character.

Number of specimens examined.—21.

Remarks.—*Productus gratiosus* is most numerously represented in the middle division of the Salt Range Productus limestone. It is rare, both in the Katta beds and in the upper Productus limestone. In the lower Productus limestone it has not been discovered. Rothplets recognised the species among the permian fossils from Timor. A variation of the true *P. gratiosus* has been collected by Schellwien in the upper carboniferous Fusulina limestone of the Carnian Alps.

The affinities of our species to *P. portlockianus*, Norw. and Pratt., *P. longispinus*, Sow., *P. griffithianus*, de Kon., and *P. costatus*, Sow., have been fully discussed by Waagen, Schellwien and Rothpletz. Regarding this subject I consequently refer to the monographs of these authors quoted above.

<h3 style="text-align:center">III. Section. SPINOSI.</h3>

f. GROUP OF PRODUCTUS CANCRINI, Verneuil.

7. PRODUCTUS CANCRINIFORMIS, Tschernyschew. Pl. IV, figs. 6 a, b, 7 a-d.

1889. *Productus cancriniformis*, Tschernyschew, Allgemeine geologische Karte von Russland. Blatt 139. Beschreibung des Central Ural und des Westabhanges, Mém. Com. Geol. Russ. St. Pétersbourg, III, No. 4, p. 372. pl. VII. figs 32, 33.

1898. *P. cancriniformis*, Diener, Pal. Indica, ser. xv. Himalayan Fossils, I pl. iv. The permian fauna of the Productus shales, etc., p. 31, Pl. I. figs, 7-10.

For a complete list of synonyms I refer to this memoir.

This characteristic species is represented in my collection from Chitichun No. I by two incomplete ventral valves, agreeing very well with Tschernyschew's type specimens, for the loan of which I am greatly indebted to Professor Th. Tschernyschew of St. Petersburg. They are of the same size as the latter, and are considerably larger than the average sized specimens from the Himalayan Productus shales.

The better preserved of my two examples particularly resembles in its shape and sculpture Tschernyschew's type specimen from the Bijas River (fig. 33). Their only difference consists in the larger number of the wrinkled, concentric folds, which extend from the wings and lateral margins across the entire shell. Otherwise the ornamentation is of exactly the same pattern, consisting of numerous delicate striae and elongated spines. A few small erect spines are placed along the hinge line. A median sinus is completely absent.

In my second specimen the surface has been more strongly weathered. In its sculpture therefore the concentric wrinkles remain as the only predominant features.

I do not hesitate to identify these specimens with *P. cancriniformis*, although, without the knowledge of the dorsal valve, it is rather difficult to make their determination with sufficient accuracy. There is especially one species in the American coalmeasures of Nebraska, which strongly resembles *P. cancriniformis*, and this is *P. pertenuis*, Meek and Hayden (Final Report of the U. S. Geol. Survey of Nebraska, p. 164, Pl. I, fig. 14, Pl. VIII, fig. 9). According to Tschernyschew, the two species can be distinguished only by the shape of the dorsal valve, which is flat and strongly geniculate in *P. cancriniformis*, whereas it is strongly curved in *P. pertenuis*. Nevertheless I am not inclined to assign my specimens from Chitichun to the American species. *P. pertenuis* is always small, the largest specimens reaching a length of about 15 mm. only, whereas in my specimens the entire length of the shell is more than 20 mm. Nor are the concentric wrinkles so strongly defined in any of the types figured by Meek and Hayden as in *P. concriniformis*.

If the larger number of concentric folds should prove a constant character in the form from Chitichun No. I, it may be considered desirable to distinguish it from the typical *Prod. concriniformis* by a varietal denomination. For the present, however, this difference appears to me of too small importance to make the shell under description a distinct variety.

The measurements of the more complete specimen (fig. 7) are the following :—

Length of the shell in a straight line . . . . 34 mm.
  "   "   " along the curve . . . . 42 "
Breadth of the shell . . . . 34 "
Thickness of the ventral valve . . . . 11 "

Number of specimens examined.—2.

Remarks.—Tschernyschew's type specimens of this characteristic species were obtained from the Artinskian horizon of the Ural Mountains. Schellwien mentions it from the upper carboniferous Fusulina limestone of the Carnian Alps in Carinthia. It is a common fossil in the permian Productus shales of the Central Himalayas. In Central Asia it has been collected by Bogdanowitsch in the brachiopod bearing limestone near the Gumass River (Western Kwen Lun) of permocarboniferous or permian age.

IV. Section. FIMBRIATI.

g. GROUP OF PRODUCTUS HUMBOLDTI, d'Orb.

8. PRODUCTUS ABICHI, Waagen. Pl. III, fig. 8 a-d.

1862. *Productus Humboldti*, (d'Orb.) Davidson, Quart. Journ. Geol. Soc., London, XVIII, p. 23, Pl. II, fig. 6.

1863. *P. Humboldti*, (d'Orb.) Davidson in L. de Koninck, Mémoire sur les fossiles paléozoiques recueillis dans l'Inde, p. 39, Pl. III, fig. 6.

1878. *P. semireticulus* (Martin) Abich, Geologische Forschungen in den Kaukasischen Ländern, Bd. I, Ueber eine Bergkalk fauna aus der Araxes-Enge bei Djoulfa, p. 33, Taf. V, fig. 3.

1879. *Strophalosia horrescens*, (Vern.) v. Möller, Ueber die bathrologische Stellung der jüngeren paläozoischen Schichtensysteme von Djoulfa in Armenien, Neues Jahrbuch, p. 233, 234.

1883. *Productus Humboldti*, (d'Orb.) Lydekker, Geology of the Kashmir and Chamba Territories and of the British district of Khágkán, Mem. Geol. Surv. Ind., XXII, Pl. II, fig. 9.

1884. *P. Abichi*, Waagen, Pal. Indica, ser. xiii, Salt Range Fossils, I Productus Limestone Fossils, p. 697, Pl. LXXIV, fig. 1-7.

1892. *P. Abichi*, Rothpletz, Die Perm- Trias- und Jura-formation auf Timor and Rotti, Palaeontographica, XXXIX, p. 76, pl. X, fig. 10.

This species, which has been excellently described by Waagen, is represented in the materials from Chitichun No. I, by two ventral valves, the smaller of which, though slightly weathered, is sufficiently well preserved to allow its identification with certainty.

The specimen under consideration is rather strongly inflated, recalling in this respect the one figured by Waagen, Pl. LXXIV, fig. 7. It is but slightly broader than long, provided with a distinct median sinus and with very small barely flattened wings. From the latter the sides ascend rather abruptly to the flattened visceral portion of the valve. The hinge line is considerably shorter than the greatest breadth of the shell. The apex is attenuated, strongly bent over, but scarcely overhanging the hinge line.

The sculpture exhibits the characteristic quincuncial arrangement of the numerous, coarse, elongated and club shaped tubercles, which is a peculiar feature of this elegant species. The difference in the ornamentation of the visceral portion and of the lateral parts of the valve is not so strongly marked in my two specimens as in the majority of Waagen's types from the Salt Range. In the larger of my specimens, which is, however, only an internal cast with its surface deteriorated by weathering, numerous and strong, imbricating striæ of growth are developed near the lateral and frontal margins. In the smaller specimen this concentric sculpture is confined to the frontal portion of the valve.

The measurements of this latter specimen are as follows :—

Length of the valve in a straight line	27	m.m.
,, ,, ,, along the rostrum	40	,,
Greatest breadth of the valve	30	,,
Thickness of the valve	14	,,
Length of the hinge line	20	,,

In my monograph of the fauna of the permian Productus shales of the Central Himálayas (Vol. I, Pt. 4) a species has been described under the denomination of *Prod. gangeticus* (Pl. I, fig. 1-3 ; Pl. II, fig. 3), the ventral valve of which most perfectly agrees with that of *P. Abichi* in shape and sculpture. As the differences between the two species consist in the shape and in the internal characters of the dorsal valve, the larger size of *P. gangeticus* is a good character, which may serve for a distinction, if one has to deal with ventral valves only. My specimens are of the average size of *P. Abichi*, and I therefore believe to be justified in identifying them with the latter species.

Number of specimens examined.—2.

Remarks.— *Productus Abichi* is a characteristic fossil of the middle and upper divisions of the Salt Range Productus limestone. It has been collected by Waagen both in the Virgal and Kalabagh beds, but not in the Katta beds of the middle Productus limestone. It is known from the permian rocks of Julfa in Armenia and of the island of Timor, and does not seem to descend into strata of upper carboniferous age. The stratigraphical position of the beds in Kashmir, where the species has been collected by Lydekker, is yet doubtful. A discussion of this subject will only be possible after the fossils from Kashmir have been described in detail.

V. Section. IRREGULARES.

A. GROUP OF PRODUCTUS STRIATUS, Fisch.

9. PRODUCTUS MONGOLICUS, nov. sp. Pl. IV, fig. 8, 9, 10.

1883. *Productus cf. Cora* (Orb). Kayser, *Obercarbonische Fauna von Loping*, Richthofen's China, IV, p. 184, Pl. XXVII, fig. 8.

A shell belonging to this species has been described by Kayser from the upper carboniferous beds of Loping in China, but has been provisionally united with *Prod. cora*, although this author himself was apparently aware of the remarkable differences which forbid an identification with d'Orbigny's well known species. In his description he correctly mentions the pointed shape of the beak, recalling that of *P. striatus*, and the presence of concentric wrinkles crossing the visceral portion of the ventral valve, as characters which have never been observed in the true *Prod. Cora*. He, however, believed, that the general features of the relief of his solitary specimen could not be much relied upon, as it seemed to have been a good deal altered by pressure, and he consequently considered his material too scanty for establishing a new species.

In 1884 Waagen described two very similar forms from the Productus limestone of the Salt Range, which had been originally united with *P. striatus* by Davidson, as *Prod. compressus* and *P. mytiloides*, and arrived at a satisfactory determination of the systematic position of the Chinese shell among the *Producti*.

Among the materials from the permocarboniferous limestone crag of Chitichun No. I, there are a few specimens, which, in my opinion, are identical with Kayser's *P. cf. Cora*. As this species must receive a new denomination, I introduce the name of *P. mongolicus*. There are fairly complete ventral valves and the fragment of a dorsal valve accessible to observation.

The general shape of the species is nearly triangular, elongated, with an acuminated apex and a rounded front margin. The ventral valve is either strongly inflated, as in the specimen fig. 8, or somewhat flattened, as in the specimen fig. 9. The curve is more regular in the transverse than in the longitudinal direction. In the latter it is less strongly vaulted, especially so in the vicinity of the apical region.

In the specimens figs. 9 and 10, the trail is marked off from the remainder of the shell by a blunt geniculation. The wings are small, but distinctly developed, though rarely preserved. They are bent down almost vertically, leaving barely any room for the development of a straight hinge line, which, properly speaking, does neither exist in this species, nor in the very nearly allied *P. compressus*, Waagen (Pal. Indica, ser. xiii, I, p. 10, Pl. LXXXI, fig. 1, 2). The apical region is strongly compressed and terminates in a distinctly incurved, slightly prominent and pointed beak. A mesial sinus is entirely absent. The median portion of the valve is always strongly convex.

The sculpture is very characteristic. It consists of delicate, radiating striæ, which are crossed by broad, prominent, concentric wrinkles.

The longitudinal striæ increase in number considerably by the interpolation of other striæ which spread out in a very regular manner, always meeting the lateral and front margins at right angles. By means of a magnifying glass a system of extremely delicate transverse striæ may be noticed, which cross the radiating sculpture, and are clustered together very closely, exactly as has been represented by Kayser in fig. 5b on Pl. XXVII of his memoir. The concentric wrinkles are by far the most conspicuous feature in the ornamentation of this valve. They are tolerably regular, obtusely rounded on their tops, and imbricating towards the apical region. They are not restricted to the median portion of the valve, as in *P. compressus*, but extend across the wings. A few scattered spines are confined to the wings and to the lateral parts in their vicinity.

The shell is extremely thin and fragile.

The dorsal valve is distinctly concave, following the curve of the opposite one and leaving but very little room between them. As far as I can judge from the fragment at my disposal, its ornamentation is similar to that of the ventral valve.

The measurements of the smallest of the specimens, which is, however, fairly complete, are the following :—

Length of the shell in a straight line	21 mm.
„ „ „ along the curve	27 „
Breadth of the shell	16 „
Thickness of the ventral valve	12 „
Approximate distance of the two valves from each other	2 „

Of the internal characters of the species nothing is known to me.

Number of specimens examined.—3.

Remarks.—The identity of the Tibetan species with Kayser's specimen from Loping may best be proved by a comparison of the drawings. They agree in every respect, even in the minor details of their sculpture.

In its general appearance this species approaches very nearly *Prod. compressus* Waagen from the middle and upper Productus limestone of the Salt Range. The points of difference between the two forms are of a comparatively small importance. The absence of any transverse sculpture on the wings and the smaller size of the concentric wrinkles in *P. compressus* are the most conspicuous ones. As a remarkable feature in *P. compressus*, the following is mentioned by Waagen. The ventral valve is always so strongly compressed in the apical region, that in a view from above the lateral margins are concealed below the overhanging lateral parts of the valve. This character has not been noticed in any of my specimens.

From *Prod. mytiloides*, Waagen (Pal. Indica, ser. xiii, I, p. 711, Pl. LXXX, fig. 4) which is very closely allied to *P. compressus*, our species differs by its larger wings and more prominent sculpture. From *P. striatus*, Fisher,[1] a common mountain limestone fossil in England and Belgium, it is distinguished by the same characters, which led Waagen to the introduction of his two new species from the Salt Range.

---

[1] L. de Koninck, Monographie des genres Productus et Chonetes, Pl. I, fig. 1, p. 90. Davidson, Monograph of the British Carboniferous Brachiopoda, p. 139, Pl. XXXIV, figs. 1-4.

Subgenus : MARGINIFERA, Waagen.

1883. *Marginifera*, Waagen, Pal. Indica, ser. xiii, Salt Range Fossils, I, Productus Limestone Fossils, p 715.

In his monograph of the Productus limestone Brachiopoda Waagen introduced the name *Marginifera* as a generic designation for such species of *Productidæ*, as are distinguished from the true *Producti* by a strange clambering of the visceral part of their shells, produced by prominent shelly ridges, placed vertically on the internal surface of the dorsal and within the wings of the ventral valve.

He admitted the occasional presence of similar internal ridges in some specimens of *Productus longispinus*, Sow., and of *P. proboscideus*, Vern., but insisted on the development of these ridges never being anything like that occurring in the shells of the Salt Range. The strong development of the ridges, by which in the latter species the visceral part of the shell is girt, seems to him " perfectly sufficient for the generic distinction of those forms. Certainly it is as well worthy of notice as the existence of an area in *Aulosteges* or the like."

Representatives of this new genus were recognised by him outside the Salt Range only in the permian beds of Julfa (*Marginifera spinosocostata*, Abich, and *M. helica* Ab.), and in the North American coalmeasures of Illinois, Missouri and Indiana (*M. splendens*, Norw. and Pratt).

Tschernyschew (Mém. Com. Géol. Russ., St. Pétersbourg, 1889, III, No. 4, p. 373—375) accepted Waagen's new genus and proved it to be rather largely represented in the upper carboniferous deposits of the Ural. Nikitin (Mém. Com. Géol. Russ., St. Pétersbourg, 1890, V, No. 5) however entirely differs from Waagen and Tschernyschew in the interpretation of the species, which the latter authors united in the genus *Marginifera*. In his opinion the development of circular ridges near the line of contact of the two valves is a character of only very small importance. This character has been mentioned, moreover, in different species of true *Producti*, as in *P. semireticulatus*, Mart., or in *P. longispinus*, Sow., and might probably have been found in many other species of *Productus* if any attention had been paid to it by previous authors. It is a very common character in the Russian shells, which have been identified with *Prod. longispinus* by Trautschold. But the most important evidence against a generic value of these ridges is the fact that they are no constant character in specimens from the same locality and geological horizon, which are perfectly identical in every other respect. Nikitin therefore believes that these ridges indicate a difference of age of the individuals, and not a generic distinction.

Schellwien, who in his monograph of the Brachiopoda from the Carnian Fusulina limestone (Palæontographica, XXXIX, 1892) had accepted Waagen's genus *Marginifera* and had introduced a new species, *M. pusilla*, has lately given up this view on the strength of Nikitin's observations.[1]

[1] E. Schellwien, Ueber eine angebliche Kohlenkalk Fauna aus der egyptisch-arabischen Wüste, Zeitschr. deutsche geol. Ges. XLVI, 1894. p. 70.

Notwithstanding Nikitin's arguments, I prefer to adopt Waagen's view and to claim at least a subgeneric rank for the species united by the latter author in his genus *Marginifera*.

It must be admitted that the peculiarities, which have been considered by Waagen as of generic value in *Marginifera*, are occasionally developed in true *Producti*, but it is likewise true that in none of them, with the possible exception of *Productus longispinus*, Sow., their development is anything like that in a typical *Marginifera*. The internal ridges in specimens of *P. semireticulatus* var. *Martini*, Davidson (Monograph of the British Carboniferous Brachiopoda, Pl. XLIII, fig. 8-10), to which Nikitin alludes, cannot be compared for a moment with those in any Indian *Marginifera*. Nor can I accept, from my own personal examination of Himálayan *Marginifera*, Nikitin's supposition that the internal ridges indicate a difference of age between single individuals. Having examined about 20 specimens of *Marginifera typica* from Chitichun No. I, I came to the conclusion that these ridges are a perfectly constant character in individuals of every size, and that in this species at least they are a feature, which marks not a certain stage of growth, but a distinction between this species and true *Producti*.

That a large number of the Russian forms, united with *Productus longispinus* by Trautschold are provided with internal ridges, as distinctly developed as in any *Marginifera* from the Salt Range or from the Ural Mountains, is no decisive argument against the generic or subgeneric rank of *Marginifera* itself. It simply proves that this genus (or rather subgenus, as I should prefer to consider it) is largely represented among the carboniferous *Productidæ* of Central Russia, as indeed has been clearly recognised by Tschernyschew, although this fact was not known to Waagen at the time when he was publishing his monograph of the *Productidæ* of the Salt Range Productus limestone.

The only important evidence against a generic or rather subgeneric value of the internal ridges in *Marginifera* is Nikitin's statement "that, having broken many specimens of the so called *Productus longispinus* from different Russian localities, he found the presence of the ridges to be a very inconstant character, even in specimens from the same locality and geological horizon, in specimens otherwise perfectly identical, in their outlines, generic and specific features."

The importance of this argument must not, however, be overrated. It must be borne in mind, that within the family of *Productidæ* the generic distinctions are altogether difficult and the limits of the different genera rather uncertain on account of the existence of transitional forms. The characters which are used for a generic distinction in this family do not make their appearance suddenly, but are developed by degrees. Thus in this family transitional forms are met with rather frequently, which barely allow a satisfactory determination of their generic position. *Productus, Aulosteges, Strophalosia, Chonetes, Productella, Daviesiella, Chonetella, Proboscidella, Etheridgina, Aulocorhynchus*, seem to bear scarcely a less intimate relation towards each other, than *Marginifera* does towards *Productus*; so that it is not wonderful that transitional forms should be met with between *Productus* and *Marginifera*, in which the subgeneric characters of the latter have not yet been sufficiently developed to mark a constant feature. But I do not think that the existence of such transitional forms ought to preclude the establishment of a separate subgenus for such species, in which the subgeneric characters of *Marginifera* have become a constant feature.

There can be no doubt that in the overwhelming majority of *Producti* no internal ridges are present similar those in a true *Marginifera*. It will therefore be found convenient to separate these forms, which bear in each valve a projecting shelly ridge, from *Productus*, and in this way to restrict the latter genus, which already includes a larger number of species than may be conveniently dealt with. Nor is the subgenus *Marginifera* difficult to recognise, if attention is paid to its leading character only, the broken-off shell margin disclosing at once the presence of the prominent internal ridges, especially in the vicinity of the wings.

The distinction between *Productus* and *Marginifera* is not based, it is true, on such conspicuous characters as in many genera of brachiopods, and there always will remain forms, whose systematic position is rather doubtful. But this case is certainly not worse than in other classes of *Molluscoidea* or *Mollusca*. Those who have had to describe *Ammonoidea* or *Gastropoda* will undoubtedly have felt the difficulty in the distinction of genera or subgenera, but notwithstanding this difficulty these genera are maintained, and often on the ground of less striking differences than those existing between *Productus* and *Marginifera*. Among the *Ammonoidea* especially many genera have been recently established, which are so closely related to each other, that they may be considered with nearly equal reason as different stages of development only of one and the same group of forms. Though the limits of the different genera may become continually more difficult to trace, these distinctions must be made so long as the tendency to a narrower restriction of single species and genera prevails among modern palæontologists.

In the permocarboniferous limestone crag of Chitichun No. I, *Marginifera* is represented by a single species, which is identical with one of the most common forms of this subgenus in the middle and upper divisions of the Salt Range Productus limestone.

I. MARGINIFERA TYPICA, Waagen. Pl. IV, fig. 11, 12, 13, Pl. V, fig. 1, 2.

1862. *Productus longispinus* (Sow.), Davidson, Quart. Journ. Geol. Soc., London, XVII, p. 31, Pl. I, fig. 10.

1863. *Productus longispinus* (Sow.), Davidson, in L. de Kœnrnh. Mémoire sur les fossiles palé-zoïques recueillis dans l'Inde, p. 37, Pl. X, fig. 19.

1884. *Marginifera typica*, Waagen, Pal. Indica, ser. xiii, Salt Range Fossils, I, Productus Limestone Fossils, p. 717, Pl. LXXVI, figs. 4-7 ; Pl. LXVIII, fig. 3.

1890. *M. typica*, Tschernyschew, Allgemeine geologische Karte von Russland, Blatt 139, Beschreibung des Central-Urals und des Westabhanges, Mém. Com. Géol. Russ., St. Pétersbourg, III, No. 4, p. 374, Pl. VII, fig. 22, 23, 24, 25.

The Tibetan specimens agree perfectly with Waagen's species both in their general shape and sculpture, and in the strongly marked development of the internal ridges.

The ventral valve is very strongly inflated, almost spirally inrolled in the majority of the specimens. The curve is, as a rule, rather irregular in the longitudinal direction, the valve being considerably flattened in the vicinity of the apex. The flattened portion of the valve unites with the remainder of the visceral part either in a regular spiral curve or in an obtusely rounded geniculation.

The apex is prominent, distinctly pointed and overhanging the hinge line, which corresponds to the greatest breadth of the shell.

The wings are rather large, spirally inrolled, and marked off from the lateral parts by a distinct furrow. The lateral parts are bent down towards the latter very steeply, in larger specimens nearly perpendicularly. If the wings are broken off, as is often the case, the general outlines of the shell appear to be almost square.

In all the specimens a median sinus is indicated, but it is rather variable in its depth and shape. If well developed, it recalls in shape the typical examples of *Productus pratiosus*. It always originates in the vicinity of the apex, and, as a rule, reaches its maximum strength near the point at which the valve appears most highly elevated above the hinge line. In most of the specimens it becomes more shallow in the frontal region, and often disappears completely near the front margin.

The trail is sometimes marked off from the remainder of the shell by a furrow or band, corresponding to the internal ridges along the margins of the opposite valve.

The ridges in the ventral valve, characteristic of the subgenus *Marginifera*, are accessible to observation when the wings have been broken off. They exhibit the peculiar crenulated appearance, which has been excellently described and figured by Waagen.

The sculpture consists of numerous, delicate, radial striæ, which originate in the apex and are nearly parallel for a considerable distance. Towards the front the radial sculpture becomes either quite indistinct, or is replaced by a few broader and more elevated ribs, especially so on the lateral portions of the valve. A distinct convergence of the radial striæ towards the median line of the sinus is rarely noticed. In this respect the sculpture of the majority of the Tibetan specimens resembles most closely that of Waagen's type from the cephalopod beds of Jabi (Pl. LXXVI, fig. 5). In the apical region a concentric sculpture is almost invariably indicated, but the concentric striæ are always but faintly marked and never equal in strength the radial sculpture.

The number of spines, which are scattered all over the surface of this valve, is very variable. They are however mostly distributed on the apical region and on the lateral parts along the furrow, which separates the wings from the remainder of the valve.

The dorsal valve (Pl. IV, fig. 12) is very deeply concave, leaving a comparatively small distance between itself and the opposite valve. Before reaching the ventral valve it suddenly flattens, presenting the appearance of a flat band, which passes around the anterior and lateral borders. This band, which is about 1 mm. in width, corresponds to the internal shelly ridges. The median fold is but very shallow. The shelly layer, which is but partially preserved, exhibits an indistinct reticulation in

the apical region and traces of a radial sculpture in its anterior portion. Small
rounded grooves are disseminated irregularly on its surface. The wings are not
accessible to observation.

The measurements of an average sized ventral valve with fairly preserved wings
are as follows:—

Length of the shell in a straight line	. . . .	23 mm.
„ „ „ „ along the curve	. . . .	30 „
Breadth of the shell, with the wings	. . . .	40 „
„ „ „ without the wings	. . . .	26 „
Thickness of the ventral valve	. . . .	17 „

The measurements of a complete specimen (Pl. IV, fig. 12), in which the two
valves have been preserved, but without the wings, are as follows:—

Length of the shell in a straight line	. . . .	30 mm.
„ „ „ „ along the curve	. . . .	37 „
Breadth of the shell without the wings	. . . .	26 „
Length of the dorsal valve	. . . .	17 „
Thickness of the ventral valve	. . . .	14 „
Distance of the two valves in the apical region	. . . .	7 „

The shelly substance of both valves is very thin.

Number of specimens examined.—23.

Remarks.— *Marginifera typica* is one of the more common species of this subgenus in the Productus limestone of the Salt Range. It occurs chiefly in the middle
and upper divisions of this formation, ranging there from the Katta beds to the
Cephalopoda (Jabi) beds. It is, however, very rare in the Katta beds, whilst in the
lower Productus limestone no characteristic specimens of the species have as yet
been found.

Tschernyschew collected a good number of types of *M. typica* in the Artinskian
horizon of the Ural Mountains and discovered a second very nearly allied species,
M. uralica (p. 374, Pl. VI, fig. 16-18) in the upper carboniferous Fusulina
limestone of the Ural. These two species are very similar and agree perfectly in
their internal characters. The only points of difference between them are the
presence of spines along the hinge line and the nearly smooth surface of the Russian
shell in the vicinity of the lateral and frontal margins.

Some of Nikitin's specimens of *Productus longispinus* from the upper carboniferous strata of Central Russia bear a great resemblance to *M. typica,* especially
the specimen figured on Pl. I, fig. 8, of his memoir. Nevertheless I dare not
identify them with the present species, on account of the presence of numerous spines,
both on the wings and along the hinge line, a character which is absent in *M.
typica,* according to Waagen's description.

The differences between the Indian shell and its nearest allies, *M. pusilla,*
Schellwien (Palæontographica, XXIX, p. 20, Pl. IV, fig. 18-21), *M. uralica*
Tschernyschew, *M. splendens,* Norwood and Pratten (Jour. Ac. Nat. Sci. Phila-
delphia, III, 1855, p. 11, Pl. I, fig. 5), and *M. exoarata,* Waagen (p. 713, Pl.
LXXVIII, figs. 2, 3) have been fully discussed by Waagen, Tschernyschew and

Schellwien. I therefore need not dwell upon them, but may refer to the memoirs of these learned authors.

Notwithstanding the great resemblance between *M. typica* and *Productus longispinus*, Waagen believes a distinction of the two species to be possible, even if founded on external characters alone. In one respect, however, Waagen's statement needs a correction, when he denies the presence of *P. longispinus* in the Productus limestone of the Salt Range. Specimens identical with Sowerby's species and different from the true *M. typica* have been recognised by Schellwien among the fossils of the Schlagintweit collection from the Punjab.

With *Productus longispinus* and *P. semireticulatus* a specimen of *Marginifera* has been identified by Stoliczka, which is most nearly related to M. typica. The differences between the two consist chiefly in the strong reticulation and in the presence of a prominent median fold in the dorsal valve of the Himálayan form from Kashmir and Spiti. The casts of the dorsal valve were mistaken by Stoliczka for *Prod. semireticulatus*, whereas he identified the ventral valves with *P. longispinus*. A detailed description of this species for which I shall introduce the name of *Marginifera himalayensis*, will be given in a special memoir on the anthracolitic rocks of Kashmir and Spiti (Pt. 2 of the present volume).

Genus: AULOSTEGES, Helmersen.

To this interesting genus belongs a new species from the permocarboniferous limestone of Chitichun No. I, which, while exhibiting the generic characters of *Aulosteges*, differs remarkably from any of the hitherto described congeneric forms. It consequently ought to be considered as type of a special group.

1. AULOSTEGES TIBETICUS, nov. sp. Pl. V, figs. 3-6.

This strange little shell is of a broadly triangular outline, with a semicircular front margin. Its ventral valve is moderately inflated, but rather variable in its convexity, which is very unequal in different directions. In its longitudinal direction it is but slightly curved in the apical region. Then follows a somewhat flattened part, which extends half way or more towards the front, when a blunt geniculation takes place, the remainder of the valve bending suddenly down to the front line. This geniculation is much more strongly marked, than in *Aulosteges medlicottianus*, Waagen (p. 663, Pl. LXII, fig. 1-1), and imparts to the present species a very peculiar and characteristic shape. Transversely the curve is more regular, being interrupted by a median sinus of variable depth and width. This median sinus originates in the immediate vicinity of the apex and reaches its greatest development near the line of geniculation of the ventral valve, whereas it is but faintly marked in the front margin.

The apex is always more or less deformed, prominent, and often incurved. It is not distinctly pointed. In some of my specimens it looks exactly, as if it had been fixed to a foreign body, whereas in others it does not show any mark of attachment.

a 2

The area is either perfectly flat, or a little concave (fig. 6). It is very variable in its height and breadth, the specimens (figs. 4 and 6) representing extreme types in this respect. It is not reclining, but either slightly overhanging (fig. 6), or forming one even plane with the dorsal valve. On its surface I noticed, in well preserved specimens, a few striæ of growth, running parallel to the hinge line, but no vertical striation. The area is interrupted in the middle by a very narrow deltidial fissure, which is of nearly equal width for its whole extent, and is closed by a prominent roof shaped pseudodeltidium.

The hinge line is always shorter than the greatest breadth of the shell, but in some specimens (fig. 3) nearly approaches the latter in its length.

No wings are present in this species.

The sculpture is very peculiar and differs from that of the majority of congeneric species in the absence of a dense cover of small spines. The ornamentation consists of very numerous concentric striæ or wrinkles of unequal strength in the apical region and of much more prominent radial plications in the geniculated marginal portion of the valve. The line of geniculation forms a very sharp boundary between these two types of sculpture. The concentric ornamentation is often interrupted by elevated roundish tubercles, which must have supported thick spines. If the shell substance is removed (fig. 6), a delicate, radial plication becomes visible on the cast, imparting to the apical region of the valve a semireticulate appearance.

The dorsal valve is either perfectly flat or slightly concave, provided with a linear area, and a pointed, flat apex. It is covered with very numerous, delicate striæ, corresponding to the tubercles in the opposite valve, which are but rarely interrupted by grooves. In the immediate vicinity of the lateral and front margins a radial plication is combined with this concentric sculpture.

Of the internal characters of this species nothing is known to me.

The measurements of two specimens (figs. 3 and 4) are as follows:—

	I (fig. 3).	II (fig. 4).
Entire length of the shell	20 mm.	19 mm.
„ „ „ „ along the curve	23 „	20 „
Length of the dorsal valve	15·5 „	14 „
Entire breadth of the shell	24·5 „	23 „
Length of the hinge line	21 „	15 „
Breadth of the pseudodeltidium at the hinge line	0·5 „	0·5 „
Entire thickness of the shell	12 „	7·5 „
Apical angle of the ventral valve	97° „	106° „
„ „ „ „ dorsal valve	180° „	

Number of specimens examined.—5.

Remarks.—The present species cannot be related to any of the forms belonging to the group of *Aulosteges Wangenheimi*, Vern. Nor does any closer relationship seem to exist between *A. tibeticus* and *A. medlicottianus*, Waagen, from the lower Productus limestone of the Salt Range.

The only shell, which can be compared with our species, is *Strophalosia popengensis* Kayser (Obercarbonische Fauna von Loping, in Richthofen's 'China,' IV, p. 196, Pl. XXVIII, fig. 9) from the upper carboniferous limestone of Loping. The

above quoted figure, which I consider as typical, strongly recalls *A. tibeticus*.
The geniculate character of the ventral valve and the strange sculpture are especially similar in the two shells. Specifically, however, the two forms are certainly
distinct, even if *Strophalosia poyangensis* should turn out to be no true *Strophalosia*
but a representative of the genus *Aulosteges*. The presence of distinct auricular
expansions and the irregular character of the radial plications in the Chinese species
make a distinction easy. Nevertheless they certainly ought to be united in the
same group, if *Strophalosia poyangensis* should be found to belong to the genus
Aulosteges.

Family : *LYTTONIIDÆ* (Waagen) Zittel.

In his monograph of the Productus limestone fossils, Waagen (1883) intro-
duced the subfamily of *Lyttoniinæ* for the reception of two very strange genera,
Oldhamina and *Lyttonia*, the nearest allies of which he found among the family of
Thecideidæ.

Of this subfamily he gave the following diagnosis : " Shell of large size, flat or
vaulted, attached by the larger valve ; hinge line straight and short, no area or
pseudodeltidium ; internally the ventral valve with a median and numerous lateral
septa ; dorsal valve rudimentary, forming together with the brachial apparatus one
strongly lobed, shelly plate, which fits between the external septa of the large valve."

The *Lyttoniinæ* have in common with *Thecidea* and *Pterophloios*, which have
been united by Waagen in the subfamily of *Thecideinæ*, the punctuate shell,
the lobed brachial apparatus, and the attached larger valve. With regard to the
persistence of these most striking characters Waagen proposed to leave these two
subfamilies, together with the subfamily of *Megathyrinæ*, in the family of *Thecideidæ*.

Œhlert (Fischer, Manuel de Conchyliologie, Brachiopoda, p. 1327) purified the
family of the *Megathyrinæ*, which are considered by him as a proper family, but
again refers the *Lyttoniinæ* to the family of *Thecideidæ*. K. von Zittel in his " Elements of Palæontology " (p. 235) elevated the *Lyttoniidæ* to the rank of a family.
His view has been adopted in the present memoir.

Genus : LYTTONIA, Waagen.

1. Lyttonia nobilis, Waagen. Pl. I, fig. 5, 6, 7.

1883. *Lyttonia nobilis* Waagen, Pal. Indica, ser. xiii, Salt Range Fossils, I, Productus Limestone Fossils,
p. 366, Pl. XXIX, XXX, figs. 1, 2, 3, 4, 5, 10, 11.

Among the material from the permocarboniferous limestone of Chitichun No. I,
there are several fragments of a large *Lyttonia*, which I believe to be identical
with the present species from the Salt Range. No complete specimen has been met
with, but the specimen figured Pl. I, fig. 5, is in a sufficiently good state of preservation, to exhibit most of the characteristic features, belonging to the shell of
Lyttonia.

This specimen is of a broadly triangular outline, flatly spread out, and firmly attached by the underside of its ventral valve for its whole extent. Laterally it is strongly bent over, shelly expansions being occasionally noticed a little outside the proper margins. In the cardinal regions small fragments of the shelly substance have been preserved. The shell is about one half millimetre in thickness. I have not been able to examine its structure.

On the inner side of the ventral valve a median, longitudinal septum is distinctly developed, but is considerably surpassed in size by the lateral septa, which form high roof-shaped crests. The septa do not reach the lateral margins, but are rather suddenly obliterated a short distance from the latter. Thus a narrow, smooth rim is formed around the entire valve along its lateral margins. The septa are quite regularly developed and are placed symmetrically to the median, longitudinal septum. The last, rudimentary pair has been considered by Waagen as replacing the cardinal teeth. Excepting this cardinal pair of septa, twelve lateral septa have been noticed in my specimen (fig. 5) on each side of the middle line.

Of the dorsal valve a few fragments only have been preserved. Its shelly substance is very thin. Of its internal character the presence of a longitudinal median septum is the only one, which I have been able to observe.

Another specimen, figured Pl. I, fig. 6, exhibits the internal characters of the cardinal portion of a ventral valve. It is likewise remarkable for the regularity of the lateral septa, which are strongly curved upwards in the vicinity of the margins.

In fig. 7, the fragment of the ventral valve of a very large specimen has been represented. In this fragment the porous character of the shelly substance is very well exhibited. The pores are distinctly arranged in two zones along both sides of each lateral septum. These zones, which are distinguished by their punctuate or grooved sculpture, are equally well visible in Kayser's figure of *Lyttonia* (*Leptodus* Kayser), *Richthofeni* (Richthofen's "China" IV, Pl. XXI, fig. 10) from the upper carboniferous limestone of Lo-Ping.

Number of specimens examined.— 5.

Remarks.—In my first report on the geology of the Chitichun crag (Denkschr. Kais. Akad. Wiss. Wien 1895, mat. nat. Cl., LXII, p. 582) I considered the present species to be rather nearly allied to *Lyttonia tenuis* Waagen (Pl. XXX, figs. 3, 4, 7, 9, p. 401). Waagen himself states *L. nobilis* and *L. tenuis* to be very similar species, which are not easy to distinguish, especially in fragments. It was only after a careful examination of my specimens, that I became convinced of their identity with *L. nobilis*. Their larger size, the massive development of the lateral septa in the ventral valve, and especially the greater thickness of the shelly substance, fragments of which I only detected after long painstaking, forbid an identification with *L. tenuis*.

The genus *Lyttonia* is a very peculiar southern type of the family *Lyttoniidæ*. It is known from China, from Sicily,[1] from Kashmir, from the Salt Range, and

¹ *G. Gemmellaro*, Pull. Soc. Sct. Nat. ed Econ, Palermo, 1891, No. I, 1892, No. III, 1894, No. 1.

from Chitichun No. I, but has not yet been found outside the subtropical portions
Europe and Asia.[1] *Lyttonia nobilis* is common in the Virgal and Kalabagh beds
of the middle Productus limestone of the Salt Range, but is absent in any other
division of the Productus limestone.

Suborder : HELICOPEGMATA, Waagen.

Family : *SPIRIFERIDÆ*, King.

Subfamily : SUESSIINÆ, Waagen.

Genus : SPIRIFERINA, d'Orbigny.

Among the family of *Spiriferidæ*, which is one of the most natural families
of this suborder, being easily recognised by very remarkable external and internal
characters, Waagen has distinguished a number of subfamilies, all of which have
their representatives in the permocarboniferous fauna of Chitichun No. I.

According to him, the first natural subfamily is formed by a group, of which
the genus *Spiriferina* is the prototype. It is most nearly allied to Davidson's
family of *Nucleospiridæ*, being provided with a transverse shelly band connecting
the primary lamellæ, as in *Uncites*, and with a punctuate shell, as in *Retzia*. In
this group, which were elevated to the rank of a subfamily, the *Suessiinæ*, by Waagen,
Spiriferina, *Suessia* (which is, however, provided with a fibrous shell,) *Cyrtina* and,
provisionally, *Mentzelia* were included. The latter genus must, however, be removed
from the *Suessiinæ* and united with the *Martiniinæ* or *Reticulariinæ*, as has been
proved by Bittner (Abhandl. K. K. Geol. Reichs-Anstalt, Wien, XIV, 1890,
p. 25).

In the Chitichun fauna only the genus *Spiriferina* is represented, by *Sp.
cristata*, Schloth., one of the most common and far spread species in carboniferous
and permian strata.

1. SPIRIFERINA CRISTATA Schlotheim *var.* OCTOPLICATA, Sowerby. Pl. VII,
fig. 5, 6, 7.

1816. *Terebratulites cristatus*, Schlotheim, Denkschr. K. Akad. Wien, München, VI. p. 26, Pl. I, fig. 3.
1827. *Spirifer octoplicatus*, Sowerby, Min. Conch., p. 190, Pl. 562, tab. 2, 3, 4.
1837. *Spirifer cristatus*, L. von Buch, Ueber Spirifer oder Delthyris, und Orthis, p. 39.
1843. *Sp. cristatus*, L. de Koninck, Description des animaux fossiles, qui se trouvent dans le terrain carbonifère de Belgique, p. 240, Pl. XV, fig. 5.
1850. *Trigonotreta cristata*, King, Monograph of the permian fossils of England, p. 127, Pl. VIII, fig. 9-14.
1851. *Spirifer octoplicatus*, L. de Koninck, Supplément de la Description des animaux fossiles, qui se trouvent dans le terrain carbonifère de la Belgique, p. 658, Pl. XV, fig. 5.
1858. *Spiriferina cristata*, Davidson, Monograph British Permian Brachiopoda, p. 17, Pl. I, figs. 37-40, 43, 46, Pl. II. figs. 43-46.
1858. *Spiriferina cristata var. octoplicata*, Davidson, Monograph British Carboniferous Brachiopoda, p. 32, Pl. VII, figs. 37-47.

---

[1] This is not the case with *Oldhamina*, which has a true representative (*O. Alicia* Keyserl.) in the carboniferous
rocks of the Urals. A. de Keyserling, Note sur la présence de l' *Oldhamina* dans la Russie. Bull. Com. Géol.,
Russ., St. Pétersbourg, 1891, p. 257.

1861. *Spirifer cristatus*, Geinitz, Dyas, II. p. 66, Pl. XVI, fig. 8-10.

1862. *Spiriferina octoplicata*, Davidson, Quart. Journ. Geol. Soc., London, XVIII, p. 29, Pl. I, fig. 12. 13, 14, (non 11)

1863. *Spiriferina octoplicata*, Davidson, in L. de Koninck, Mémoire sur les fossiles paléozoïques recueillis dans l' Inde, p. 36, Pl. X, figs. 12, 13, 14 (non Pl. IX. fig. 11).

1863. *Spiriferina cristata*, Sp. octoplicata, Davidson, Monograph British Carboniferous Brachiopoda Appendix, p. 267, Pl. LIV, figs 10-12.

1865. *Spirifer cristatus*, Beyrich, Ueber eine Kohlenkalkfauna von Timor, Abhandl K. Akad. Wiss Berlin, 1867, p. 79, Pl. I, fig 4.

1876. *Spirifer cristatus*, Trautschold, Die Kalkbrüche von Mistschkowo, p. 79, Pl. VIII, fig. 5.

1877. *Spiriferina cristata*, White, in Wheeler's Report upon the U. S. Geograph. Surveys west of the one hundredth Meridian, Vol. IV, Palaeontology, p. 130, Pl. X, fig. 8.

1883. *Spiriferina cristata*, Waagen, Pal. Indica, ser. xiii, Salt Range Fossils. I, Productus Limestone Fossils, p. 509, Pl. XLIX, figs. 3-7.

1884. *Spiriferina cristata*, Walcott, Palaeontology of the Eureka District. Mon. U. S. Geol. Survey, VIII. p. 215, Pl. XVIII, figs. 12-13.

1887. *Spiriferina octoplicata*, L. de Koninck, Faune du calcaire carbonifère de la Belgique, Ann. Mus Royal d'hist. nat. Belgique, 6 ème ptie., p. 100, Pl. XXII, fig. 33-39.

1889. *Spiriferina cristata*, Tschernyschew, Allgemeine Geologische Karte von Russland, Bl. 139, Geologische Beschreibung des Central Urals und des Westabhanges. Mém. Com. tidal. Russ., et. Petersbourg. III, No. 4, p. 273.

1890. *Spiriferina cristata*, Tschernyschew, Travaux exécutés au Timane en 1889, Bull. Com. Géol. Russ. p. 83.

1892. *Spiriferina cristata*, Rothpletz, Die Perm- Trias- and Juraformation auf Timor and Rotti, Palaeontographica, XXXIX. p. 81

The identification of the Tibetan specimens with the present species can only be maintained, if the latter is accepted in the extension which has been adopted by Davidson and Waagen. They agree perfectly with Waagen's types from the Productus limestone of the Salt Range, but deviate slightly from the permian *Spiriferina cristata*, especially from Schlotheim's German Zechstein types.

All are of a comparatively large size, and of variable outlines. The greatest breadth of the shell is either situated at the hinge line, or as in the two specimens, figs. 5 and 6, a little towards the front. The area is of variable width and shape, more or less strongly reclining, and provided with a moderately large triangular fissure. As a rule, the lateral margins meet the hinge line in a sharp angle, at least in those specimens, in which the greatest breadth of the shell coincides with the hinge line. In some specimens, however (Pl. VII, fig. 6), the cardinal angles are almost rounded, not prolonged with acute terminations.

The ventral valve is as strongly inflated as the dorsal one, and is very regularly curved in both directions. The beak is small, pointed and strongly incurved. In the area of the specimen fig. 5, a distinct horizontal striation has been noticed.

The number of folds varies from eight to ten. The sinus is deep, more than twice as broad than the depressions between the adjoining folds, and extending from the extremity of the beak to the front. In the two specimens, figs. 5 and 6, it is considerably produced beyond the frontal level of the lateral portions of the valve, whereas in other specimens (fig. 7) this is not the case. It is more or less sharply rounded in its bottom, and in one of the specimens (fig. 6) shows a well marked tendency to develop a rudimentary median fold. The crests of the folds are comparatively high and acutely rounded, exactly as in Waagen's figures.

The dorsal valve is provided with a linear area only. The number of its folds varies from nine to eleven. The median fold is always larger than those situated on the lateral portions. In specimens with a strongly protracted sinus this fold is elevated considerably above the general convexity of the valve. It is but slightly flattened on the top, especially so in the vicinity of the front margin, but never assumes the obscurely triplicated appearance, which has been observed by Davidson in some of his British types of *Spiriferina octoplicata*.

In the front of the two valves a few more or less strongly marked, imbricating striæ of growth are developed in the majority of my specimens.

The coarsely punctuate character of the shelly substance can easily be observed, even with the naked eye.

A concentric ornamentation, as in *Spiriferina insculpta* Phill., or in *Spiriferina kentuckensis* Shum., is completely absent.

The measurements of a specimen (fig. 6) with eleven folds on the dorsal valve are, as follows :—

Entire length of the shell	.	.	.	.	.	.	.	.	30	mm.
Length of the dorsal valve	.	.	.	.	.	.	.	.	15	„
Entire breadth of the shell	.	.	.	.	.	.	.	.	24	„
Length of the hinge line	.	.	.	.	.	.	.	.	22½	„
Thickness of both valves	.	.	.	.	.	.	.	.	13½	„
Apical angle of the ventral valve	.	.	.	.	.	.	.	108°	„
„ „ „ dorsal „	.	.	.	.	.	.	.	app. 125°	„

Number of specimens examined.—5.

Remarks.—Davidson was inclined to consider *Spiriferina cristata*, Schlotheim and *Sp. octoplicata*, Sow. as specifically identical, declaring, that "the latter form cannot claim to be considered more than a variety of *Sp. cristata*." In this view the majority of palæontologists agree with Mr. Davidson, with the exception, however, of L. de Koninck and Schellwien, who think the differences between the two forms sufficient for a specific distinction.

L. de Koninck lays a special stress on the following characters of difference – the permian *Spiriferina cristata* is always smaller, the number of its folds is less, rudimentary ribs along the median fold of the dorsal valve are absent, the cardinal angles are rounded.

Both in the Tibetan and Salt Range types the number of folds is extremely variable (from 8 to 14 in the ventral, from 8 to 10 in the dorsal valve), but is never so small as in some specimens from the German Zechstein. The cardinal angles are either acute or indistinctly rounded. Secondary ribs, which occur at the sides of the median fold in the dorsal valve, have never been observed. In their dimensions they certainly exceed the typical form of Schlotheim's species, and approach more nearly the British types of *Spiriferina octoplicata*.

A more important difference between *Spiriferina cristata* and *Sp. octoplicata* than those, which were enumerated by L. de Koninck, has been signalized by Schellwien (Die Fauna des Karnischen Fusulinenkalks, Palæontographica, XXXIX, 1892, p. 50). He draws attention to the peculiar shape of the sinus in specimens from the German Zechstein, for which the name of *Sp. cristata* was first introduced by Schlotheim, and which ought consequently to serve as prototypes of this species. In a very large number of specimens, which he was able to examine, he invariably found the bottom of the sinus forming an even plane and marked off by sharp borders from the adjoining lateral portions. Having myself no sufficient material at hand for comparison, I would hardly consider myself warranted in offering any decided opinion as to the specific claims of *Spiriferina cristata* and *Sp. octoplicata*. If their specific distinction should be found to be maintainable, the Tibetan specimens ought to be referred to *Sp. octoplicata*, not to Schlotheim's species.

Spiriferina cristata, in Davidson's definition of the species, is a form of very large geographical and geological distribution. It is not only common in the carboniferous and permian rocks of Western Europe, but has also been found in the Moscovian stage of Koroptschews in Central Russia, in the upper carboniferous and permocarboniferous rocks of the Ural and Timan Mountains, in the permian deposits of Timor, and in the Salt Range, where it reaches through the entire thickness of the Productus limestone series. All the Asiatic types of this species are more nearly allied to *Spiriferina octoplicata*, than to the true *Sp. cristata*.

To the present species, if taken in the wide extension of Davidson, Waagen, Beyrich and Rothpletz, some American shells have been correctly attributed by White and Walcott. I am not, however, inclined to accept all the synonyms in the list of the latter author, and must object to the identification of the shells in question with *Spiriferina spinosa*, Norw. and Pratt., from the Kaskaskia limestone, and with *Sp. kentuckensis* from the coalmeasures of the central and western parts of the United States. From the two last mentioned American species the Tibetan specimens differ in the absence of spines and of a regular concentric ornamentation.

From Höfer Island (Barents Islands, N.W. Novaya Semlya) *Spiriferina cristata* var. *octoplicata* has been mentioned by Toula (Eine Kohlenkalk-Fauna von den Barents Inseln, Sitzungsb. K. Akad. Wiss. Wien, LXXI, 1875, p. 20), but the identification was based on incomplete casts only.

The shell from Spitzbergen, which has been described by L. de Koninck (Nouvelle notice sur les fossiles du Spitzberge, Ac. Roy. Belg., XVI) as *Sp. cristata*, can scarcely be united with Schlotheim's species, but deserves at least a varietal denomination.

Subfamily : DELTHYRINÆ, Waagen.

Genus : SPIRIFER, Sowerby.

L. von Buch divided the forms, which constitute the genus *Spirifer* proper, i.e. the radially plicated forms with a fibrous shell—his group of *Sp. alatus*—into two sections, the *Ostiolati* (type, *Spirifer ostiolatus*, Schloth.) with a smooth mesial sinus, and the *Aperturati* (type, *Sp. aperturatus*, Schloth.) with a plicated mesial sinus. The three species, by which this genus is represented in the permocarboniferous fauna of Chitichun No. I, belong exclusively to the section of

A perturati. Of these species two occur also in the Productus limestone of the Salt Range and in Europe; these are *Spirifer musakheylensis,* Dav., and *Sp. Wynnei,* Waag. The third species, *Sp. tibetanus,* nov. sp., is peculiar to the Chitichun fauna, but rather nearly related to *Sp. rajah,* Salter, from the upper carboniferous rocks of Kashmir and Spiti.

These species may be grouped most conveniently in the following manner :—

> I. GROUP OF SPIRIFER FASCIGER, Keyserl.
> 1. *Sp. musakheylensis,* Davidson.
> II. GROUP OF SP. DUPLICICOSTA, Phill.
> 2. *Sp. Wynnei,* Waagen.
> III. GROUP OF SPIRIFER RAJAH, Salter.
> 3. *Sp. tibetanus* nov. sp.

Spirifer Wynnei and *Sp. tibetanus* are among the more common fossils of the material from Chitichun No. I, but *Sp. musakheylensis* is a very rare species.

I. GROUP OF SPIRIFER FASCIGER, Keyserl.

1. *Spirifer musakheylensis,* Davidson. Pl. VI, fig. 8.

1862. *Spirifer musakheylensis,* Davidson. Quart. Journ. Geol. Soc., London. XVIII, p. 28, Pl. II, fig. 2.

A complete list of synonyms has been given in my monograph of the permian fauna of the Productus shales (Pt. 4 of this volume).

Among the material from Chitichun No. I, there is only a single dorsal valve, which I refer to this species, one of the most common forms of the genus, both in the permian Productus shales of the Central Himalayas and in the upper carboniferous rocks of Kashmir and Spiti.

The present valve is transversely fusiform, with a hinge line, which is as long as the greatest breadth of the shell. The apex is a little more prominent, than is usually the case in specimens of a similar size, overhanging slightly the very narrow, almost linear area. The strongly elevated median fold is narrowly rounded on its top and is moderately curved in the longitudinal direction.

The fasciculation of the ribs is the most prominent feature in the ornamentation of this valve. Six or seven bundles of narrowly rounded ribs can be distinguished on each side of the median fold. Lamellose striæ of growth have not been noticed.

The measurements of this specimen are, as follows :—

Length of the dorsal valve	.	.	.	.	.	.	.	.	.	.	.	15	mm.
Breadth	„	„	.	.	.	.	.	.	.	.	.	85	„
Thickness	„	„	.	.	.	.	.	.	.	.	.	6	„
Apical angle	„	„	.	.	.	.	.	.	.	.	abt.	135°	„

Number of specimens examined.—1.

Remarks.—The affinities of *Spirifer musakheylensis* to the congeneric species of the group of *Sp. fasciger,* Keys., have been fully discussed in my memoir on

F 2

the Productus shales fossils, to which I refer for further explanation. In this memoir I have explained the reasons, why, in my opinion, a special stress must be laid on the rounded shape of the folds in *Sp. musakheylensis*, but not on the presence of the lamellose striæ of growth, which are sometimes missing. Notwithstanding their absence, I deemed it preferable to identify the present specimen with Davidson's species, rather than *Sp. fasciger*, on account of the character of its lateral folds, which are very flat and wavy, but not acute, as in Grünewaldt's and Tschernyschew's type specimens of Keyserling's species.

The rarity of these shells, so common at other localities of the Himálayas, in the Chitichun crag, is a rather astonishing fact.

II. Group of SPIRIFER DUPLICICOSTA, Phillips.

2. *Spirifer Wynnei*, Waagen. Pl. VII, figs. 1-4.

1883. *Spirifer Wynnei*, Waagen, Palæontologia Indica, ser. xiii, Salt Range Fossils. 1, Productus Limestone Fossils, p. 517, Pl. XLIV, figs. 6. 7.

1889. *Sp. Wynnei* (?), Tschernyschew, Mém. Com. Géol. Russ., St. Pétersbourg, III, No. 4, p. 571, Pl. V, fig. 7, 8.

The differences between this species and the nearly allied *Spirifer duplicicosta* Phill. (Davidson, Mon. Brit. Carb. Brach., p. 24, Pl. III, figs. 7-10, Pl. IV, figs. 3, 5-11) from the mountain limestone of Western Europe have been clearly defined by Waagen. They consist chiefly in the very small development of the median fold and in the presence of a broader, more strongly reclining area in the dorsal valve of the median shell.

The specimens, from Chitichun No. I, agree entirely with Waagen's types from the Salt Range. The peculiar characters, so well displayed in the Punjab examples of *Sp. Wynnei*, can also be observed in the Tibetan shells. Owing to the larger number of specimens available for examination, the shape has been found more variable than was admitted by Waagen. Although the general outline is transversely oval in the majority of specimens, the length and width of the shell are equal in a small number. The hinge line is always shorter than the greatest breadth of the shell. The cardinal angles are distinctly rounded.

The ventral valve is always deeper than the opposite one. It is provided with a broad, concave, strongly reclining area and with a rounded mesial sinus, which is very shallow as a rule, but exceptionally is comparatively deep and narrow, as in the specimen fig. 2.

The radiating ribs are narrowly rounded at their tops, either single, or dichotomous, but never arranged in bundles. Their number is rather variable. Strong and erect, concentric lamellæ of growth have frequently been noticed.

In the dorsal valve of this species the most characteristic feature is the flatness of the median fold, which, as a rule, is barely marked off from the lateral parts. Waagen's specimen, Pl. XLIV, fig. 7, most nearly approaches the typical shape of the dorsal valve in the Tibetan specimens. Even in shells like the one figured on Pl. VII, fig. 2, of this memoir, which are provided with a comparatively deep sinus, the mesial fold in the dorsal valve is very flat. Specimens, in which the mesial fold projects a little above the frontal wave of the ventral valve, as the one figured by Waagen on Pl. XLIV, fig. 6, are exceptional among the material from Chitichun No. I. The flatness of the median fold in the dorsal valve of *Sp. Wynnei* consequently appears to be a constant feature and may appropriately serve for a distinction between this species and *Sp. duplicicosta.*

The sculpture is exactly the same as in the larger valve.

The presence of strong dental plates in the ventral valve has been noticed. Otherwise I can say nothing of the internal characters of this species.

The measurements of my largest specimen (fig. 1) are, as follows :—

Entire length of the shell	.	.	.	.	.	.	.	.	.	40	mm.
Length of the dorsal valve	.	.	.	.	.	.	.	.	.	33	„
Entire breadth of the shell	.	.	.	.	.	.	.	.	.	61	„
Length of the hinge line	.	.	.	.	.	.	.	.	.	36	„
Thickness of both valves	.	.	.	.	.	.	.	.	.	23	„
Apical angle of ventral valve	.	.	.	.	.	.	.	.	106°	„
„ „ „ dorsal valve	.	.	.	.	.	.	.	.	131°	„

Number of specimens examined.—27.

Remarks.—In the Salt Range, *Spirifer Wynnei* is a very rare fossil, entirely restricted to the Virgal beds of the middle Productus limestone. A ventral valve, collected in the Artinskian deposits of Russia by Professor Tschernyschew, has been provisionally assigned to this species. Perrin Smith quotes *Sp. Wynnei*, from the argillites above the McCloud limestone in California, which he considers to be homotaxial with the topmost carboniferous limestones of the Ural Mountain and with the lower part of the Artinskian stage.[1]

III. Group of SPIRIFER RAJAH, Salter.

3. *Spirifer tibetanus,* nov. sp. . Pl. VI, fig. 1-7.

This characteristic and beautiful species recalls *Spirifer ovalis*, Phillips (Davidson, Mon. Brit. Carb. Brach., p. 53, Pl. IX, fig. 20-26) or *Sp. integricosta* Phill. (ibidem, p. 55, Pl. IX, fig .18-19) in its general shape and outlines. It is longitudinally oval, or, especially in young specimens, nearly semicircular. The hinge line is always less than half the width of the shell, with rounded cardinal angles.

The ventral valve is considerably deeper than the opposite one. Its beak is prominent and strongly incurved. A mesial sinus of variable width and depth extends from the extremity of the beak to the front margin. It invariably reaches its maximum near the front, which is not produced beyond the lateral portions in any of the specimens. The area is of nearly equal height and width, and so narrow that the large triangular fissure occupies more than half of its entire surface.

[1] Perrin Smith. Mesozoic changes in the faunal geography of California : Journal of Geology, Chicago, III. 1895, p. 373.

The sculpture is both very elegant and prominent. The mesial sinus, in the centre of which there exists a narrow, thread-like rib, is bordered on each side by a large, broadly rounded rib. These two ribs become dichotomous in the apical region and are again subdivided into two smaller ones in the vicinity of the front. The lateral portions of the valve are ornamented with six to eight similar ribs on each side of those central ribs, which are, however, always the largest as well as the most prominent. Beyond the apical region all the lateral ribs are subdivided into two or three smaller ones of irregular strength and width. In the vicinity of the cardinal edges the lateral sculpture becomes gradually more indistinct.

The dorsal valve is less strongly convex than the ventral valve. It is provided with a very small, but distinctly developed area. The mesial fold is but slightly elevated above the regular convexity of the valve, but nevertheless sufficiently well defined from the lateral plications. It is composed of a single rib at its origin, and continues so to some distance, when it becomes dichotomous. In large specimens each of these two main ribs is again subdivided, before reaching the front. The intercostal depressions between the median fold and the lateral plications are ornamented with a smaller rib, which is either single or dichotomous, but always thin and far inferior in strength to the other ribs. The lateral folds, which occur to the number of seven or eight on each side of the median fold, are all broadly rounded, simple at their origin, and increase in height and width as they approach the margins of the valve. In young examples they continue as simple ribs to the very margin of the shell ; in larger specimens, however, they become dichotomous or produce smaller ribs on each of their lateral portions.

In both valves the surface is almost invariably covered with numerous concentric lines of growth of irregular strength.

The internal characters of this species are not known to me.

The measurements, taken from an average sized example (fig. I), are as follows :—

Entire length of the shell	. . . .	55 mm.
Length of the dorsal valve	. . . .	39 „
Greatest breadth of the shell	. . . .	53 „
Length of the hinge line	. . . .	14 „
Thickness of both valves	. . . .	21 „
Apical angle of the ventral valve	. . . .	80°
„ „ „ dorsal valve	. . . .	118°

The species can attain very considerable dimensions, as specimens, measuring 60 mm. in length, have been noticed.

Number of specimens examined.—32.

Remarks.—The species, which is beyond doubt most nearly related to the present one, is *Spirifer rajah*, Salter (Palæontology of Niti in the Northern Himalayas, p. 59) from the upper carboniferous rocks of Kashmir and Spiti. According to the excellent description and figures, given by Davidson (Quart Journ. Geol. Soc., London, XXII, 1866, p. 40, Pl. II, fig. 3), the sculpture of *Sp. rajah* is of a very similar type. The only important difference consists in the triplicate division of the main ribs in *Sp. rajah*, while they are, as a rule, dichotomous in *Sp. tibetanus*. This difference is most distinctly marked in the character of the median fold in the dorsal valve. In Salter's species this median fold is composed of a prominent median rib, which produces smaller ones on each of its lateral portions, but continues as main rib to the front. In *Sp. tibetanus* the original median rib is not continued to the front but becomes subdivided in two ribs of equal width and strength. With regard to their general shape and outline the two species stand in a similar relationship, as *Sp. ovalis*, Phill. to *Sp. pinguis*, Sow. *Spirifer tibetanus* is easily distinguished from *Sp. rajah* by the shortness of its hinge line and area, the last being much more triangular and higher in proportion to its width.

Spirifer Keilhavii, von Buch (Ueber *Spirifer Keilhavii*, über dessen Fundort und Verhältniss zu ähnlichen Formen, Abhandl. K. Akad. Wiss. Berlin, 1876, p. 65), which is nearly allied to *Sp. rajah*, differs from the present species in the same characters, by which it is distinguished from the former, and by the absence of a median rib in the sinus of the ventral valve.

Another species, which must be compared with *Spirifer tibetanus*, is *Sp. parryanus*, Toula (Permocarbon Fossilien von der Westküste von Spitzbergen, Neues Jahrbuch, 1875, p. 256, Pl. VII, fig. 8) from Spitzbergen (Hinlopen Straits). Both from Toula's figures and description and from a personal examination of his type specimens, I have been convinced of the close relationship existing between these two species. *Sp. parryanus* approaches more nearly the *Sp. rajah* by its comparatively longer hinge line and area, but agrees with the present species in the dichotomous character of the median fold of the dorsal valve. The ornamentation is of the same pattern, as in *Sp. tibetanus* and in *Sp. rajah*, being composed of broadly rounded ribs, which become subdivided in the vicinity of the front. The development of a well defined median rib in the sinus of the ventral valve is also an important character, which is common to these three species.

Among the representatives of this group of the genus *Spirifer*, characterised by large, broadly rounded, subdivided radial ribs, *Spirifer Wilotski*, Toula, *Sp. tasmaniensis*, Morris, and *Sp. interplicatus*, Rothplets, may be mentioned. Their relationship to our species is, however, only a rather distant one, and will be more fully discussed in the description of *Sp. rajah* in the second part of the present volume.

This group seems to be entirely absent from the Productus limestone of the Salt Range.

Subfamily : *MARTINIINÆ*, Waagen.

Genus : MARTINIA, M'Coy.

The name *Martinia* has been proposed by M'Coy as a generic denomination for such forms, as belong to the relationship of *Anomites glaber*, Martin. Davidson and L. de Koninck, however, rejected it and classed these forms in the genus

Spirifer, proving M'Coy's diagnosis of his new genus to be insufficient and in part even erroneous. Generic rank was restored to *Martinia* by Waagen, who was the first to draw attention to some characters of its typical species, deserving of a distinct generic designation. Such characters are the existence of a punctured surface of the epidermis, and the absence of septa in the ventral valve.

Martinia, if Waagen's definition of this genus is adopted, is rather richly represented in the permocarboniferous fauna of Chitichun No. I, both in number of species and individuals. But most of the specimens are too incomplete to allow a satisfactory determination. Nevertheless six species at least can be distinguished.

The typical *Martinia glabra* Mart., which in Europe is the most common and widely spread type of this genus in carboniferous and permocarboniferous deposits, is barely represented. I can only attribute a single specimen to *M. glabra*, and even this does not approach the typical shape of that form, but is distinguished by the absence of a distinct sinus in the ventral valve.

All the rest of the specimens differ from *M. glabra* by a considerably shorter hinge line, terminating occasionally on both sides in little wing-shaped prominences, and by a smaller area.

Among the species, which I am able to distinguish, *Martinia nucula*, Rothpletz, *M. semiplana*, Waagen, and *M. contracta*, Meek and Worthen, are identical with previously described species. The two remaining species are new ones and will receive the names of *M. acutomarginalis* and *M. elegans*. The first is nearly allied to *M. carinthiaca*, Schellwien, from the Carnian Fusulina limestone of upper carboniferous age. The second one seems to bear a close relationship both to *M. nucula*, Rothpl. and to *M. Warthi*, Waagen.

Waagen arranged the Salt Range species of the genus *Martinia* in three different groups. I did not think it advisable to follow his example, as these groups are based on external characters only. As has been proved by Tschernyschew's remarks regarding the internal characters of *Martinia semiplana* (?) and *M. corculum* (Mém. Com. Géol. Russ., St. Pétersbourg, 1880, III, p. 369), very different types may easily be united in one single group, if their internal structure is not accessible to observation.

The rich development of the genus *Martinia* in the fauna of the Chitichun limestone contrasts sharply with its rarity in the fauna of the Salt Range Productus limestone. In the latter the few species, which have been described by Waagen, " occur in rather sporadic and isolated specimens, thus clearly indicating, that they are either stragglers from a territory, where the genus is far more plentifully developed, or that they are the last representatives of a group of forms, which is on the verge of becoming extinct, and which had been more copiously developed in a former period." The number of specimens examined, as quoted in the diagnosis of each species described in this memoir, does not give an adequate idea of the actual number of representatives of this genus in the Tibetan collection. But among 76 specimens of *Martinia*, 19 only were found to be sufficiently well preserved to allow a specific determination.

1. MARTINIA CF. GLABRA, Martin. Pl. IX, fig. 4.

1809. *Conchyliolithes Anomites glaber*, Martin, Petrificata Derbyensia, p. 11.

For further synonyms see Waagen, Pal. Indica, ser. xiii, Salt Range Fossils, I. Productus Limestone Fossils, p. 531, and my memoir on the fauna of the Himalayan Productus shales (Pt. IV of the present volume).

The only specimen available for description resembles in its circular outline, and in the absence of a well defined sinus, the example from Yorkshire figured by Davidson on Pl. XII, fig. 10, of his monograph of the British carboniferous brachiopoda, and only referred to *M. glabra* with considerable hesitation. The present specimen, however, differs considerably from the British fossil in the remarkably small disproportion in the depth of the two valves.

The ventral valve is moderately vaulted and of almost equal length and breadth. The beak is but slightly prominent beyond the hinge line, incurved, and of moderate dimensions. The hinge line is considerably longer, than in all the rest of congeneric species, occurring in the permocarboniferous limestone of Chitichun No. I. It is more than one half the entire breadth of the shell. The area is comparatively large, elongated, and marked off from the remainder of the shell by sharply defined lateral margins. A sinus is but very indistinctly indicated in the vicinity of the frontal region. Nevertheless the frontal line is strongly produced, forming a tongue-shaped wave, which causes the front margin of the dorsal valve to be elevated above its general convexity.

The dorsal valve is but partly preserved. It is very regularly curved in both directions, but slightly flattened in the proximity of the hinge line, where the latter unites with the lateral margins. A median fold seems to have been present in the frontal region only.

The ornamentation of the shell is of the same pattern as in the majority of specimens of the common *M. glabra*.

The measurements of this specimen are, as follows:—

Entire length of the shell	.	.	.	.	.	.	.	.	.	29	mm.
Length of the dorsal valve	.	.	.	.	.	.	.	.	.	23	„
Entire breadth of the shell	.	.	.	.	.	.	.	.	.	25	„
Length of the hinge line	.	.	.	.	.	.	.	.	.	19·5	„
Thickness of both valves	.	.	.	.	.	.	.	.	.	17·5	„
Apical angle of the ventral valve	.	.	.	.	.	.	.	app.	90°	
„ „ „ „ dorsal	.	.	.	.	.	.	.	.	108°	

Number of specimens examined.—1.

Remarks.— Waagen (p. 537) considers the shells, figured by Davidson on Pl. XII, figs. 9 and 10, of his monograph, and referred with hesitation only to *M. glabra*, as very near allies to the Russian *M. corculum*, Kutorga (Verhandl. K. Russ. Mineralog. Ges., St. Pétersbourg, 1872, p. 25, Pl. V, fig. 9.) As regards its external shape the present specimen may likewise hold an intermediate position between the true *M. glabra* and *M. corculum*. I do not, however, think, that it should be identified with the latter species, which is characterised by a very unequal depth of its two valves.

o

In their internal characters a remarkable difference between *M. glabra* and *M. corculum* has been noticed by Tschernyschew (Mém. Com. Géol. Russ., St. Pétersbourg, 1889, III, No. 4, p. 369), who observed the existence of a large median septum in the ventral valve of Kutorga's type specimens, whereas in *M. glabra* both septa and dental plates are absent.

In my specimen the internal characters are unfortunately not accessible to observation. Externally, however, it seems to be much more nearly allied to Martin's species than to the Russian form. I consequently deemed it preferable to leave it with *M. glabra*, without venturing, however, to identify it with the typical form of this latter species.

Although *M. glabra* is most abundant in carboniferous rocks, specimens, which only differ from the typical form by a shorter hinge line, are found in permian strata (Productus shales of Kiunglung encamping ground, Painkhánda). No geological importance can therefore be attributed to the presence of the specimen under consideration in the permocarboniferous limestone of Chitichun No. I.

2. MARTINIA NUCULA, Rothpletz. Pl. VIII, figs. 5, 6.

1892. *Martinia nucula*, Rothpletz, Die Perm- Trias- and Jura-formation auf Timor and Rotti, Palæontographica, XXXIX, p. 85, Pl. IX, fig. 3, 7.

This species, which is not at all rare in the permocarboniferous limestone of Chitichun No. I, is distinguished from the true *Martinia glabra* Mart., according to Rothpletz, by the strongly inflated apical region, by its very short hinge line, and by its small, indistinctly defined area.

In young specimens the sinus is only indicated by a median depression in the frontal part of the ventral valve. In larger specimens it is, however, more strongly developed as a rule. Its continuation towards the apical region is sometimes marked by a narrow, longitudinal impression or furrow. The front line ascends in a highly elevated tongue-shaped curve. In young specimens this frontal wave is almost acutely rounded, as in the figure given by Rothpletz on Pl. IX, fig. 7, of his memoir, but in later stages of growth it forms a more regular curve.

Rothpletz states that the apices of the two valves touch each other in full grown individuals. None of my specimens exhibit this character, although in some of them the two apices are rather approximate. But even my largest example does not attain the dimensions of the type specimen from Timor (loc. cit. Pl. IX, fig. 3), and, moreover, the shell is but partially preserved in its apical region (cf. Pl. VIII, fig. 5). When the beak is perfectly preserved, as is the case in the smaller specimen, fig. 6, it appears to be strongly incurved and pointed.

In spite of the strong inflation of the apical region in the ventral valve, the two valves do not differ in depth very conspicuously. This is a good character of distinction from *M. semiplana*, Waagen, in which the dorsal valve is but slightly vaulted.

In the shape of the very small and obscurely defined area *M. nucula* agrees with *M. contracta*, Meek and Worthen (Palæontology of Illinois, p. 298, Pl. 23, fig. 5), as has been remarked by Rothpletz. To the differences of these two species, enumerated by that author, the shape of the beak may be added, which is very prominent in the Asiatic fossil, but barely projects beyond the hinge line in the American species

The surface of the shell is smooth and covered with very numerous puncta, where its epidermis has been preserved. In the specimen, figured on Pl. VIII, fig. 5, a number of indistinct, rounded ribs may be observed near the lateral margins of the dorsal valve.

Rothpletz noticed the presence of two dental plates in the ventral valve of his species, a character, which is absent in the typical *Martinia glabra*. I have not succeeded in making out anything of the internal structure of my specimens.

The measurements of the largest specimen (Pl. VIII, fig. 5) are, as follows :—

Entire length of the shell	.	.	.	.	.	.	.	.	.	35 mm.
Length of the dorsal valve	.	.	.	.	.	.	.	.	.	24 „
Entire breadth of the shell	.	.	.	.	.	.	.	.	.	38·5 „
Length of the hinge line	.	.	.	.	.	.	.	.	.	9 „
Thickness of both valves	.	.	.	.	.	.	.	.	.	24 „
Apical angle of the ventral valve	.	.	.	.	.	.	.	.	90°
„ „ „ „ dorsal	.	.	.	.	.	.	.	.	app. 110°

Number of specimens examined — 5.

Remarks — Rothpletz believes his species to be identical with the Artinskian form, which has been identified with Waagen's *M. semiplana*, by Tschernyschew (Mém. Com. Géol. Russ., St. Pétersbourg, 1889, III, No. 4, p. 369, Pl. V, figs. 1, 3). I am not inclined to follow this view, in which no sufficient strength has been laid on the fact, that the Artinskian species is provided internally with a large median septum, whereas two dental plates are present in the ventral valve of *M. uncula*.

From the Indian *Martinia semiplana*, Waagen (Pal. Indica, ser. xiii, I, p. 580, Pl. XLIII, fig. 4) the present species differs considerably by its larger size, by the absence of ridges bordering the area, and by the nearly equal depth of the two valves.

Martinia uncula has been collected by Wichmann in the permian rocks of the island of Timor.

3. MARTINIA CONTRACTA, Meek and Worthen. Pl. IX, fig. 3.

1866. *Spirifer glaber* (Mart.) var. *contractus*, Meek and Worthen, Geological Survey of Illinois, Vol. III, Palæontology, p. 298, Pl. 23, fig. 5.

The specimens from the permocarboniferous limestone of Chitichun No. I, which ought, in my opinion, to be identified with the present species, agree in every respect with the description and figures as given by Meek and Worthen in the memoir quoted above.

Although in my monograph of the Productus Shales Fossils (Pt. iv of this volume) I have quoted *Martinia contracta* as a mere variety of *M. glabra*, Mart. sp., I now perfectly agree with Rothpletz (Palæontographica, XXXIX, p. 80) in considering it to be a true species. It belongs to the same group of forms as *M. semiplana*, Waagen, its two valves being of very unequal depth.

o 2

It differs from *M. glabra* "in having a much smaller and more obscurely defined ventral area. Indeed the sides of the beak of its ventral valve round in so regularly to the foramen, that it is often difficult to see where the margin of the area is." (Meek and Worthen, *loc. cit.*, p. 209.)

From *M. semiplana* the present species is easily distinguished by its larger size, by the absence of any ridges, bordering the small, triangular area, and by the presence of a shallow sinus in the ventral valve in the proximity of the front.

The differences between *M. contracta* and *M. nucula* have been enumerated by Rothplotz. They chiefly consist in the strongly inflated apical region of the ventral valve in the latter species, and in the different shape of the frontal wave which is sharply rounded or tongue-shaped in *M. nucula*, whereas it forms a shallow curve only in *M. contracta*. Nor does the small, well incurved beak project beyond the hinge line in the present species.

No median fold or crest is developed in the dorsal valve.

The figured specimen is of nearly the same size as the American one, figured by Meek and Worthen. Its measurements are, as follows :—

Entire length of the shell	 22	mm.
Length of the dorsal valve	. . . . 17·5	„
Entire breadth of the shell	. . . . 22·5	„
Length of the hinge line	. . . . . . app. 10	„
Thickness of both valves	. . . . 14·5	„
Apical angle of the ventral valve	. . . . 81°	
„ „ „ „ dorsal „	. . . . . . app. 110°	

The ornamentation of the nearly smooth shell is of the same pattern, as in *M. glabra*. The puncta are of the epidermis have been observed in a few places of the ventral valve.

Number of specimens examined.— 3.

Remarks.— *M. contracta* has been collected in the Chester group of the lower carboniferous series in Illinois.

4. MARTINIA SEMIPLANA, Waagen. Pl. VIII, fig. 7.

1883. *Martinia semiplana.* Waagen, Palæontologia Indica, ser. xiii, Salt Range Fossils. I, Productus Limestone Fossils. p. 556, Pl. XLIII, fig. 4.

1889. *Martinia*(?) *semiplana* ?. Tschernyschew, Geologische Beschreibung des Central-Urals und des Westabhanges, Mém. Com. Géol. Russ., St. Pétersbourg, III, No. 4, p. 369, Pl. V, fig. 1, 2.

1892. *Martinia semiplana*, Schellwien, Die Fauna des Karnischen Fusulinenkalks. I. Th., Palæontographica, XXXIX, p. 39, Pl. IV, fig. 13-15.

This species belongs to a group of forms, which are characterised by the unequal depth of the two valves, of which the dorsal valve is considerably flatter.

The ventral valve is of a circular outline, strongly vaulted, and provided with a broad beak, which projects considerably beyond the hinge line. The small and concave area is pierced by a comparatively large triangular fissure and marked off from the remainder of the shell by distinct, though very low, ridges. In the specimens under description a sinus is entirely absent, but replaced by a thread-like mesial furrow. The frontal part is but slightly produced and ascends in a very flat curve.

The dorsal valve is much flatter than the opposite one, and more strongly curved in the transverse than in the longitudinal direction, whereas in the ventral valve the curve is almost equally strong in both directions. No median fold corresponds to the longitudinal impression in the ventral valve, but the median part of the dorsal valve is shaped into a sort of an obtusely rounded crest, from which on both sides the valve flatly slopes down in a roof-like manner.

Martinia semiplana is a rather small species. The measurements of the figured specimen are as follows :—

Entire length of the shell	.	.	.	.	.	.	.	.	14	mm.
Length of the dorsal valve	.	.	.	.	.	.	.	.	13	„
Entire breadth of the shell	.	.	.	.	.	.	.	.	14	„
Length of the hinge line	.	.	.	.	.	.	.	.	8	„
Thickness of both valves	.	.	.	.	.	.	.	.	8	„
Apical angle of the ventral valve	.	.	.	.	.	.	.	85°
„ „ „ „ dorsal „	.	.	.	.	.	.	.	116°

Number of specimens examined.—2.

Remarks.—*Martinia semiplana* has been quoted by Waagen from the Virgal beds of the middle Productus limestone in the Salt Range, and by Schellwien from the Carnian Fusulina limestone of the Krone, which corresponds in its stratigraphical position to the Gshelian stage of Central Russia and to the upper carboniferous Fusulina limestone of the Ural Mts.

Tschernyschew identified an Artinskian *Martinia* with Waagen's Salt Range species, but grave doubts have been raised as to the correctness of this identification. Tschernyschew gives the following description of the internal characters in his Artinskian form :

The internal structure of this species "differs considerably from that in a typical *Martinia*, which is distinguished by the absence of any septa or dental plates in its ventral valve. A median septum distinctly shows through the transparent shells of the Artinskian specimens, from the apex of the ventral valve for the third part of the length of the latter. A septum of exactly the same shape is visible in the ventral valves of *M. corculum*, as I have been able to state from my personal examination of Kutorga's type specimens. In this respect the upper carboniferous and Artinskian species recall the triassic genus *Mentzelia*, Quenstedt, from which they differ, however, in the absence of dental plates. I think that these forms might rightly serve as prototypes of a distinct genus, although I prefer to abstain from giving a name to the latter, in view of our insufficient knowledge of its internal characters."

Waagen has observed nothing of the internal characters of his type specimen from the Salt Range. But Schellwien, who examined some of his Carnian examples, did not succeed in finding any trace of a median septum in their ventral valves. That the specimens from the Fusulina limestone of Carinthia are identical with the Indian *M. semiplana*, cannot be questioned, as his identification has been confirmed by Waagen himself on the ground of his personal examination of Schellwien's types. It is therefore very doubtful, if the Artinskian shell, described by Tschernyschew, ought to be united with the present species.

Rothplets (Palæontographica, XXXIX, p. 81) likewise believes Tscherny-schow's Artinskian species to be different from *Martinia semiplana*, but identifies it, although with some hesitation, with his *M. nucula* from Timor. I think this should not be done, because in *M. nucula* the ventral valve is internally provided with two dental plates, but not with a median septum.

5. MARTINIA ELEGANS, nov. sp.   Pl. VIII, figs. 1, 2; Pl. IX, figs. 1, 2.

The specimens, for which this new denomination is introduced, recall the group of *Martinia Warthi*, Waagen (Pal. Indica, ser. xiii, I, p. 530) by the development of little wings on both sides of the very short hinge line.

M. elegans is a large and beautiful species. Its general outline is either slightly elongated or as long as it is broad. The valves are not equally inflated, the dorsal valve being, as a rule, less deep than the ventral one, but the disproportion in their depth is not very considerable. It is more conspicuous in the type figured on Pl. IX, fig. 1, than in all the rest of my specimens.

The ventral valve is strongly vaulted and about equally curved in both directions. The apical region is not so much inflated as in *M. nucula*. The beak is rather small, well incurved, and scarcely projects beyond the hinge line. It bears on its dorsal side a very small, concave area of a nearly equilateral shape, which is almost entirely occupied by the triangular fissure. This area is separated from the remainder of the shell by sharp and narrow, often considerably elevated ridges. In some of the specimens these ridges are bordered by a flatly rounded furrow along their exterior margins.

The hinge line is very short, reaching one third, or, exceptionally, two fifths of the entire breadth of the shell only, and terminates on both sides in little wing-shaped prominences.

A mesial sinus is always distinctly indicated, originating either in the apical region or reaching up at least half way from the line. It causes the frontal part of the shell to ascend rather strongly into a frontal wave. The latter is much more strongly elevated in full grown than in adolescent individuals. In the largest specimen (Pl. VIII, fig. 1) it is distinctly tongue-shaped and produced into a sort of a ridge, which overhangs the general convexity of the dorsal valve. This character is chiefly observable in a lateral view of the shell (fig. 1. c).

In the dorsal valve there is no median fold or crest, corresponding to this sinus. It is much more strongly vaulted in the transverse than in the longitudinal direction. Longitudinally it is distinctly curved in the apical region only, but flattened or even slightly hollowed in the vicinity of the front line. The apex is little prominent and slightly bent over. The narrow concave area is bordered by similar ridges, like in the ventral valve.

In one of the specimens (Pl. VIII, fig. 2) the surface of the shell is sufficiently well preserved to show the punctation. As in a typical *Martinia* it consists of very numerous irregularly clustered granulations and impressions, giving to the epidermis of the shell the appearance of shagreen. Besides this very faint ornamentation, both valves are covered with numerous thin, concentric lines of growth. Traces of a radial sculpture are occasionally indicated in the proximity of the lateral margins.

Of the internal characters of this species nothing is known to me.

The measurements of the largest specimen are as follows:—

Entire length of the shell	.	.	.	.	.	.	.	.	.	.	61	mm.
Length of the dorsal valve	.	.	.	.	.	.	.	.	.	.	41½	"
Entire breadth of the shell	.	.	.	.	.	.	.	.	.	.	47	"
Length of the hinge line	.	.	.	.	.	.	.	.	.	.	19	"
Thickness of both valves	.	.	.	.	.	.	.	.	.	.	37	"
Apical angle of the ventral valve	.	.	.	.	.	.	.	.	.	84°	
" " " dorsal "	.	.	.	.	.	.	.	.	.	102°	

Number of specimens examined.—6.

Remarks.—This species can be easily distinguished from all the congeneric forms which have been hitherto described.

From *Martinia nucula*, Rothpletz, which it resembles in external shape, especially in the development of the sinus and of the tongue-shaped frontal wave, it differs in its less strongly inflated apical region, its sharply limited area, and the development of little wings on both sides of the hinge line. Nor are the beaks of the two valves ever so conspicuously approximate.

From *M. semiplana*, Waagen, it is distinguished by its much larger size, by the lesser disproportion in the depth of the two valves, by the presence of a well developed median sinus, and by its hinge line terminating in little wings.

The differences between our species and all the Salt Range forms belonging to the group of *M. Warthi*, Waag., are so conspicuous that I need not dwell upon them.

6. MARTINIA ACUTOMARGINALIS, nov. sp.   Pl. VIII, figs. 3, 4.

This is a small species, with a slightly elongate and pentagonal outline.

The ventral valve is strongly but very regularly vaulted. The apical region is barely less inflated than in *Martinia nucula*, Rothpletz. The beak is thick, little prominent beyond the hinge line, and well incurved. The area is small, indistinctly defined, and occupied almost entirely by the equilateral triangular fissure. The hinge line is very short, with rounded terminations, and without any wing-shaped prominences. The sinus is very broad, perfectly flat, and entirely restricted to the strongly produced frontal part. It is limited on both sides by a short, rounded fold, which is followed again by a rounded depression or furrow. This sort of sculpture gives to the lateral margins of the valve a zigzag shape. In the frontal region the middle part of the shell, corresponding to the sinus, is much produced, and causes the frontal wave to ascend very strongly.

The dorsal valve is also strongly vaulted, especially so in the apical region, whereas it gradually flattens in the longitudinal direction, when approaching the front line. The beak is slightly bent over, and overhangs an extremely narrow, indistinct area. The apices of the two valves are rather approximate.

In the frontal portion a broad, indistinct median fold corresponds to the sinus in the opposite valve. It is limited on both sides by rounded depressions which are followed again by very low and short folds. This sort of sculpture is entirely restricted to the proximity of the lateral and front margins, which are slightly elevated and shaped into sharp edges.

The ornamentation consists of very thin and numerous concentric striæ of growth. In the depressions on both sides of the median fold of the dorsal valve, thin radial striæ have been noticed.

The measurements of the smaller, but more complete, specimen (Pl. VIII, fig. 3) are as follows:—

Entire length of the shell	. . . .	13 mm.
Length of the dorsal valve	. . . .	10 „
Entire breadth of the shell	. . . .	11·5 „
Length of the hinge line	. . . .	4·5 „
Thickness of both valves	. . . .	10 „
Apical angle of the ventral valve	. . . .	75°
„ „ „ dorsal „	. . . .	110°

Number of specimens examined.—2.

Remarks.—This species seems to be very nearly related to *M. carinthiaca*, Schellwien (Palæontographica, XXXIX, 1892, p. 41, Pl. VIII, fig. 15, 16) from the upper carboniferous beds of the Krone in the Carnian Alps. It is only distinguished from the latter by very subordinate characters,—namely, by its thicker, more strongly incurved beak of the ventral valve, and by the trapezoidal shape of the frontal wave. In this respect *M. acutomarginalis* is very similar to *M. Warthi*, Waagen (p. 533, Pl. XLIII, figs. 2, 8), from which species it differs, however, in the absence of any ridges or furrows separating the area from the remainder of the ventral valve, and in the rounded, wingless terminations of its hinge line. In spite of the external similarity to *M. Warthi*, the present species therefore seems to be more nearly allied to *M. carinthiaca*. Their relationship is certainly a very close one.

Subfamily : RETICULARIINÆ, Waagen.

Genus : RETICULARIA, M'Coy.

This genus, which was introduced by M'Coy for *Reticularia reticulata* and its allies, is represented in the permocarboniferous fauna of Chitichun No. I by one single species only, *R. lineata*, one of the most common and far-spread forms in the carboniferous and permian rocks of the Old World.

1. RETICULARIA LINEATA, Martin. Pl. IX, figs. 5, 6, 7, 8.

1809. *Conchyliolithus Anomides lineatus*, Martin, Petrificata Derbyensia, II, XXXVI, fig. 8.
1821. *Terebratula lineata*, Sowerby, Mineral Conchology, Vol. IV, p. 39, Pl. CCXXXIV, figs. 1, 2.
1836. *Spirifera lineata*, Phillips, Geology of York-shire, p. 219, Pl. X, fig. 17.
1844. *Spirifera lineata*, v. Buch, Essai d'une classification des Delthyris, Mém. Soc. Géol. France, IV, p. 199, Pl. X, fig. 24.

1843. *Spirifer lineatus*, L. de Koninck, Description des animaux fossiles, qui se trouvent dans le terrain carbonifère de Belgique, p. 270, Pl. XVII, fig. 8.

1844. *Reticularia lineata*, M'Coy, Synopsis of the characters of the carboniferous fossils of Ireland, p. 143.

1845. *Spirifer lineatus*, de Verneuil, Géologie de la Russie d'Europe, etc., Vol. II, Paléontologie, p. 147, Pl. VI, fig. 8.

1847–1863. *Spirifera lineata*, Davidson, Monograph of the British Carboniferous Brachiopoda, p. 63, Pl. XIII, figs. 4–10.

1865. *Spirifer lineatus*, Beyrich, Ueber eine Kohlenkalkfauna von Timor, Abhandl. K. Akad. Wiss. Berlin p. 76, Pl. I, fig. 13.

1873. *Spirifer lineatus*, L. de Koninck, Monographie des fossiles carbonifères de Bleiberg en Carinthie, p. 45, Pl. II, fig. 11.

? 1876. *Spirifer lineatus*, Trautschold, Die Kalkbrüche von Mjatschkowa, p. 79, Pl. VIII, fig. 7.

1878. *Spirifer lineatus*, Abich, Geologische Forschungen in den kaukasischen Ländern I Th. Eine Bergkalk-Fauna aus der Araxesenge bei Djoulfa, p. 79, Pl. VI. fig. 6, 7, 8, Pl. VII, fig. 10; Pl. IX, fig. 5.

1883. *Retzularia lineata*, Waagen, Palaeontologia Indica, ser. xiii. Salt Range Fossils, I, Productus Limestone Fossils, p. 540, Pl. XLII, figs. 6–8.

1883. *Reticularia lineata*, Kayser, Obercarbonische Fauna von Loping, Richthofen's China. IV, Bd., p. 174, Pl. XXII, figs. 6–7.

1889. *Reticularia lineata*, Tschernyschew, Mém. Com. Géol. Russ., St. Pétersbourg, III, No. 4, p. 364.

1890. *R. lineata*, Nikitin, Mém. Com. Géol. Russ., St. Pétersbourg, V. No. 5, p. 66.

1892. *R. lineata*, Schellwien, Die Fauna des Karnischen Fusulinenkalks, I Th., Palaeontographica. XXXIX, p. 38, Pl. VI, fig. 10–13.

1892 *R. lineata*, Rothpletz, Die Perm.-Trias- und Juraformation auf Timor und Roti, Palaeontographica, XXXIX. p. 81 Pl. IX, fig. 8.

1894 *R. lineata*, Stuckenberg, Beiträge zur Stratigraphie Central Asiens, Denkschr. K. Akad. Wiss. math. nat. Cl., LXI, p. 458.

1895. *R. lineata*, Tornquist, Das fossilführende Untercarbon am östlichen Rothliegendsgebiet in den Südvogesen, Abhandl. zur Geologischen Spezialkarte von Elsass Lothringen. V, pt. 4, p. 591, Pl. XVI, fig. 6.

1896. *Martinia lineata*, Julien, Le terrain carbonifère marin de la France Centrale, p. 96, Pl. II, fig. 9–12, Pl. VIII, fig. 6

Among the collections from Chitichun No. I, this species is rather richly represented by specimens of very different size, but varying little in their general shape and outlines. The majority come very near the typical form from the English mountain limestone, especially the type from the Craven district, figured by Davidson on Pl. XIII, fig. 9, of his monograph.

My specimens are suborbicular or elongately oval, although the difference in length and breadth is somewhat small. They are strongly inflated, the two valves being, as a rule, almost equally deep. The hinge line is always considerably shorter than the greatest breadth of the shell, and provided with rounded cardinal angles.

In all my specimens the ventral valve is distinguished by a high and strongly incurved beak. In this respect they recall the variety of *R. lineata*, from the permian rocks of Timor, which has been figured by Beyrich. In the Tibetan specimens, however, the beak never projects so strongly as in Beyrich's figure (Pl. I, fig. 13b), but the apices of the two valves are always rather distant, like in the specimen, fig. 10, on Pl. XIII, of Davidson's monograph. In well preserved examples the beak is distinctly pointed. The area is small and only obscurely defined by lateral margins. The triangular fissure, in which traces of a pseudo-deltidium have been occasionally noticed, is of the same shape and size as in Davidson's British examples.

H

In the majority of the specimens the sinus is very shallow and confined to the proximity of the front. In young individuals (Pl. IX, fig. 7) it is barely indicated at all. The only specimen among the materials from Chitichun No. 1, provided with a deeper sinus, is the one figured on Pl. IX, fig. 6. In this specimen the frontal part of the shell is slightly produced and elevated.

The dorsal valve is very evenly convex and without any median elevation. The shape of the apical region and of the area agree with the description given of these characters in the Salt Range specimens by Prof. Waagen.

The sculpture is exactly the same as in European specimens from the mountain limestone of England or Belgium. The specimen, fig. 8, in which the shelly layer has been well preserved, exhibits the characteristic ornamentation, marked by numerous, very thin, radiating striæ, which are arranged between the more distant concentric lines. The reticulation is never so distinct as in the Indian specimens, which have been figured by Waagen.

The measurements of two specimens (Pl. IX, figs. 5 and 8) are as follows :—

	I (fig. 5).	II (fig. 8).
Entire length of the shell	20 mm.	23 mm.
Length of the dorsal valve	23 „	18·5 „
Entire breadth of the shell	24 „	19·5 „
Length of the hinge line	17 „	13 „
Thickness of both valves	20·5 „	15 „
Apical angle of the ventral valve	90°	75°
„ „ „ „ dorsal „	?	130°

I have not been able to ascertain the internal characters in any of the Tibetan specimens.

Number of specimens examined.—7.

Remarks.—*Reticularia lineata* is a very common and widespread species ranging from the lowest carboniferous into permian strata. In Great Britain it ranges from the lower limestone shales into the millstone grit, whilst in Belgium it is most abundant in the Calcaire de Visé. It is known from the mountain limestone of Alsatia, Central France and Silesia. Möller, de Verneuil, Grünewaldt, Trautschold and Nikitin quote it from the carboniferous deposits of Russia. Tschernyschew discovered it in the Artinskian marls of permocarboniferous age, Schellwien in the upper carboniferous Fusulina limestone of Carinthia. In the eastern region it has been mentioned from the permian strata of Julfa by Abich; from the upper carboniferous rocks of Loping by Kayser, and of Tongitar in the Koktan Range by E. Suess, who discovered it in the collections made by Dr. Stoliczka; from the lower Productus limestone of the Salt Range by Waagen from the permian rocks of Timor by Beyrich and Rothplets.

A somewhat doubtful fragment (ventral valve) from Barents Island (Novaya Semlya) has been described as *Spirifer lineatus*, Mart. var., by Toula (Sitzgsber. K. Akad. Wiss. Wien, LXXI, I Abth., 1875, p. 10, Pl. II, fig. 3). L. de Koninck (Recherches sur les fossiles paléozoiques de la Nouvelle-galles du Sud, Bruxelles, 1876-77, 3 ptie, p. 224, Pl. XI, fig. 9) figures a ventral valve from the carboniferous strata of New South Wales, which he identifies with *Spirifer lineatus*, Mart., but it is impossible to decide from the figure whether the Australian form is identical with the European and Indian types of *Reticularia lineata*.

According to Waagen the presence of the true *R. lineata* in America is very doubtful. The smaller species, which has been called *R. perplexa* by McChesney (Descr. New Palæozoic Fossils, 1860, p. 43), must at all events be kept separate.

A species, which I am likewise inclined to consider as specifically distinct from the present form, is *R. conularis*, Grünewaldt (Beiträge zur Kenntniss der sedimentären Gebirgsformationen, etc., Mém. Acad. Imp. Sci. St. Pétersbourg, sér. vii, II, No. 7, 1860, p. 102, Pl. IV, fig. 2) although Trautschold quotes it partially among the synonyms of *R. lineata* and of *Martinia glabra*. According to Grünewaldt's description, the most important feature in the species from the Ural Mountains is the extraordinary disproportion in the size of its valves, the ventral valve being about three times as deep as the dorsal one. Neither in this character nor in the shape of the strongly elongated beak do any of my specimens agree with Grünewaldt's type specimen from Artinsk.

Family : *ATHYRIDÆ*, Phillips.

Sub-family : ATHYRINÆ, Waagen.

Genus : ATHYRIS, M'Coy. (SPIRIGERA, d'Orbigny.)

1. ATHYRIS ROYSSII, Léveillé.   Pl. X, figs. 1, 2, 3, 6.

1833. *Spirifer de Royssii*, Léveillé, Mém. Soc. Géol. France, II, p. 39, Pl. II, fig. 19—20.
1843. *Terebratula Royssii*, de Koninck, Description des animaux fossiles qui se trouvent dans le terrain carbonifère de Belgique, p. 300, Pl. XXI, fig. 1.
1858. *Athyris Royssii*, Davidson, Monograph British Carbon. Brachiopoda, p. 87, Pl. XVIII, fig. 1—11.
1863. *A. Royssi*, Davidson, Quart. Journ. Geol. Soc., London, XVIII, p. 27, Pl. I, fig. 6.
1863. *A. Royssii*, Davidson in L. de Koninck, Mém. sur les fossiles Paléozoïques, recueillis dans l'Inde, p. 38, Pl. IX, fig. 6.
1861. *Spirigera Royssii*, Deyrich ex parte, Ueber eine Kohlenkalkfauna von Timor, Abhandl. K. Akad. d. Wissensch., Berlin, 1864, p. 74, Pl. I, fig. 5 (á—e 2).
1883. *Athyris Royssii*, Waagen, Palæontologia Indica, ser. xiii, Salt Range Fossils, I, Productus Limestone Fossils, p. 475, Pl. XL, fig. 6—12, Pl. XXXIX, fig. 10.
1897. *A. Royssii*, Diener, Himalayan Fossils, Palæontologia Indica, ser. xv, I, Pt. 4, Pl. V, figs. 3, 7.

In my memoir on the fauna of the Himálayan Productus shales I have fully explained my intention in giving this specific name to the Himálayan specimens. Waagen has distinguished among his group of *Athyris Royssii* quite a number of forms, to which he applied specific denominations. Although I agree with Rothpletz (Palæontographica, XXXIX, 1892, p. 81) in considering these distinctions, on rather insignificant characters, of no great importance, I do not object to accepting Waagen's species, as I can easily distinguish them among the Tibetan material.

The overwhelming majority of specimens, which have been collected in the permocarboniferous limestone of Chitichun No. I, belong to Waagen's *Athyris*

Roysii typica. They agree very well both with Waagen's types from the Salt
Range and with the permian specimens from the Productus shales of Johár and
Painkhánda. The three figures given in this memoir show the amount of variability
among the Tibetan types. Their general outline is either circular or slightly
subpentagonal (Pl. X, fig. 1). As a rule the transverse diameter is a little larger
than the longitudinal one. The valves are comparatively flat. Specimens which
are more strongly vaulted than that figured on Pl. X, fig. 3, are rare.

Neither the sinus in the ventral and the corresponding median fold in the dorsal
valve are ever very conspicuous, and in some of the specimens (Pl. X, fig. 2) even
entirely absent. But even in such specimens a flat frontal wave is indicated.
The foramen of the beak is rather variable in size, but, as a rule, larger than in
Beyrich's specimen from the Island of Timor.

The ornamentation of the shell is of exactly the same pattern, as has been
noticed by Waagen in his Indian specimens.

In the example figured on Pl. X, fig. 0, the internal characters are partly
accessible to observation, a portion of the spinal appendages having been laid bare.

My specimens do not attain as large dimensions as those which have been
figured by Davidson from the British mountain limestone. The largest example
nearly attains the same size as the one figured by Waagen on Pl. XI., fig. 8, of
his memoir, but is considerably flatter. The measurements of two smaller specimens (Pl. X, figs. 2 and 3) are as follows : —

	I (fig. 2).	II (fig. 3).
Entire length of the shell	34·5 mm.	29 mm.
Length of the dorsal valve	27 „	27·5 „
Entire breadth of the shell	31 „	31 „
Thickness of both valves	16 „	19 „
Apical angle of the ventral valve	115°	107°
„ „ „ dorsal	121°	119°

Number of specimens examined.—117.

Remarks.—*A. Royssii* is by far the most common shell in the permocarboniferous limestone of Chitichun No. I. In the Salt Range it is likewise largely
distributed through the middle and upper divisions of the Productus limestone,
while it is very rare in the lower Productus limestone. Waagen considers his
Indian specimens as absolutely identical with the typical forms from the mountain
limestone of Europe. L. de Koninck (Faune du calcaire carbonifère de la Belgique,
Ann. Mus. Roy. d'Hist. Nat. Bruxelles, XIV, 1887, pt. 6, p. 85) does not share
in this view, but believes all the shells from upper carboniferous and permian strata,
which have been classed by palæontologists among the synonyms of *A. Royssii*, to
be specifically different from Léveillé's species.

The Indian *A. Royssii* is certainly identical with the types from the permian
Productus shales, from Timor and from Chitichun No. I. Whether the specimen
from the permian rocks of Julfa, which has been described and figured by Abich
(Geologische Forschungen in den kaukasischen Ländern, I. Th. Eine Bergkalk fauna aus der Araxesenge bei Djoulfa, p 62, Pl. VII, fig. 8) as *Athyris*

Royssii, belongs to this species, even in the wider definition adopted by Waagen, is, to say the least, very doubtful.

2. ATHYRIS SUBEXPANSA, Waagen. Pl. X, fig. 4.

1883 *Athyris subexpansa*, Waagen, Palæontologia Indica, ser. xiii. Salt Range Fossils, I, Productus Limestone Fossils, p. 476, Pl. XXXIX, figs. 1—6.

The only specimen among the material from Chitichun No. I, which I am able to identify with the present species, agrees perfectly well with Waagen's description and figures. In its transversely oval outlines and in its comparatively flat shape it especially recalls Waagen's specimen from Musakheyl (loc cit. Pl. XXXIX, fig. 4), but in the characters of the sinus it approaches more nearly his example from Kálábagh (fig. 3). The sinus is very shallow and indicated only in the proximity of the front, causing the middle portion of the frontal part to be slightly bent up. In this respect my specimen is most nearly related to *A. expansa* Phill. (Davidson, Mon. Brit. Carb. Brach, p. 82, Pl. XVI, figs. 14, 16, 18; XVII, figs. 1—5).

I am really at a loss how to distinguish the latter species, from the mountain-limestone of Western Europe, from Waagen's *A. subexpansa*. The two characters which, according to Waagen's diagnosis, ought to serve for their distinction, namely, the presence of a well developed sinus in the ventral valve and the fringed expansions of the Indian form, do not seem to be constant features. The two specimens figured by Waagen on Pl. XXXIX, figs. 2 and 3 of his memoir, have only a very shallow sinus. On the other hand, Davidson in his description of *Athyris expansa* states the ventral valve to be occasionally provided with a gentle median depression. Nor do the fringed lamelliform expansions in the sculpture of the shell seem to be entirely absent in the European form, so far as we may judge from Sowerby's description and figures in his Mineral Conchology (Pl. DCXVI, fig. 1). It is, however, true, that the median depression is extremely shallow in the specimens figured by Davidson. Thus a distinction of the two species may possibly be based on this insignificant feature

Of the ornamentation of the shell I can say but little. The numerous, imbricating striæ of growth are present, as in most of the congeneric forms, but I have not been able to verify Waagen's observation regarding the shelly fringes, the uppermost layer of the shell in my specimen having been injured by weathering.

The measurements are as follows :—

Entire length of the shell	.	.	.	.	.	.	.	.	25	mm.
Length of the dorsal valve	.	.	.	.	.	.	.	.	23·5	„
Entire breadth of the shell	.	.	.	.	.	.	.	.	33	„
Thickness of both valves	.	.	.	.	.	.	.	.	16	„
Apical angle of the ventral valve	.	.	.	.	.	.	.	125°	
„ „ „ „ dorsal „	.	.	.	.	.	.	.	134°	

Number of specimens examined.—1.

Remarks.- Waagen quotes *A. subexpansa* from the middle and upper divisions

of the Salt Range Productus limestone. It is most abundant in the Virgal and
Káiabagh beds, but very rare in the Katta beds of the middle Productus limestone.
Whether it is also present in the upper carboniferous or permocarboniferous deposits of Russia cannot be determined, as no figures of *A. Royssii* have been given
by Tschernyschew and Nikitin, who quote the latter species from Central Russia and
from the Ural Mountains.

3. ATHYRIS CAPILLATA, Waagen. Pl. X, fig. 5.

1883. *Athyris capillata*, Waagen, Palæontologia Indica, ser. xiii, Salt Range Fossils, I, Productus Limestone
Fossils, p 479, Pl. XXXIX, figs 6—9; Pl. XL. figs 1—5; Pl. XLII, figs 1—8.
1892. *Spirigera Royssii var capillata*, Rothpletz, Die Perm- Trias- and Juraformation auf Timor und
Rotti, Palæontographica, XXXIX, p. 82, Pl. X fig 8

There is only one fragmentary specimen in the Tibetan collection, which exhibits the characteristic ornamentation of this species, and may consequently be kept
separate from Waagen's *Athyris Royssii typica*. The shell is of a nearly circular
outline. Its valves are more strongly inflated than in the majority of the specimens of *A. Royssii*. The mesial depression in the ventral valve is shallow, but
distinctly indicated. The very large number of lamellar striæ of growth, which
are covered with thickly set hair-like fringes, is a good character of this species.

The measurements of my specimen are as follows :—

Entire length of the shell	23 ½ mm.
Length of the dorsal valve	24 „
Entire breadth of the shell	28 „
Thickness of the two valves	16 „

Number of specimens examined.— 1.

Remarks.— *Athyris capillata* has been quoted by Waagen from the middle and
upper divisions of the Salt Range Productus limestone, where it is very common.
It has not been collected either in the lower Productus limestone, or in the
Katta beds of the middle Productus limestone. Rothpletz discovered the species in
the collections, which have been made by Wichmann in the permian rocks of the
island of Timor.

Subgenus : SPIRIGERELLA, Waagen.

The genus *Spirigerella* was introduced by Waagen (Pal. Indica, ser. xiii,
I, p. 450) for such species of *Athyrinæ*, as differ externally from *Athyris*, M'Coy
(*Spirigera*, Orb.) in the shape of their apical region, the beak being bent over
so strongly and appressed to the apex of the dorsal valve, that the small foramen of
the beak is entirely concealed. Internally the shape of the cardinal process and the
mode of attachment of the primary lamellæ mark a character of distinction of
Athyris. Rothpletz (Palæontographica, XXXIX, 1892, p. 82) does not consider
these characters of generic importance. Oehlert (in Fischer, Manuel de Conchyliologie, III, p. 1290) and Zittel (Grundzüge der Palæontologie, 1895, p. 240) do,
however, accept *Spirigerella* as a subgenus of *Athyris*. In the present memoir I
have followed the views of these two learned authors.

In the permocarboniferous limestone of Chitichun No. I, *Spirigerella* is represented by three species. Two among them are identical with forms, which have been previously described from the middle and upper divisions of the Salt Range Productus limestone.

1. Spirigerella Derbyi, Waagen. Pl. XI, fig. 4.

1883. *Spirigerella Derbyi*, Waagen, Palæontolog's Indica, ser. xiii, Salt Range Fossils, I, Productus Limestone Fossils, p. 753, Pl. XXXV, figs. 4-7, 9-13; Pl. XXXVII, figs. 11-13; *var. acuteplicata*, p. 754, Pl. XXXV, figs. 10, 11, Pl. XXXVII, fig. 11.

For further synonyms *vide* my monograph of the fossils of the Productus shales (Pl. iv of the present volume).

There is a solitary specimen among the material from Chitichun No. I, which, according to my opinion, must be identified with this species. In its dimensions it approaches most nearly the types, figured by Waagen under the varietal denomination *acuteplicata* (especially Pl. XXXV, fig. 10), or the specimens from the permian Productus shales (Pl. iv of the present volume, Pl. V, figs. 6, 8). It is of the same size, a little longer than broad, and provided with a broad median fold in the dorsal valve, to which an equal sinus in the ventral valve corresponds. This sinus originates in the frontal portion of the latter valve only, and is only as deep, as in the typical form of the Salt Range species. The beak is entirely appressed to the apical region of the dorsal valve, its foramen being therefore absolutely concealed. A tongue-shaped process is formed in the frontal region by the different breadth of the shelly zones in the frontal and lateral portions of the ventral valve, as it has been likewise noticed by Waagen in his types from the Salt Range.

The surface of the well preserved shell is entirely smooth.

The measurements of this specimen are as follows : —

Entire length of the shell	. . . .	16 mm.
Length of the dorsal valve	. . . .	13 5 „
Entire breadth of the shell	. . . .	14 „
Thickness of both valves	. . . .	10 „
Apical angle of the ventral valve	. . . .	88°
„ „ „ dorsal	. . . .	110°

Number of specimens examined.—1.

Remarks.—*Spirigerella Derbyi* is peculiar to permocarboniferous and permian strata. It is one of the most common fossils in the Virgal and Kálábagh beds of the middle Productus limestone and in the upper Productus limestone of the Salt Range. In the Himálayas it has been collected by Griesbach in the permian Productus shales of Painkhánda. In America it occurs in the coalmeasures of Itaituba (Brazil), which, according to Waagen, ought to be correlated with the upper divisions of the middle Productus limestone.

The shape of the sinus does not allow an identification of the present species with Waagen's *Sp. Derbyi* var. *acuteplicata*.

2. SPIRIGERELLA GRANDIS (Davidson), Waagen.   Pl. XI, fig. 3.

1862. *Athyris subtilita*, Hall, *var. grandis*, Davidson, Quart. Journ. Geol. Soc., London, XVIII, p. 28 Pl. I, fig. 8 (7).

1863. *Athyris subtilita var. grandis*, Davidson, in L. de Koninck, Mémoire sur les fossiles paléozoïques recueillis dans l'Inde, p. 32, Pl. IX, fig. 8.

1883. *Spirigerella grandis*, Waagen, Palæontologia Indica, ser. xiii, Salt Range Fossils, I. Productus Limestone Fossils, p. 461, Pl. XXXVI, figs. 1-7 ; XXXVII, fig. 1

The Tibetan specimens do not vary so much, in shape and dimensions, as the representatives of this species in Salt Range Productus limestone, which have been figured by Waagen. The reason may, however, be found in the fact, that I had to deal with a rather small number and with adult individuals only.

All are distinguished by a suborbicular shape, strongly inflated valves, a distinctly appressed beak, the foramen of which is very small and entirely concealed, and by the presence of a very shallow sinus. In the latter character they agree with the examples from Musakheyl, figured by Waagen on Pl. XXXVI, figs 3, 4. Both in the longitudinal and in the transverse direction the curve of the two valves is very equal. The frontal part is not produced.

The ventral valve is about as deep as the dorsal one, and provided with a comparatively small, strongly incurved beak, which is bordered on both sides by a very indistinct false area. The shallow depression, which corresponds to the sinus, is confined to the proximity of the front. In the dorsal valve no median fold has been observed. Waagen describes the lateral margins of this valve as "hanging down laterally over the margins of the ventral valve in full grown specimens, thus enveloping them more or less." I have noticed this character in several of my specimens, though only in a slight degree.

The ornamentation of the two valves consists of numerous imbricating striæ of growth, which occur at irregular distances. In places, where the fibrous shell has been injured by weathering, very numerous but extremely thin, radial striæ make their appearance within the zones, which are marked off by two lines of growth.

In its general shape this species recalls *Athyris globulina*, Waagen (p. 467, Pl. XLI, figs. 1-3), but it is considerably larger and the rounded foramen, as it is exhibited in the latter form, is not visible in any of my specimens.

The dimensions of my largest specimen are as follows:—

Entire length of the shell	.	.	.	.	.	.	.	.	.	.	.	.	30 mm.
Length of the dorsal valve	.	.	.	.	.	.	.	.	.	.	.	.	27 „
Entire breadth of the shell	.	.	.	.	.	.	.	.	.	.	.	.	26 „
Thickness of the two valves	.	.	.	.	.	.	.	.	.	.	.	.	21 „
Apical angle of the ventral valve	.	.	.	.	.	.	.	.	.	.	.	.	95°
„ „ „ dorsal „	.	.	.	.	.	.	.	.	.	.	.	.	107°

Number of specimens examined.— 4

Remarks.—In the Salt Range *Spirigerella grandis* is almost entirely restricted to the middle division of the Productus limestone. Among 46 specimens examined by Waagen, there was only a single one from the upper division. In the middle Productus limestone it is common both in the Virgal and Kálábagh beds, but does not descend into the Katta beds.

3. Spirigerella pertumida, nov. sp. . Pl. XI, figs. 1, 2.

This species, which is represented by two specimens only, stands in a similar relationship to *Spirigerella grandis*, Waagen, or to *Sp. media*, Waagen, as does *Retioularia conularis*, Grünewaldt, to *R. lineata*, Mart. It differs from all the hitherto described species of the subgenus *Spirigerella* in the very strong development of the apical region in the ventral valve. The greatest thickness of the two valves coincides with the cardinal region.

The ventral valve is as deep or a little deeper than the opposite one. It is almost equally curved in the longitudinal direction, perhaps slightly more strongly in the apical than in the frontal region. Transversely the bend is very steep in the proximity of the margins, but becomes rather flat in the central portion of this valve.

The apex is enormously inflated and strongly bent over. The beak is pointed and firmly appressed to the apex of the dorsal valve, its small foramen being thus entirely concealed. It is bordered on both sides by a comparatively large false area, which is not, however, marked off distinctly from the remainder of the shell. Laterally it is not overlapped by the margins of the dorsal valve, as is often the case in full grown individuals of *Sp. grandis*. The frontal part of the valve is strongly produced. Of the two type specimens the one, which is distinguished by its more elongated shape, is provided with only a very shallow median sinus, whereas in the second the sinus is well developed and originates in the apical region, extending for about three quarters of the entire length of the valve.

The dorsal valve is more strongly curved in the transverse than in the longitudinal direction. The median fold is developed as a sort of an obtusely rounded crest, in which the lateral parts of this valve unite, enclosing a right angle. This median crest originates in the very apex, but is only developed into a fold in the frontal region. This fold is neither very prominent nor rectangular. A small part of the apical region only is concealed below the beak of the ventral valve.

The ornamentation of the two valves is the same as in *Sp. grandis*, or in *Sp. Derbyi*. The shell is fibrous in its structure, which is well exhibited in a portion of the dorsal valve of the specimen figured Pl. XI, fig. 1.

Of the interior characters nothing, unfortunately, is known to me. The only feature, I have been able to observe, is the enormous thickness of the ventral valve in the apical region, reminding me of Waagen's description of *Spirigerella grandis* (loc. cit, p 462).

The measurements of the two type specimens are as follows :—

	I (fig. 1).	II (fig. 2).
Entire length of the shell	33 m.m.	28 mm.
Length of the dorsal valve	24 „	24·5 „
Entire breadth of the shell	30 „	27 „
Thickness of both valves	23 „	20 „
Apical angle of the ventral valve	?	app. 70°
„ „ „ dorsal „	130°	130°

Number of specimens examined.—2.

Remarks.—It appears to me, that the characters, which have been enumerated in the preceding diagnosis, especially the extraordinary inflation of the apical region in the ventral valve, are sufficient to distinguish this form from the hitherto described congeneric species.

This species is certainly nearly allied to *Spirigerella grandis*, Waagen, or perhaps still more nearly to *Sp. timorensis*, Rothpletz (Die Perm- Trias- und Jura-formation auf Timor und Rotti, Palæontographica, 1892, XXXIX, p. 82, Pl. X, figs. 4, 5). My specimens differ, however, from the figures of *Sp. timorensis*, given by Rothpletz, not only in the more strongly inflated apices of the ventral valve, but also in the fact, that a considerably smaller part of the apical region in the dorsal valve is concealed by the beak of the opposite one. Among the numerous types of *Spirigerella* from the permian rocks of Julfa, which have been united with true *Athyris* in one single species, *Spirigera protea*, by Abich (Geologische Forschungen in den Kaukasischen Ländern, I. Th. Eine Bergkalk-Fauna aus der Araxesenge bei Djoulfa, Wien, 1878, p. 52) there is especially the one figured on Pl. VII, fig. 7 of his memoir, which recalls *Spirigerella pertumida*. It has been described by Abich (loc. cit., p. 58) as *Spirigera protra var. globularis*, Phill. That it must be kept separate from the true *Spirigera (Athyris) globularis*, Phill., has been proved by von Möller (Neues Jahrbuch, 1879, p. 225). It is probably identical with *Spirigerella timorensis*, Rothpletz, a species, which this author himself believes to be represented among the variations of Abich's *Sp. protea*. An identification of this Armenian form with *Sp. pertumida*, is, however, impossible in spite of their general resemblance. In the specimen from Julfa the greatest thickness of the two valves is situated about half way between the front and the cardinal region, and the difference between the entire length of the shell and of the dorsal valve is only one sixth of the former, whereas it is more than one quarter in the Tibetan species.

Suborder ANCISTROPEGMATA, Zittel.

Family : *PORAMBONITIDÆ*, Davidson.

Subfamily: ENTELETINÆ, Waagen.

Genus: ENTELETES, Fischer v. Waldh. (SYNTRIELASMA, Meek.)

The genus *Enteletes* is represented in the permocarboniferous fauna of Chiti-chun No. I, by a new species, belonging to the section of *centrisinuati*, Waagen, in which the ventral or smaller valve is provided with the sinus, and the dorsal or larger valve bears the corresponding median fold. To this series of forms, the typical one of the genus, belong *E. Lamarcki*, Fischer v. Waldh., *F. Kayseri*, Waagen, *E. hemiplicatus*, Hall, *E. lævissimus*, Waagen, *E. carnicus*, Schollwien. My new species, which will receive the name of *E. Tschernyscheffi*, is very nearly related to *E. Kayseri*, and belongs to a group of forms of which *E. hemiplicatus*, Hall, is the prototype.

1. ENTELETES TSCHERNYSCHEFFI, nov. sp.   . Pl. V, figs. 7-11.

This is a rather large species of slightly transversely oval or nearly circular outlines, and strongly inflated valves, of which the dorsal one is considerably larger and more strongly vaulted.

The ventral valve is provided with a comparatively short, little prominent, moderately incurved apex, and with a high and concave area, which is cut open in the middle by a large triangular fissure. The median sinus originates in the apical region. In full grown specimens (fig. 9), it may be traced almost to the very apex of the valve. It is angular and bordered on both sides by high and tolerably sharp folds, which are followed on each side by three other folds. In some specimens a fourth fold is slightly indicated. These lateral folds likewise originate in the proximity of the apex.

The dorsal valve, which is the larger one, is of a nearly semicircular shape. Its strongly incurved apex distinctly overhangs the hinge line, which equals only one half the entire breadth of the shell in length. The area is considerably lower than in the opposite valve, but likewise cut open in the middle by a large deltidial fissure. In this valve a strong median fold, originating in the apical region, corresponds to the sinus of the ventral valve. It is followed on each side by three or occasionally even by four lateral folds. The median fold is the broadest and highest of all, the lateral folds gradually diminishing in height and breadth. The crests of the folds as well as the median edges of the angular valleys between them are tolerably sharp.

The front margin of both valves forms a deeply and sharply zig-zag shaped line. Numerous striæ of growth, which are bent up and down in similar zig-zag lines, cover the surface of the two valves in the vicinity of the front. In well preserved specimens (fig. 11) a delicate, radial striation may be noticed.

Of the internal structure of this species I have only been able to make out traces of the three septa in the ventral valve.

The present species attains very considerable dimensions. As far as we may judge from the size of the smaller valve, figured on Pl. V, fig. 8, it is not much surpassed in this respect by *E. pentameroides*, Waagen, or by *E. Suessi*, Schellwien, the largest hitherto known species of this genus.

The measurements of a moderately sized but fairly complete specimen are as follows :—

Entire length of the shell	20 mm.
Length of the dorsal valve	18 „
Entire breadth of the shell	22.5 „
Length of the hinge line	11 „
Thickness of both valves	19.5 „
Height of the area of the smaller valve	1.5 „
Apical angle of the smaller valve	116°
„ „ „ „ larger „	102°

1 2

Number of specimens examined.—11.

Remarks.—The present species seems to be very closely related to *E. Kayseri*, Waagen (= *E. hemiplicatus*, Kayser *non* Hall, Obercarbonische Fauna von Loping, Richthofen's China, IV, p. 179, Pl. XXIV, figs. 2, 3) from the upper carboniferous rocks of Loping. As may be seen from the figures and descriptions given by Kayser, Waagen (Pal. Indica, ser. xiii, I, p. 553), and Schellwien (Palæontographica, 1892, XXXIX, p 35), these two species bear a strong resemblance, and must certainly be united in the same group of forms. Nevertheless the Tibetan species can be easily distinguished from *E. Kayseri* by its more globular profile and by the strong plication of its valves. In *E. Kayseri* the number of lateral folds is always smaller, and both the folds and the median ridges in the intercostal valleys are neither so sharp nor do they extend so near the apical region, as in the present form. The remarkably flatter shape of the dorsal valve in *E. Kayseri* is also a good character of distinction.

There is no other species yet known from the section of the *centrisinuati*, to which *E. Tschernyscheffi* could be more particularly compared. It is, however, very similar, if not actually identical, with a yet undescribed species from the permian rocks of Sicily, which the palæontological museum of the Vienna University has quite recently been able to acquire.

This beautiful species is dedicated to Prof. Tschernyschew of St. Petersburg, who by his geological work in the Ural Mountains contributed so considerably to our knowledge of the palæozoic faunas of Russia.

Family : *RHYNCHONELLIDÆ* Gray.

Subfamily : RHYNCHONELLINÆ, Waagen.

Genus : UNCINULUS, Bayle.

Although representatives of the genus *Rhynchonella*, Fisch., are not altogether absent from the permocarboniferous limestone of Chitichun No. I, they are extremely rare. Among the extensive material I have only found two specimens, which may be attributed to this genus, and they are in a too fragmentary state of preservation to allow any specific determination. It is a rather astonishing fact that the family of *Rhynchonellidæ* is almost exclusively represented, in the Chitichun limestone, by the genus *Uncinulus*, which has been met with hitherto but very rarely, and in rather sporadic specimens only, in beds of a similar geological age. To this genus a species belongs, which I believe to be identical with Beyrich's *Rhynchonella timorensis* from the island of Timor, and which is not at all rare in the Chitichun limestone.

1. UNCINULUS TIMORENSIS, Beyrich. Pl. X, figs. 7, 8, 9, 10.

1845. *Rhynchonella Timorensis*, Beyrich, Ueber eine Kohlenkalk-Fauna von Timor, Abhandlgn K. Ak. Wiss., Berlin. 1844. p. 71, Pl. I, fig. 10.

1883. *Uncinulus Theobaldi*, Waagen, Palæontologica Indica, ser. xiii, Salt Range Fossils, I, Productus Limestone Fossils. p. 425. Pl. XXXIV, fig. 1

1892. *Rhynchonella (Uncinulus) Timorensis*, Rothpletz, Die Perm- Trias- und Jura-formation auf Timor und Rotti, Palæontographica, XXXIX, p. 87, Pl. X, fig. 6.

Beyrich based his diagnosis of this interesting species on a single specimen from the Ajermati river near Kupang on the island of Timor. Among Wichmann's collections from the same locality Rothpletz likewise discovered a single specimen of this species. Waagen introduced the name *U. Theobaldi* for two specimens from the Productus limestone of the Salt Range, which, (though resembling *U. timorensis* very much, seemed to differ from the latter in some minor details. Rothpletz suggested that both *U. Theobaldi* and *U. jabiensis*, Waagen (Pl. XXXIV, fig. 2, p. 427) ought to be united with *U. timorensis* until their characters of difference, which are all within the limits of variations of a species, have been proved constant by the comparison of more material. My examination of a large number of specimens has convinced me, that *U. timorensis* and *U. Theobaldi* at least ought no longer to be maintained as separate species, as they do not differ by constant characters.

The points of difference, which Waagen considered to be of specific value, are the following: *Uncinulus timorensis* is distinguished by the greater smoothness of the shell, the smooth part in the cardinal region of *U. Theobaldi* being considerably smaller, by the far broader median fold and sinus in comparison to the lateral parts and by the greater flatness or the nearly impressed form of the ventral valve.

Among my specimens both the typical forms of *U. timorensis* and of *U. Theobaldi* are represented, but linked together by a considerable number of intermediate types uniting one or more of the above mentioned characters, which Waagen believed to be of specific importance. The specimen figured on Pl. X, fig. 8, perfectly agrees with the typical *U. Theobaldi*. The lateral parts and the median portion of the shell are of nearly equal breadth, the ventral valve is flatly arched, not impressed, the ribs cover nearly half the extent of the valve. In the specimen fig. 7, the ribs are considerably shorter, the smooth part occupying more than one half of the entire length of the ventral valve. But this distinction, which is based on the length of the ribs, loses all its importance from the fact, that in some of my specimens the ribs, although thinning very gradually, extend even as far into the apical region, as in *Uncinulus posterus*, Waagen (loc. cit., p. 428).

It will also be found that in different examples the breadth of the median portion, corresponding to the mesial fold and sinus, varies considerably. In the specimen fig. 10, the median part of the shell is more than twice as broad as the lateral portions, a proportion, which perfectly agrees with that noticed in Beyrich's type specimen. Nor is the shape of the ventral valve a more constant character. In the specimen fig 9, it is barely less distinctly impressed than in the typical *U. timorensis*.

Of other variations the following may be mentioned: The shelly rectangular lobe, into which the median part of the ventral valve is prolonged, is very variable in its height, and bordered on both sides by either vertical or very steeply inclined margins. The number of ribs is also variable. I counted from eight to twelve ribs in the sinus or corresponding fold and an equal or larger number in the lateral portions of the shell.

My examination of the fossil material in the Tibetan collection has convinced me, that Waagen's illustration of *Uncinulus Theobaldi*, is only one of the numerous modifications, assumed by this variable species. After the discovery of so many intermediate shapes it has appeared to me impossible to arrive at any other satisfactory conclusion, than that they are all variations of *U. timorensis*. Both species apparently merge into each other so completely, that I cannot help considering them as specifically inseparable.

I am not, however, inclined to add *Uncinulus jabiensis*, Waagen (loc. cit., p. 427, Pl. XXXIV, fig. 2) to the synonyms of *U. timorensis*, as has been advocated by Rothpletz. All my specimens of *U. timorensis* are transversely oval and much shorter than broad. I have never met with a single type among them, representing the triangular outlines of *U. jabiensis*, or *U. posterus*. I therefore believe the transversely oval shape of *U. timorensis*, to be a good and constant character of specific importance. The number of ribs in *U. jabiensis* is also considerably smaller than in any of my examples of the present species.

It is barely necessary to add anything else to Waagen's excellent description of *U. Theobaldi*. Regarding the internal structure of the species, I may only remark, that besides the tolerably large dental plates in the ventral valve a median septum has been noticed in the dorsal valve.

My largest specimen attains a diameter of 28 mm. The measurements of a median sized specimen are as follows:—

Entire length of the shell	.	15 mm.
Length of the dorsal valve	.	13·5 ,,
Entire breadth of the shell	.	20 ,,
Thickness of both valves	.	12 ,,
Apical angle of the ventral valve	.	113°
" " " dorsal "	.	app. 134°

Number of specimens examined.—28.

Remarks.—The present species has been described both from the Kálábagh beds of the middle Productus limestone and from the Jabi beds of the upper Productus limestone in the Salt Range by Waagen, and from the permian rocks of Timor by Beyrich and Rothpletz.

Lóczy (Wissenschaftliche Ergebnisse der Reise des Grafen Béla Széchenyi, I Th. Wien, 1893, p. 723) quotes the species from beds of a probably permocarboniferous age near Yar-ka-lo in South-Eastern China.

Subfamily : CAMAROPHORIINÆ, Waagen.

Genus : CAMAROPHORIA, King.

This characteristic and easily recognisable genus is represented in the permo-carboniferous fauna of Chitichun No. I, by three species, all of which belong to the group of *Camarophoria crumena*, Mart. Of these three species only one can be identified with a well known Salt Range form *C. Purdoni*, Davidson. The second, although very nearly related to *C. Purdoni*, must be considered as new species distinguished by its unusually large size and by the larger number of ribs, covering both valves. The third species is too imperfectly preserved to allow an exact determination.

1. CAMAROPHORIA PURDONI, Davidson.   Pl. XII, figs. 6, 8, 9.

1862. *Camarophoria Purdoni*, Davidson, Quart. Journ. Geol. Soc. London, XVIII, p. 30, Pl. III, fig. 4.

1863. *Camarophoria Purdoni*, Davidson, in L. de Koninck, Mémoire sur les fossils paléozoiques recueillis dans l'Inde, p 36, Pl. XII, fig. 4.

1882. *C. Purdoni*, Waagen, Salt Range Fossils, ser. xbi, Palæontologica Indica, I, Productus Limestone Fossils, p 497, Pl XXXII, figs. 1-7.

1890. *C. Purdoni*, Nikitin, Mém. Com. Géol. Russ., St. Pétersbourg, V, No. 5, p. 71, Pl. III, figs. 6-7.

This species is very common in the permocarboniferous limestone of Chitichun No. I.

The specimens do not vary greatly in their general shape and outlines. The majority among them are transversely oval, but examples are not at all rare in which the length and breadth of the shell are nearly equal. The valves are either strongly or moderately inflated. Flat specimens, like the one figured on Pl. XII, fig. 9, are quite an exception. Both valves are convex, but the dorsal valve is always much more strongly vaulted. The sinus in the ventral valve is rather broad, always broader than the lateral parts, but not deeply sunk in. The beak is well incurved and pointed, but it is very rarely sufficiently well preserved to exhibit the triangular slit at its lower side.

The ventral valve is covered with sharply rounded ribs of a rather irregular strength and number. They originate in the proximity of the very apex, leaving a very small portion of the shell only entirely smooth. This is a good character of distinction between the present species and *C. crumena*, Mart, or *C. pinguis*, Waag., in which the ribs are always restricted to the frontal, marginal and central portions of the shell, whereas the cardinal region is quite smooth. In average sized specimens there are from ten to eighteen ribs present in this valve. There are from five to seven ribs within the sinus. On the lateral parts of the valve the number of ribs is rarely equal on both sides of the sinus. Like Waagen's types from the Salt Range the Tibetan specimens show a remarkable tendency to become unsymmetrical.

The augmentation of ribs in full grown specimens is always due to the inter-calation of new ribs, not to a bifurcation of the original ones, which remain undivided from the apical region to the front.

The dorsal valve is distinctly subdivided into three portions, of which the median one is very broad, but not much elevated, rounded and slightly flattened on its top. The sculpture is of the same pattern as in the opposite valve, but the number of folds is, as a rule, a little larger.

Traces of expansions have been noticed on the frontal and lateral margins of several specimens.

The internal characters of the Tibetan examples agree perfectly with Waagen's and Nikitin's descriptions and figures. They are, as a rule, better exhibited in the ventral than in the dorsal valve. The strong median septum is fixed to the dental plates so as to leave an ogival interval between the latter. In the dorsal valve both the high median septum and the spathulate, trilobate plate have been observed.

The measurements of an average sized specimen are as follows :—

Entire length of the shell	.	.	.	.	.	.	.	.	.	26	mm.
Length of the dorsal valve	.	.	.	.	.	.	.	.	.	23·5	„
Entire breadth of the shell	.	.	.	.	.	.	.	.	.	23	„
Breadth of the median fold	.	.	.	.	.	.	.	.	.	16	„
Thickness of both valves	.	.	.	.	.	.	.	.	.	18	„
Apical angle of the ventral valve	.	.	.	.	.	.	.	.	96°	
„ „ „ dorsal „	.	.	.	.	.	.	.	.	118°	

Number of specimens examined.—51.

Remarks.—There cannot, I think, be much doubt as to the identity of the Tibetan specimens with Davidson's species, as they agree with the latter not only in their more important characters, but also in their minor details.

The only other species, with which they might be compared, is *Camarophoria alpina*, Schellwien (Palæontographia, XXXIX, 1892, p. 51, Pl. VIII, figs. 4-8), which only differs from *C. Purdoni* in very minor details, but the number and arrangement of ribs in the Tibetan examples is more in accordance with Davidson's Indian species.

In the Salt Range *Camarophoria Purdoni* is very common in the Virgal and Kálábagh beds of the middle Productus limestone, but very rarely extends into higher strata. Nikitin quotes the species from the Gshelian stage of the upper carboniferous deposits of Central Russia. He likewise refers *C. plicata*, Tschernyschow (Mém. Com. Géol. Russ., St. Pétersbourg, III, No. 4, p. 309) from the Artinskian horizon of the Ural Mountains to *C. Purdoni*.

2. CAMAROPHORIA GIGANTEA, nov. sp.   . Pl. XII, figs. 5, 7, 10.

From *Camarophoria Purdoni* a species must be separated, which, although very nearly related to the former, is readily distinguished by its very remarkable size and by the larger number of ribs in both valves. This species attains gigantic dimensions, surpassing largely in size all the types of *C. Purdoni*, which have been figured by Davidson, Waagen and Nikitin. Nor can any intermediate types be traced between the two forms.

In their general shape and outlines the two species agree perfectly with each other. The only difference consists in the sculpture, which is formed by a larger number of ribs in *C. gigantea*. The ventral valve bears from eight to twelve ribs within the sinus, and from six to eight ribs in each of the lateral parts. In the latter the same tendency to become unsymmetrical, is exhibited, which is a remarkable character in *C. Purdoni*. In the dorsal valve the number of ribs amounts from twenty to thirty, of which always more than a third lie on the flatly rounded median fold. The ribs are highly rounded but not acute. Their augmentation is due to intercalation only. All the ribs remain undivided from the apical region to the front. No bifurcate ribs have been noticed in any of my specimens.

Similar traces of lateral expansions, as in *C. Purdoni* have been observed on the frontal and lateral margins of several specimens.

So far, as I have been able to make out the arrangement of the dental plates and septa, the internal structure is the same as in the preceding species. The figure Pl. XII, fig. 10, gives a lateral view of the ventral and dorsal septa, which agrees almost entirely with Waagen's figure of these characters in *C. Purdoni* (Pl. XXXII, fig. 6).

The measurements of my type specimen (fig. 5) are as follows:—

Entire length of the shell	45 mm.
Length of the dorsal valve	43 "
Entire breadth of the shell	51·5 "
Breadth of the median fold	29 "
Thickness of both valves	24 "
Apical angle of the ventral valve	107°
" " " dorsal "	115°

Number of specimens examined.—25.

Remarks.—The specific claims of this form may perhaps be considered uncertain by some palæontologists, but the distinctions suggested in the preceding diagnosis are, in my opinion, of sufficient importance to allow it to be maintained as a separate species. Its remarkable size and the larger number of ribs constitute appreciable differences between the present species and *C. Purdoni*, although both are certainly very nearly related to each other.

By its unusually large dimensions this species recalls *C. Plicata*, Kutorga, from which it is, however, distinguished by the character of its ornamentation. In the Russian species the ribs are augmented by bifurcation, which is never the case in the Tibetan types.

3. CAMAROPHORIA SP. IND. AFF. C. CRUMENA, Mart. . Pl. XI, fig. 6.

A third species of the genus *Camarophoria* is represented among the material from Chitichun No. I by three specimens, attaining a length of 8 to 12 mm. only. The internal arrangement proves that they belong to the present genus. In their outlines and sculpture they recall *C. crumena*, Mart., and *C. pinguis*, Waagen. They differ from *C. purdoni*, in which the shell is almost entirely covered by ribs, even in quite young individuals (*vide* Pl. XII, fig. 8), by their ribs being restricted to the frontal portion of the shell and being acute on their tops. The number of ribs within the sinus varies from two to four.

K

As apparently no full grown specimens are among the Tibetan examples, I must leave it undecided, whether they should be attributed to one of the two above quoted forms, or to a separate species.

Suborder: ANCYLOPEGMATA, Zittel.

Family: *TEREBRATULIDÆ*, King.

Subfamily: TEREBRATULINÆ Waagen.

Subgenus: DIELASMA, King.

1. DIELASMA BIPLEX, Waagen. Pl. XI, figs. 6, 7, (?) fig. 5.

1883. *Dielasma biplex.* Waagen, Palæontologica Indica, ser. xiii. Salt Range Fossils, I. Productus Limestone Fossils, p. 349, pl. XXV, figs. 3, 4, 5.

Among a small number of specimens, by which the subgenus *Dielasma* is represented in the Chitichun fauna, the two examples figured (figs. 6, 7) may, in my opinion, be safely identified with the present species. The rest I must leave undetermined. My reasons for doing so will be easily understood by any one who has had to deal with incomplete forms belonging to a genus, in which the specific distinctions have been based on very subordinate details alone. A species of so large an extension, as was *Dielasma sacculus*, Martin, as defined by Davidson, may be of little use for stratigraphical or for biological purposes, but it can scarcely be denied, that species of so narrowly restricted, as those which have been introduced by L. de Koninck in *Dielasma*, are likewise practically useless.

The two figured specimens exhibit the elongately pentagonal outline, which is peculiar to Waagen's specimen Pl. XXV, fig. 5, from the middle Productus limestone of Musakheyl. They are, however, somewhat variable in their general shape, the specimen fig. 6 being more strongly elongated and having its greatest transverse diameter shifted considerably nearer towards the frontal part. The most conspicuous character, which is especially well developed in the specimen fig. 7, is the presence of two converging folds in the dorsal, and of a straight mesial fold in the ventral valve. In my second specimen, which has been slightly deformed by pressure, these characters are less distinctly indicated.

The ventral valve is provided with a tolerably large and thick beak, which is pierced by a broad, oval foramen. The apex is slightly bent over but not appressed to the apex of the dorsal valve. The small deltidium is clearly visible in the specimen fig. 6. From both sides of the foramen indistinct ridges descend for a short distance to the lateral parts of the valve, bordering the broad false area, whose lower part is not, however, distinctly marked off from the remainder of the shell. The frontal line is distinctly biplicate and produced into two slightly bent up lappets, corresponding to the two folds of the dorsal valve. The obtusely rounded, low ridge originates in the middle of the length of the valve, and gradually increasing in width, reaches down to the front line. It is bordered on both sides by low, rounded depressions.

The dorsal valve is much more strongly vaulted in the transverse than in the longitudinal direction. In the apical region two indistinct, rounded folds originate, which diverge towards the front line leaving a low, triangular depression between them. In the specimen fig. 6 the marginal slope of these folds is marked by a distinct bend in the transverse profile of the valve.

Of the internal structure the presence of strongly developed dental and septal plates has alone been determined.

The measurements of the specimen figured on Pl. XI, fig. 7, are as follows :—

Entire length of the shell	.	.	.	.	.	.	.	.	.	**24**	mm.
Length of the dorsal valve	.	.	.	.	.	.	.	.	.	**3·5**	,,
Entire breadth of the shell	.	.	.	.	.	.	.	.	.	**26**	,,
Thickness of both valves	.	.	.	.	.	.	.	.	.	**16·5**	,,
Apical angle of the ventral valve	.	.	.	.	.	.	.	.	.	**75°**	
,, ,, ,, dorsal ,,	.	.	.	.	.	.	.	.	.	**105°**	

Number of specimens examined.—2.

Remarks.—It is only with great hesitation, that I refer the adolescent specimen Pl. XI, fig. 5, to the present species. In this specimen the peculiar characters of *Dielasma biplex* are not yet distinctly indicated.

In the Salt Range *D. biplex* is restricted to the middle Productus limestone. Waagen quotes it from the Virgal and Kálábagh beds, but not from the Katta beds of this division.

The specimens of *Dielasma* are too incomplete to state whether other species besides the present form, are represented among them.

Subgenus: HEMIPTYCHINA, Waagen.

The genus *Hemiptychina* has been introduced by Waagen for such species of *Terebratulina fimbriata*, in which distinct septal but no dental plates have been developed. Rothpletz (Palæontographica, XXXIX, 1886, p. 72) rejected Waagen's genus, which he considers to be merely a stage of development in different series of the genus *Terebratula*. Oehlert (in Fischer's Manuel de Conchyliologie, III, p. 1315) likewise classes *Hemiptychina* among the synonyms of *Terebratula*. Dealongchamps (1884) however elevated the species united by Waagen in his genus *Hemiptychina* at least to the rank of a distinct group (a) of *Terebratula*, and Geheimrath von Zittel in his recently published Grundzüge der Palæontologie (p. 275) advocates the retention of the name *Hemiptychina* as a subgeneric denomination. In the present memoir the latter view has been followed.

In the permocarboniferous fauna of Chitichun No. 1 the subgenus *Hemiptychina* is represented by three species, all of which are identical with Salt Range forms.

1. HEMIPTYCHINA HIMALAYENSIS, Davidson.   Pl. XII, fig. 4.

1862. *Terebratula himalayensis*, Davidson, Quart. Journ. Geol. Soc. London, XVIII, p 27, Pl. II, fig. 1.
1863. *T. himalayensis*, Davidson, in L. de Koninck, Mémoire sur les fossils paléozoïques recueillis dans l'Inde, p. 32, Pl. IX, fig. 1.
1879. *T. himalayensis*, Waagen, Rec. Geol. Surv. Ind., XI, p. 188.
1883. *Hemiptychina himalayensis*, Waagen, Pal Indica, ser. xiii, Salt Range Fossils, I, Productus Limestone Fossils, p. 368, Pl. XXVI, figs. 6-10.

This species is represented in the collection from Chitichun No. I by two specimens, agreeing perfectly with the figures and descriptions given by Davidson and Waagen.

The specimens are strongly elongated, truncated at their front, and provided with ten or eleven short folds   The two valves are equally vaulted.   The beak is pierced by an oval foramen.   The sculpture of both valves is smooth up to within a few millimetres of the margins, where the short rounded ribs are developed.   In my examples, which have been slightly deteriorated by weathering, only the four or five folds occupying the front, are distinctly visible.

The measurements of my larger specimen are as follows :—

Entire length of the shell	.	.	.	.	.	.	.	.	22	mm
Length of the dorsal valve	.	.	.	.	.	.	.	.	19·5	„
Entire breadth of the shell	.	.	.	.	.	.	.	.	16	„
Thickness of both valves	.	.	.	.	.	.	.	.	14	„
Apical angle of the ventral valve	.	.	.	.	.	.	.	90°	
„ „ „ „ dorsal „	.	.	.	.	.	.	app.	115°	

Number of specimens examined.—2.

Remarks.— The present species is very common in the middle and upper Productus limestones, but very rare both in the Katta beds and in the lower Productus limestone of the Salt Range.   Mr. Hughes collected it in a white crinoidal limestone in the Central Himalayas near one of the passes leading from Milam into Hundes.

2. HEMIPTYCHINA SPARSIPLICATA, Waagen.   Pl. XII, figs. 1, 2.

1883. *Hemiptychina sparsiplicata*, Waagen, ser. xiii, Salt Range Fossils, Pal. Indica, I, Productus Limestone Fossils, p. 366, Pl. XXVII, figs. 4, 5, 6.
1892. *Terebratula himalayensis var. sparsiplicata*, Rothpletz, Die Perm- Trias- und Juraformation auf Timor und Rotti, Palæontographica, XXXIX, p. 63, Pl. X, fig. 10.

This species, which is a little more numerously represented among the collections from Chitichun No. I is only distinguished from *Hemiptychina himalayensis* by the smaller number of its folds.

The two species are linked together, according to Waagen's own statement, by a number of transitional forms.   Nevertheless Waagen "thought it expedient" to note the typical form of *H. sparsiplicata* by a special name " by reason of its geological importance.   To *H. sparsiplicata* a more varietal rank has been attributed by Rothpletz.   I deemed it preferable to follow Waagen's view, without overrating the geological importance of the two species, the number of which among my material is altogether too small to allow any conclusion in this respect.

In all the examples under description the number of folds is considerably smaller, than in the two figured types of *H. himalayensis*. The species is rather variable in its outlines. The specimen figured on Pl. XII, fig. 2, exhibits the roundish shape and the large apical angle, which, according to Waagen, chiefly characterize the typical form of *H. sparsiplicata*. Fig. 1, on the contrary, represents a specimen of unusually elongated outlines.

The measurements of a specimen, agreeing with the typical form of *H. sparsiplicata* (fig. 2) are as follows :—

Entire length of the shell	.	20 mm.
Length of the dorsal valve	.	17 ,,
Entire breadth of the shell	.	15½ ,,
Thickness of both valves	.	12½ ,,
Apical angle of the ventral valve	.	83°
,, ,, ,, dorsal ,,	.	114°

Number of specimens examined.— 8.

Remarks.—The present species is abundant in the lower Productus limestone and in the Katta beds, but rare in the Virgal and Kálábagh beds of the Salt Range. From the permian rocks of Timor the species has been quoted by Rothpletz.

3. HEMIPTYCHINA INFLATA, Waagen. Pl. XII, fig. 3.

1882. *Hemiptychina inflata*, Waagen, Palæontologia Indica, ser. xiii, Salt Range Fossils, I, Productus Limestone Fossils, p. 375, Pl. XXVII, figs. 7, 8, 9.

To this characteristic species I can attribute only a single specimen, which is distinguished from the other specimens of *Hemiptychina* by its large size, its strongly inflated valves, its globular outlines, and its distinctly developed plications, which extend across the larger portion of the shell to the front line.

My specimen has been considerably damaged by weathering. In the dorsal valve especially the sculpture has been almost entirely destroyed. But the considerable inflation of the shell and the character of the folds which in the ventral valve originate in the proximity of the apical region, make a distinction from the two preceding species an easy matter. There are about ten folds present, four of which occupy the frontal portion of the ventral valve. The absence of any dental plate in this valve does not allow the present specimen to be confounded with *Dielasmina plicata*, Waagen.

The measurements of this specimen are as follows :—

Entire length of the shell	.	24 mm.
Length of the dorsal valve	.	20 ,,
Entire breadth of the shell	.	19 ,,
Thickness of both valves	.	18 ,,
Apical angle of the ventral valve	.	app. 71°
,, ,, ,, dorsal ,,	.	app. 130°

Number of specimens examined.—1.

Remarks.—*Hemiptychina inflata* is quoted by Waagen from the Virgal and Kálábagh beds of the middle Productus limestone and from the lowest beds of the upper Productus limestone in the Salt Range.

Genus : NOTOTHYRIS, Waagen.

1. NOTOTHYRIS TRIPLICATA, nov. sp.,   Pl. XIII, figs. 1, 2.

It is only on the analogy of its external shape, that I place the present species in Waagen's genus *Notothyris*, as I have not been able to make out the internal characters, although I have sacrificed a good number of specimens for their investigation. Nevertheless the reference of this species to *Notothyris* is barely more doubtful than Waagen's identification of *Terebratula djoulfensis*, Abich, with one of his species of *Notothyris* from the Salt Range Productus limestone.

In their general shape and outlines *Notothyris* and *Hemiptychina* exhibit a remarkable similarity. According to Waagen's statement, however, there is in *Hemiptychina* always a certain tendency to form a slightly vaulted frontal line, whilst in *Notothyris*, on the contrary, the tendency prevails to bend the frontal line in the opposite direction. In my specimens a very distinct sinuation takes place in the frontal line, so much so, that the indentations corresponding to the folds of the frontal region are situated at a lower level than those of the lateral parts. This peculiar feature marks the present species as forming part of the genus *Notothyris*.

The general outline of the present species is elongately oval or indistinctly pentagonal. The two valves are of very unequal depth, the ventral valve being much more strongly inflated than the opposite one.

The ventral valve is strongly curved both in the longitudinal and transverse directions, but even more so transversely. Longitudinally the curvature is very equal, but transversely it is slightly flattened in the middle portion of the shell. The beak is moderately thick, prominent, but not strongly bent over. It is pierced by a small foramen just behind its apex. An indistinctly limited off false area extends for a short distance from both sides of the beak.

The most characteristic feature in this valve is its peculiar sculpture, which remarkably differs from the ornamentation in all the hitherto described species of *Notothyris*. The valve is smooth for about half its length from the beak. Then there appear two very strong, sharp folds, which extend down to the front and very distinctly separate the frontal portion of the valve from the lateral parts. In the rounded valley, which is enclosed by these two folds, a third median fold rises, which is, however, slightly inferior in strength. On each side of the frontal portion of the valve one or two short lateral folds are developed. The margin of the valve is distinctly sinuated in the frontal region, so that the indentations corresponding to the three folds of the frontal part are situated at a lower level than those of the lateral parts.

The three high and sharp frontal folds, of which the middle one is slightly less strongly marked, are a constant character in all the numerous specimens, which I have been able to examine.

A great many imbricating striæ of growth are crowded together in the proximity of the margins, whereas few only are developed in the middle portion of the valve.

The dorsal valve is rather flat and always remarkably less strongly vaulted than the ventral one. In the longitudinal direction it is, however, strongly deflected in the vicinity of the front, frequently imparting to the entire shell a slightly truncated shape. In this valve the folds are much shorter than in the opposite one, and three quarters of its entire surface remain almost perfectly smooth, except for the imbricating striæ of growth. Four comparatively strong folds are situated on the frontal part. The lateral folds are but indistinctly developed. They occur to the number of one or two on each side of the front.

The front margin is slightly prolonged, corresponding to the sinuation of the ventral valve.

Of the internal characters of this species nothing is known to me.

The measurements of one of my type specimens are as follows :—

Entire length of the shell	20	mm.
Length of the dorsal valve	17	„
Entire breadth of the shell	16	„
Thickness of both valves	15	„
Apical angle of the ventral valve	67°	
„ „ „ dorsal „	105°	

Number of specimens examined.—28.

Remarks.—This species of *Notothyris* is easily distinguished from all the other congeneric forms by its peculiar, triplicate sculpture. Nor is any species of *Hemiptychina* known to me, to which it might advantageously be compared.

In my preliminary report, on the geological results of the Himálayan expedition to Johár, Hundés and Painkhánda in 1892, I have quoted *N. simplex*, Waagen, among the species of this genus, represented in the permocarboniferous fauna of Chitichun No. I. After a careful examination of the specimens in question, I prefer however to abandon this view, as those examples are too imperfectly preserved to allow any satisfactory determination.

2. Notothyris cf. subvesicularis, Davidson.

1862. *Terebratula subvesicularis*, Davidson, Quart. Journ. Geol. Soc., XVIII, p. 27, Pl. II, fig. 4.
1863. *Terebratula subvesicularis*, Davidson, in L. de Koninck, Mémoire sur les fossiles paléozoïque s recueillis dans l'Inde, p. 22, Pl. IX, fig. 4.
1882. *Notothyris subvesicularis*, Waagen, Palæontologia Indica, ser. xiii, Salt Range Fossils, I. Productus Limestone Fossils, p. 376, Pl. XXVIII, figs. 3, 4.

I am unfortunately unable to give an adequate figure of the Tibetan representatives of this species, the only two available specimens having been sacrificed in fruitless attempts to make out their internal structure. I must consequently abstain from giving a detailed description but shall confine myself to the statement, that among the permocarboniferous fauna of Chitichun No. I. a few examples have been noticed, which in their external characters entirely agreed with Davidson's

species. Prof. Waagen who examined my specimens, likewise declared them to be
identical with *Notothyris subvesicularis.*

In the Salt Range this species is characteristic of the upper region of the
middle Productus limestone (Virgal and Kálábagh beds). In the upper Productus
limestone it is extremely rare. Outside the Punjab it has been mentioned by
Waagen from a white crinoidal limestone, which was picked up by Mr. Hughes
near one of the passes leading from Milam into Tibetan territory.

Class : BRYOZOA.

Order : GYMNOLÆMATA, Allm.

Suborder : CYCLOSTOMATA, Busk.

Family : *FENESTELLIDÆ,* King.

Genus : FENESTELLA, Lonsdale.

1. FENESTELLA SP. IND. . Pl. XIII, Fig. 3.

Among the material from Chitichun No. I a few small fragments of colonies
may be safely attributed to this genus. All the rest of *Bryozoa* in the collection
are too badly preserved to allow any diagnosis. Nor did I dare to attempt a
specific determination of the figured specimen of a *Fenestella,* as of the two
faces of the colony the non-poriferous is alone accessible to observation. I have
entirely failed in preparing the poriferous side, which is not only firmly attached to
the rock, but also completely destroyed by having been altered into crystallized
calcite.

The largest fragment of a colony available for description is fan-shaped, without the presence of any marked axis. It recalls *F. virgosa,* Eichwald (Lethæa
Rossica I, Pl. 23, fig. 9, p. 359), but the meshes of the network are a little
larger. The branches are rather thin, of equal thickness for their whole extent, and
bifurcate in a somewhat irregular manner. Otherwise they run tolerably parallel
to each other.

The fenestrules are oval or rectangular, with rounded off corners, and considerably longer than broad. There are about three fenestrules within the space of
5 mm. in the direction of the extension of the branches, and four in the transverse
direction. The branches and the dissepiments are of nearly equal thickness.

The presence of an indistinct longitudinal striation of the branches cannot be
positively asserted, as the surface in all the specimens appears to have been slightly
injured by weathering.

Number of specimens examined.—1.

ECHINODERMATA.

Class: CRINOIDEA.

In the permocarboniferous fauna of Chitichun No. 1 only stem joints of representatives of this class have been preserved. Although very abundant, they are insufficient for generic determination. I must consequently abstain from giving descriptions or figures, and merely record their presence in the Chitichun limestone.

CŒLENTERATA.

Class: SPONGIÆ.

Subclass: CALCISPONGIÆ.

Order: SYCONES, Hæckel.

Suborder: SPHINCTOZOA, Steinmann.

Family: *SPHÆROSIPHONIDÆ*, Steinmann.

Genus: AMBLYSIPHONELLA, Steinmann.

In placing *Amblysiphonella* among the *Sycones*, I am following the view of Rauff (Palæospongiologie, I. Theil, Palæontographica, XL, 1803, p. 102), who does not share the opinion expressed by Steinmann (Pharetronen-Studien, Neues Jahrbuch, 18·2, II. Th., p. 160) and Waagen, regarding the systematic position of this genus among the *Pharetrones*.

I regard a specimen, which, notwithstanding its rather fragmentary state of preservation, appears to be nearly related to *A. vesiculosa*, de Kon., but differs by its comparatively smaller size, as a typical representative of the genus.

1. AMBLYSIPHONELLA, APP. VESICULOSA, de Koninck. Pl. XIII, Fig. 4.

The fragment available for description consists of five segments, of which only the lower three are tolerably perfect. The segments are shortly cylindrical with convex sides, and separated from each other by deep furrows. The outer walls of the segments are thin and perforated all over by numerous, very small pores.

Both longitudinal and transverse sections exhibit a moderately wide central tube, bordered by a rather thick wall. It is surrounded by the circular chambers, corresponding to the segments, in the shape of broad rings.

L

I have not succeeded in making out the details of the internal structure although the presence of numerous vesicles within the chambers has been noticed. The measurements of this fragment are as follows :—

Entire length of the fragment	.	.	.	.	.	.	.	.	35 mm.
Diameter of the third segment	.	.	.	.	.	.	.	.	15 „
„ „ „ lowest „	.	.	.	.	.	.	.	.	13 „
Height of the third segment	.	.	.	.	.	.	.	.	6 „
Width of the central tube	.	.	.	.	.	.	.	.	3·5—5 „

Number of specimens examined.—1.

Remarks.—This fragment is rather nearly allied to *Amblysiphonella cericulosa*, de Koninck (Waagen, Pal. Indica, ser. xiii, I, p. 972, Pl. CXXII, fig. 1), from which it differs by its considerably smaller size. In this respect it agrees with *A. Borroisi*, Steinmann (Neues Jahrbuch, 1882, II Th., p. 169, Pl. VI, fig. 1) from the carboniferous rocks of Sebargas in Spain. From the latter species it is however readily distinguished by its cylindrical shape, the angle of emergence being considerably smaller than in *A. Borroisi*.

Class: ANTHOZOA.

Order: TETRACORALLIA, Haeckel.

Family: *CYATHOPHYLLIDÆ*, Haeckel.

Genus: LONSDALEIA, M'Coy.

1. LONSDALEIA INDICA, Waagen and Wentzel.   Pl. XIII, figs. 5, 6.

1888.   *Lonsdaleia indica*, Waagen and Wentzel, Palæontologia Indica, ser. xiii, Salt Range Fossils, I.
        Productus Limestone Fossils, p. 897, Pl. CI, figs. 1—3, Pl. CIV, figs. 8-4.

This is by far the commonest species of coral in the Chitichun limestone. Part of the south-eastern slope of the crag of Chitichun No I is made up almost entirely of a coral rock, composed of corallites which in every respect agree with *Lonsdaleia indica* from the Salt Range Productus limestone.

The compound corallum is composed of cylindrical, tolerably straight corallites which are rather far apart and are only in very few cases blended together. Their surface is covered with numerous transverse wrinkles, and exhibits a delicate longitudinal striation. The diameter of the calices varies from 3 to 7 mm. No traces of a secondary wall are indicated. The columella occupies nearly a third of the diameter of the entire calix. There are about twenty primary septa, and an equal number of secondary ones.

In transverse thin sections (Pl. XIII, fig. 6e) the transverse lamella, which forms the centre of the columella, is very distinctly exhibited. This character makes a distinction of *L. indica* and *L. virgalensis*, Waagen and Wentzel (loc. cit. p. 900, Pl. CXVI, fig. 2 ; Pl. CI, fig. 4) from all the congeneric species an easy matter. It is the presence of this central transverse lamella, and the exceptional arrangement of the tabulae, which in Waagen's opinion, might even be considered sufficient for the introduction of a distinct generic denomination for *L. indica* and its allies.

From the central transverse lamella radiating lamellae originate, which are crossed by a concentric lamellation. The columella is marked off very sharply from the remainder of the calyx by a distinct wall. This wall is reached by the majority of the primary septa. A few of the latter are even brought into contact with one of the radiating lamellae within the columella. The interseptal dissepiments are very numerous, but are not arranged regularly so as to form concentric accessory walls.

Remarks.—I think there can be scarcely any doubt of the identity of the specimens from the Chitichun limestone with *Lonsdaleia indica*. The larger diameter of the branches and the irregular arrangement of the interseptal dissepiments forbid an identification with *L. virgalensis*, which is, however, very nearly allied to *L. indica* and is only distinguishable by very subordinate characters. I know of no other species of the genus *Lonsdaleia*, with which my specimens could be advantageously compared.

In the Salt Range *L. indica* occurs in very large numbers in the coral beds of the middle Productus limestone (Virgal beds), but extends also, though more rarely, into the Kálábagh beds and Khundghát beds. It has neither been found in the Amb beds nor in the Katta beds.

Family : ZAPHRENTIDÆ, Haeckel.

Genus : AMPLEXUS, Sowerby.

1. AMPLEXUS, SP. IND. AFF. ABICHI, Waag. and Wentzel. Pl. XIII, fig. 7.

Among the corals from Chitichun No. 1 there are a few specimens belonging to this genus, but their state of preservation is unsatisfactory, and the matrix so spathic that their structure cannot be made out well enough to permit of specific determination. I must consequently while drawing attention to their remarkable similarity to *A. Abichi*, Waagen and Wentzel (Pal. Indica, ser. xiii. I, p. 903) abstain from any identification with previously described species of *Amplexus*.

The corallum is simple, subcylindrical, straight or slightly curved, and provided with strongly prominent, rounded or imbricating wrinkles of growth, completely encircling the corallum. The epitheca is comparatively thin and covered with very numerous and delicate longitudinal striæ. The outer walls of the septa show through the epitheca in the shape of thin, straight lamellæ. To a diameter of 14 mm. of the corallum 24 septa correspond. They are short, but their exact length cannot be made out in consequence of the spathic character of the matrix. The tabulæ are flatly curved. The shape of the calix is unknown to me.

The species, to which the present one seems to be most nearly related, are *Amplexus Abichi* (*A. coralloides*, Abich., Geologische Forschungen in den Kaukasischen Ländern, I. Th. Eine Bergkalkfauna aus der Araxes-Enge bei Djoulfa, p. 57, Pl. XI, fig. 10) and *A. coralloides*, Sowerby. The similarity with other congeneric species, described by L. de Koninck in his monograph of the carboniferous fossils of Belgium and by Stuckenberg in his memoir on the corals and bryozoa of the carboniferous rocks of the Ural and Timan (Mém. Com. Géol. Russ., St Pétersbourg, 1895, X, No. 3) is less close. I do however not venture to identify my specimens any of the two above quoted species, as my insufficient knowledge of their internal characters forbids an exact determination.

PROBLEMATICA : Gen. et sp. ind.   . Pl. XII, fig. 8.

Among the material from Chitichun No. I there is a very strange fossil, whose determination has puzzled me for a long time. I have tried in vain to make out to which class of invertebrates it might belong. Nor have any of my learned friends, who examined this curious fossil, arrived at a more satisfactory conclusion. Perhaps its true character will be revealed by the discovery of more complete specimens. I, therefore, give its description and figure, notwithstanding the impossibility of ascertaining its zoological position.

The specimen in question consists of a system of root-like appendages, uniting with a central ring-shaped body. The root-shaped appendages, which occur to the number of five, exhibit a semicircular arrangement and are fastened both to the central main-body and to each other. By reason of their semicircular shape, however, they touch each other for a short distance only, and afterwards strongly diverge. Triangular spaces are left between the places, where two neighbouring appendages are attached to the body and to each other. The appendages terminate in sharp points. The largest diameter is about 5 mm. The diameter of the central, ring-shaped body attains nearly the same size. The largest diameter of the slightly elliptical inner portion, encircled by this body, is about 16 mm.

The internal structure of the fossil is very peculiar. Both the main body and the appendages are divided by deep transverse furrows into numerous segments of irregular size. These furrows correspond in the ring-shaped body and in the appendages, and unite in the places where the latter are fastened to each other. To these deep furrows a very large number of delicate transverse furrows is added, which are more distinctly preserved near the outer margin of the fossil and thus at a superficial glance give the impression of a proper marginal zone, distinguished by their presence. This is, however, not the case, but the delicate transverse furrows are scem to continue throughout the entire thickness of the appendages and main body wherever the latter are fairly preserved.

I first thought that this strange fossil might belong to the *Calcispongiæ*, and ought perhaps to be placed in the relationship of *Amblysiphonella radicifera*, Waagen and Wentzel (Pal. Indica, ser. xiii, I, Pl. CXXIII., CXXIV., fig. 1. p. 975). This was also Prof. Waagen's view when he first examined the specimen. Geheimrath von Zittel, however, one of our most competent authorities on this subject, after a close examination of the Tibetan fossil refused to acknowledge the possibility of its belonging to the *Calcispongia*, and suggested a similarity, though rather distant, to crinoid roots. The fact of the specimen being unique forbade the use of the grindstone. I could therefore only prepare a single section of its outlying portions for microscopical examination, but this section revealed nothing of the characteristic structure of *Echinodermata*, although this structure may only have been destroyed by the process of fossilisation.

The fossil in question must consequently still be treated as a problematicum.

STRATIGRAPHICAL RESULTS.

The fauna of the limestone crag of Chitichun No. I is composed of the following species:—

CRUSTACEA.

1. *Phillipsia Middlemissi*, Diener.
2. *Cheiropyge himalayensis*, Diener.

CEPHALOPODA.

3. *Stachoceras Trimueti*, Diener.

LAMELLIBRANCHIATA.

4. *Aviculopecten* aff. *jabiensis*, Waagen.

BRACHIOPODA.

5. *Productus lineatus*, Waag.
6.   ,,   *Cora*, d'Orb.
7.   ,,   *semireticulatus*, Mart.
8.   ,,   *boliviensis* var. *Chitichunensis*, Diener.
9.   ,,   cf. *calcostatus*, Waag.
10.   ,,   *gratiosus*, Waag.
11.   ,,   *cancriniformis*, Tschernyschew.
12.   ,,   *Abichi*, Waag.
13.   ,,   *mongolicus*, Diener.
14. *Marginifera typica*, Waag.
15. *Aulosteges tibeticus*, Diener.
16. *Lyttonia nobilis*, Waag.
17. *Spiriferina octoplicata*, Sow.

18. *Spirifer musakheylensis*, Davidson.
19.   „   *Wynnei*, Waag.
20.   „   *tibetanus*, Diener.
21. *Martinia cf. glabra*, Mart.
22.   „   *elegans*, Diener.
23.   „   *uncata*, Rothpletz.
24.   „   *semiplana*, Waag.
25.   „   *acutomarginalis*, Diener.
26.   „   *contracta*, Meek and Worthen.
27. *Reticularia lineata*, Mart.
28. *Athyris (Spirigera) Royssii*, Lév.
29.   „   „   *capillata*, Waag.
30.   „   „   *subexpansa*, Waag.
31. *Spiriferella Derbyi*, Waag.
32.   „   *grandis*, Waag.
33.   „   *peritumida*, Diener.
34. *Eulolicus Pachtanscheffi*, Diener.
35. *Uncinulus timorensis*, Beyrich.
36. *Camarophoria Purdoni*, Dav.
37.   „   *gigantea*, Diener.
38.   „   *aff. crumena*, Mart.
39. *Hemiptychina himalayensis*, Dav.
40.   „   *sparsiplicata*, Waag.
41.   „   *inflata*, Waag.
42. *Dielasma biplex*, Waag.
43. *Notothyris triplicata*, Diener.
44.   „   *cf. subvesicularis*, Dav.

BRYOZOA.

45. *Fenestella*, sp. ind.

SPONGIÆ.

46. *Amblysiphonella, aff. vesiculosa*, Koo.

ANTHOZOA.

47. *Amplexus* p. ind *aff. A. Abichi*, Waag. and Wentzel.
48. *Lonsdaleia indica*, Waagen and Wentzel.

Altogether 48 species, among which the *Brachiopoda* numbering 40 species by far predominate, both in species and individuals, and comprise five-sixths of the entire fauna.

I ought now to enter into a discussion of the relations which exist between the fauna of the Chitichun limestone and the faunas of homotaxial beds in other countries. As however for a comparison of these faunas among each other the determination of the upper limit of the carboniferous system is of fundamental importance, I shall defer such a discussion till the present state of the so-called permocarboniferous problem has been explained. This will be found the more necessary, as some very important memoirs have thrown new light on the subject since the geological results of Waagen's monograph of the Productus limestone fossils were published, in consequence of which the views of this author regarding the correlation of the permocarboniferous strata will have to be partly modified.

The difficulty of correlating the series of strata intermediate between the typical carboniferous and permian deposits chiefly consists in the fact that an uninterrupted sequence of marine beds of both systems was for a long time unknown. In Europe the discovery of this sequence is due to the united efforts of the Russian geologists, whose works have made Central Russia and the Ural Mountains a classic ground for the study of the marine development of upper carboniferous and permocarboniferous beds.

In Central Russia the upper carboniferous series is divided into two well defined groups. The lower or Moscovian stage (horizon of *Spirifer mosquensis*) contains the fauna of Mjatchkowa, described by Trautschold; the upper or Gshelian stage (horizon of *Chonetes uralica*, "ourallien") contains the fauna of Gshel, which has been studied by Nikitin. In the upper carboniferous rocks of the Ural Mountains two faunistically different horizons may, according to Tschernyschew, be likewise distinguished: a lower stage (C₃) with *Spirifer cf. mosquensis*, and an upper one (C₄ horizon of *Productus Cora*) which in itself is rather manifold but consists chiefly of *Fusulina* limestones. The topmost limestones (Schwagerina horizon) must be placed on a somewhat higher level than the Gshelian stage, corresponding to the Schwagerina dolomites of the Oka-Kljasma and Oka-Wolga-basins, which have been proved by Sibirzew[1] to overlie the horizon of *Chonetes uralica* (Nikitin's Gshelian stage). The Artinskian marls and sandstones conformably overlie the carboniferous rocks. There is no break in the sequence, as has been proved by Tschernyschew and Krasnopolsky in a most convincing manner.

To the Artinskian sandstone and marls (CPg) and to the following limestone-dolomite horizon (CPe) the name permocarboniferous has been applied by Russian geologists. It is in this sense only that the term "permocarboniferous" will be used in this memoir, as it is the only one which does not inevitably lead to misunderstanding. The denomination Artinskian, which has been proposed by Munier-Chalmas and A. de Lapparent[2] may perhaps be preferable, as it excludes all misunderstandings, but the term permocarboniferous in the meaning of the Russian geologists has been accepted lately by so many authors (e.g., Waagen, Credner, Kayser, Oldham) that I should not like to drop it altogether.

It is only above these permocarboniferous strata, viz., above the Artinskian sandstones and marls and the limestone-dolomite horizon, that the true "Permian" as it had been understood by Sir Roderick Murchison, follows.

Karpinsky[3] and Tschernyschew,[4] two authors, to whom the most detailed studies

[1] N. Sibirzew,—Mém. Com Géol. Russ. St. Pétersbourg, 1896, XV, No. 2, pp. 227—242.

[2] Munier-Chalmas et A. de Lapparent,—Note sur la nomenclature des terrains sédimentaires, Bull. Soc. Géol. France, sér. iii, XXI, 1893, p. 752.

[3] A. Karpinsky.—Ueber die Ammoneen der Artinsk-stufe, Mém. Acad. Imp. Sci. St. Petersbourg, sér. xxxvii, p. 90—101.

[4] Mém. Com. Géol. Russ. St. Pétersbourg, III, No. 4, 1889, pp. 362—840.

of the Artinskian fauna are due, strongly advocate the distinction of the permocarboniferous from carboniferous and permian systems, and are decidedly averse to uniting it with either the one or the other. Tschernyschew especially strongly combats the view of the majority of geologists who proposed to unite the permocarboniferous with the permian, as a lower division of the system. According to him a separation of the permocarboniferous from the permian system is demanded by the general aspect of the fauna, in which the carboniferous types greatly predominate, chiefly among the brachiopoda. If it ought to be united either with the carboniferous or permian system, in spite of its distinctly intermediate position, it must necessarily be placed in the former, on the strength both of the carboniferous character of its fauna and of historical priority, since the Artinskian sandstone had been correlated with the carboniferous millstone-grit of Western Europe by Sir Roderick Murchison, who first introduced the name of permian.

Against the first argument the objection may be raised that notwithstanding the prevalence of carboniferous types in the Artinskian fauna, the latter "marks a very important moment in the history of development of organic remains, namely, the first appearance of true ammonites with complicated sutures."[1] Nor is the large percentage of carboniferous types in the Artinskian fauna an astonishing fact, in view of the absence of any break in the sequence of marine beds from the upper carboniferous to the true permian strata. Even in beds, which must be placed very high in the permian system, in the upper Productus limestone of the Salt Range and in the Otoceras beds of Julfa, the fauna contains a proportionately large number of carboniferous forms. It is to the faunas of these deposits, the normal representatives of the pelagic permian, not to the local fauna of the Zechstein, that the permocarboniferous fauna must be compared, if we want to get a clear idea of its relationship to those of the upper carboniferous and permian. Bearing in mind the gradual passage from an upper carboniferous to a permian fauna through the intermediate group of rocks, the question to be answered is, which consideration is of the greater importance in defining the boundary between the two systems, the appearance of a new group of cephalopoda, which become of an unparalleled stratigraphical value in mesozoic times, or the presence of a belated fauna, composed of forms which are generally not well adapted for the characterisation of narrowly limited horizons.

The majority of geologists have decided in favour of the first alternative. Gümbel,[2] Krasnopolsky,[3] Kayser,[4] Waagen,[5] Credner,[6] Munier-Chalmas and A. de Lapparent, Frech—to enumerate only a small number among them,—are unanimous in regarding the permocarboniferous as the lowest division of the permian system.

A discussion of the permocarboniferous problem from a historical point of view

[1] A. Karpinsky.—loc. cit. p. 101.
[2] C. W. Gümbel,—Geologie von Bayern, Kassel 1888, I. Th. p. 634.
[3] A. Krasnopolsky,—Mém. Com. Géol. Russ. St. Pétersbourg, XI. No. 1, p. 506.
[4] E. Kayser,—Lehrbuch der geologischen Formationskunde, Stuttgart, 1891, p. 157.
[5] W. Waagen,—Pal. Indica, ser. xiii, Salt Range Fossils, IV. Geological Results, p. 238.
[6] H. Credner,—Elemente der Geologie, 7. Aufl., Leipzig, 1891, 504.

leads to a similar result. This side of the question has been especially treated by Frech,[1] whose reasoning I consider to be entirely justified.

Sir Roderic Murchison, it is true, did not include the Artinskian deposits in his permian system, but on the other hand, his correlation of these deposits with the carboniferous millstone-grit is decidedly erroneous, and its priority cannot be respected. In the Rhenish regions, where the sequence of terrestrial and lacustrine, plant-bearing strata of this epoch is most complete, the true coal measures come to an end with the Ottweiler Schichten, whereas the following series of rocks comprising the Cuseler and Lebacher Schichten, have been united in a lower division of the permian system by Gümbel. In the Carnian Alps plant-bearing beds, containing a rich flora of the Ottweiler Schichten' alternate with Fusulina limestones, which have been proved by Schellwien[2] to be homotaxial with Nikitin's Gshelian stage in Central Russia. As has been noticed by Geyer,[3] this alternating series of dark Fusulina limestones and plant bearing beds is conformably followed by a compact mass of white Fusulina limestones (Trogkofelkalk) containing *Spirifer supramosquensis*, Nikitin, which must be corelated with the topmost carboniferous Fusulina limestones (Schwagerina horizon) of the Ural Mountains. The homotaxis of the Ottweiler Schichten and of the Carnian Fusulina limestone, which itself corresponds in age to the uppermost carboniferous beds of Central Russia (Gshelian stage) and of the Ural (Cora horizon, and Schwagerina horizon), apparently requires the boundary line between the two systems to be drawn immediately above the Schwagerina limestone of the Ural and Timan, and below the Artinskian stage.

A particular view of the permocarboniferous problem has been taken by Rothpletz[4] who hints at the probability, "that the permocarboniferous may not represent a distinct stratigraphical stage, but merely a particular facies, which is entirely wanting in Western Europe, was confined to the commencement of the permian epoch in Russia, but lasted throughout the whole of this epoch in some parts of Asia and North America."

I quite agree with Rothpletz in his statement that the normal sediments of the ocean, which covered parts of North-Western India, of Central Asia and of Russia, in the commencement of the permian epoch (Artinskian stage) must for the later stages of this epoch be looked for in the upper Productus limestone of the Salt Range and not in the Zechstein nor in Murchison's "permian system" of Russia, but I cannot understand why this view should be inconsistent with the character of the Artinskian deposits as a distinct stratigraphical horizon. A correlation of the permocarboniferous horizon proper with higher stages of the permian system cannot be

[1] F. Frech.—Die Karnischen Alpen. Halle, 1894, p. 567.

[2] E. Suess.—Das Antlitz der Erde II. Bd., p. 334 ; E. Schellwien, Die Fauna des Karnischen Fusulinen kalks, I. Th. Palaeontographica XXXIX, 1892, p. 1 ; F. Frech, Die Karnischen Alpen, Halle, 1894, pp. 309-324.

[3] E. Schellwien.—Zeitschr. Deutsch. Geol. Ges., 1893, p. 70.

[4] G. Geyer.—Ueber die Geologischen Verhältnisse im Pontafeler Abschnitt der Karnischen Alpen, Jahrb K. K. Geol. Reichs-Anstalt, 1896, 46. Bd., p. 156.

[5] A. Rothpletz, Die Perm Trias und Jura-formation auf Timor und Rotti Palaeontographica, XXXIX, 1892, p. 64.

M

admitted, seeing the difference of their cephalopod faunæ. Ammonites with ceratitic sutures are absent in the Artinskian stage. Their presence is barely less characteristic of the younger horizons of the permian system, than the first appearance of more highly developed *Ammonoidea*, is characteristic of the permocarboniferous stage. Rock-specimens like the one described by E. v. Mojsisovics (Denkschr. K. Akad. Wiss. Wien, math nat. Cl., LXI, 1894, p. 458) from Stoliczka's collections near Woábjilga (Karakorum Pass), in which ammonites with goniatitic, ceratitic and ammonitic sutures are mixed together, have up to now never been found in strata geologically older than the pelagic permian of India or Julfa.

In Russia the marine deposits of the Artinskian stage are overlaid by Murchison's " permian system," a sequence of brackish, freshwater and marine sediments, with an impoverished fauna,[1] which cannot be considered as the normal sediments of the permian epoch. This remark applies equally to the local development of the German Zechstein or of the British magnesian limestone. The pelagic equivalents of these deposits must be looked for in the permian rocks of the Mediterranean region in South-eastern Tyrol, Sicily and Armenia, of the Pamirs, of India and the Malayan Archipelago, of North America and probably also of Spitzbergen and the neighbouring islands. These sediments are proved both by their pelagic faunas and by their wide distribution within the Jhetys[2] and the Arctic-Pacific region to be the deposits of the great permian oceans and consequently the *normal* sediments of the permian epoch.

In the Mediterranean region three different rock groups have yielded fossil remains of this pelagic development of the permian epoch. These rock groups are the Fusulina limestone of the valley of Sosio in Sicily, the Bellerophon limestone of South-eastern Tyrol and Friaul, and the Otoceras beds of Julfa in Armenia. All of them are of a rather isolated occurrence and, as far as one may judge from their faunas, of different age.

The lowest position is apparently held by the Fusulina limestone of Sicily. Its cephalopod fauna seems to be more nearly related to the Artinskian one than to those of the Jabi beds of the Salt Range or of the Otoceras beds of Julfa. Ammonites with ceratitic sutures are yet absent. According to Karpinsky's statement, one species of *Medlicottia* is identical with an Artinskian form, ten more species are very nearly allied. On the other hand Karpinsky and Waagen noticed the first appearance of *Waagenoceras* and *Hyattoceras* in Sicily, two genera, which show a much more complicated sutural line than any of the Artinskian *Ammonea*. Waagen consequently places the Fusulina limestone of Sicily on a higher level than the permocarboniferous stage, but on slightly lower level than the Jabi beds of the upper Productus limestone. A more exact comparison of the Sicilian and Salt Range faunas will only be possible after the publication of Gemmellaro's monograph of the *Brachiopoda* from the Sosio limestone, as no cephalopod fauna is known in the Salt Range below the Jabi beds.

[1] *W. Amelisky*—The permian system in the Oka-Wolga basin. 1887 ; *N. Sibirzew*, Mém. Com. Géol. Russ. St. Pétersbourg. XV, No. 2, pp. 247—568.
[2] *E. Suess*—Natural Science, Vol. II, No. 13, March 1893.

The Otocerss beds of Julfa with their strongly marked triassic affinities must certainly be higher in the upper palæozoic series than the Fusulina limestone of Sosio. They cannot be much different in age from the Otoceras beds of the Himálayas, although the latter certainly hold a somewhat higher stratigraphical position, and they may consequently be placed on a level with the upper Productus limestone or with the Chidru group of the Salt Range.

The youngest of the three rock groups is probably the Bellerophon limestone of South-eastern Tyrol. Its fauna is a very peculiar one, species identical with those known outside this rock group being almost completely absent. The predominence of palæozoic types induced Stache[1] to fix the homotaxis of these beds as upper permian, whereas Gümbel supposed them to be of lowest triassic age. I have quite recently succeeded in discovering an interesting fauna of these beds in the valley of Suxten, containing the first ammonites and Orthoceras hitherto known in this horizon, which are rather in favour of Stache's view.

In none of these three permian rock groups of the Mediterranean region is a normal sequence of marine beds exposed, with the possible exception of the Bellerophon limestone of the Carnian Alps, which, however, is underlaid by an enormous mass of unfossiliferous limestones and dolomites. Their correlation must consequently be based on palæontological evidence alone.

The second region, in which a gradual transition of marine beds of the carboniferous and permian systems has been observed, is the central and south eastern portions of the United States of North America. In the typical sections of Kansas and Nebraska a gradual passage from the upper coal measures to the higher beds, with a fauna of a permian aspect, has been noticed, but the question, where the boundary between the two systems ought to be drawn, still remains to be settled. A very able paper by H. S. Williams,[2] in which the present state of the permo-carboniferous problem in America is exposed, has demonstrated that this question is still an open one. "In the first place the formations themselves are not delimited on the same basis in different provinces, and secondly the fossils have been reported under so many different names, that a thorough revision of the several biologic groups is necessary, before the various lists prepared can be scientifically correlated."

A cephalopod fauna of a permian type has been described by C. White[3] from the Baylor and Archer counties in Western Texas. Of the five species of ammonites comprising this fauna, Medlicottia Copei, Hyalcloceras Cumminsi, and Popanoceras Walcotti are very nearly allied to Sicilian forms, whereas Popanoceras Parkeri, Heilprin,[4] and Paralegoceras baylorense[5] exhibit a near relationship to Artinskian species.

[1] G. Stache.—Beiträge zur Fauna der Bellerophonkalke Südtirols, Jahrb. K. K. Geol. Reichs-Anst., Wien 1877, XXVII. pp. 271-318; 1878, XXVIII. pp. 93-168.

[2] Correlation Papers, Devonian and Carboniferous, Bull. U. S. Geol. Survey, Washington, 1891, No. 80. pp. 193-212.

[3] C. White.—The Texan Permian and its mesozic types of fossils, Bull. U. S. Geol. Surv., Washington, 1891, No. 77.

[4] Proc. Acad. Nat. Sciences. Philadelphia, 1887, VXXXVI pp. 63-68.

[5] Gonatites Baylorensis. White. From its general shape and to the character of the external line this form probably belongs to Hyatt's genus Paralegoceras.

A third region, distinguished by an uninterrupted sequence of marine beds reaching from upper carboniferous strata through the permian system, is the Salt Range in the Punjab.

This sequence begins with the Amb beds or lower Productus limestone. The stratigraphical relations of the Amb beds to the glacial boulder beds and to the *Eurydesma* sandstones with their Australian affinities are yet doubtful, as has been stated by Noetling in a recent paper on this subject.[1] The Productus limestones however are quite conformable to each other, and it is only between their topmost beds (Chidru group) and the next overlying lower Ceratite limestones, that a break in the sequence has been suggested by Waagen.

The following classification of the Productus limestone series has been adopted by Waagen :—

$$\text{Upper Productus Limestone.} \begin{cases} \text{Chidru beds.} \\ \text{Jabi beds.} \\ \text{Khund Ghat beds.} \end{cases}$$

$$\text{Middle Productus Limestone} \begin{cases} \text{Kalabagh beds.} \\ \text{Virgal beds.} \\ \text{Katta beds.} \end{cases}$$

Lower Productus Limestone.   Amb beds.

This classification has been slightly modified by R. D. Oldham,[2] who proposed a separation of the Chidru group from the rest of the series and a separation of the Katta beds from the middle Productus limestone. I did not however think it advisable to follow Oldham's second proposal, as the different meaning of the terms "lower" and "middle" Productus limestone in two memoirs of the Palæontologia Indica would be confusing. I consequently deemed it preferable to accept Waagen's classification, but to unite the Virgal and Kalabagh beds into an upper division of the middle Productus limestone, thus marking the faunistic contrast between this rock group and the underlying Katta beds.

There can be no doubt as to the permian age of the Chidru group with its peculiar bivalve fauna, and of the upper Productus limestone. The homotaxis of the lower divisions of the Productus limestone series is however uncertain, and some difference of opinion exists regarding the correlation with upper palæozoic strata outside India.

Waagen believed the entire Productus limestone series to be permian. He correlated the lower Productus limestone and the Katta beds with the Artinskian stage of Russia, considering them as permocarboniferous. To this correlation he was chiefly led by the supposition of a stratigraphical break between the upper carboniferous and permocarboniferous beds of the Ural Mountains, filled up, as he believed, in the Salt Range by the glacial boulder bed and by the *Eurydesma* sandstones. This view has become untenable, since in the Ural and Timan the con-

[1] F. Noetling, Beiträge zur Kenntniss der Glacialen Schichten permischen Alters in der Salt Range, Neues Jahrbuch, 1898, Bd. II, pp. 61—86.

[2] A Manual of the Geology of India, 2nd edition, by R. D. Oldham, Calcutta, 1893, pp. 121, 122.

tinuity of the sequence of marine beds from the horizon of *Productus giganteus* (mountain limestone of Western Europe) to the Artinskian stage has been proved by the geologists of the Russian survey in a most convincing manner. On the other hand, the recent memoirs of Nikitin, Schellwien, Tschernyschew, Suess, and Sibirzew tend to show that the number of carboniferous types in the Productus limestone is considerably larger than had been anticipated by Waagen.

The species from the Productus limestone, identical with those of upper carboniferous strata in other countries, are the following :—

Productus Cora, d'Orb.	Spirifer alatus, Schloth.
„ lineatus, Waag.	Reticularia lineata, Mart.
„ semireticulatus, Mart.	Martinia glabra, Mart.
„ indicus, Waag.	„ semiplana, Waag.
„ opuntia, Waag.	Spiriferina octoplicata, Sow.
„ spiralis, Waag.	„ cristata, Waag.
Orthothetes semiplanus, Waag.	Athyris Royssii, Lev.
Orthis Pecosii, Marcou.	Retzia (Eumetria) grandicosta, Waag.
Derbya grandis, Waag.	Camarophoria Purdoni, Dav.
Euteletes Kayseri, Waag.	Richthofenia sinensis, Waag.
Spirifer moosakhylensis, Dav.	Fusulina longissima, Müll.
„ striatus, Mart.	Stenopora ovata, Lonsd.
„ Marcoui, Waag.	Fenestella perelegans, Meek.

To these 26 species the following nine may be added as doubtful ones :—

Productus gratiosus, Waag.	Athyris pretiosifera, Sow.
„ Humboldti, d'Orb.	Dielasma elongatum, Schloth.
Streptorhynchus pelargonatus, Schloth.	Lyttonia Richthofeni, Kayser.
Orthis indica, Waag.	Orthoceras cyclophorum, Waag.

Aviculopecten derajatensis, Waag.

In all 35 species. It only needs a comparison of this list with that given by Waagen on p. 163 of his "Geological Results" (Palæontologia Indica, ser. xiii, Vol. IV) to prove the carboniferous affinities to be much more strongly marked in the Salt Range Productus limestone, than could be anticipated by this learned author.

The affinity of the fauna of the lower Productus limestone to those of the Gshelian stage of Russia, of the Cora and Schwagerina horizons of the Ural Mountains and of the Carnian Fusulina limestone induced Rothpletz (loc. cit. p. 63), Oldham (loc. cit. p. 125) and Frech (loc. cit. p 372) to correlate the lower Productus limestone with upper carboniferous rather than with permocarboniferous beds. E. Suess,[1] after a careful examination of the upper carboniferous fauna of Tongitar (southern slopes of Tian-shan) likewise agrees with the above mentioned authors in the opinion, "that the entire Productus limestone forms a series, which cannot be separated from the carboniferous system, and that the *Fusulina* bearing Amb beds certainly belong to the latter, although their correlation with the Gshelian stage may yet remain questionable." Noetling in his preliminary note on the glacial

---

[1] E. Suess, Beiträge zur Stratigraphie Central-Asiens, Denkschr. K. Akad. Wiss. Wien, math. nat. Cl. LII, pp. 438, 439.

boulder beds of the Salt Range takes a different view (*loc. cit.* p. 84), considering these beds as well as the entire Productus limestone series as permian. But from a palæontological point of view a correlation of the Amb beds with upper carboniferous strata appears to be more adequate to their faunistic character.

The fauna of the lower Productus limestone (Amb beds) consists of 58 species, allowing an exact determination, 42 of which do not extend into higher strata but give to this division its peculiar aspect. It contains altogether sixteen species identical with forms found also in upper palæozoic rocks of other countries. All of them occur in upper carboniferous beds, but only twelve extend into permocarboniferous strata. This proportion is in favour of a correlation of the Amb beds with deposits of an upper carboniferous age.

The middle Productus limestone of the Salt Range has been considered as homotaxial with the permocarboniferous beds of Russia (Artinskian stage) by Tschernyschew. Waagen however admits this homotaxis for the Katta beds only, but correlates the Virgal beds with the permian limestones of Kostroma, "which are on a lower level than the great bulk of permian beds as exposed in Perm and thus are intermediate between the permocarboniferous beds of the Ural and the typical permian strata" (*loc. cit.* p, 194). The palæontological evidence is more in favour of Tschernyschew's view, which has been accepted by Rothpletz (*loc. cit.* p. 63).

Among the brachiopoda of the upper division of the middle Productus limestone (Virgal and Kálábagh beds) seven species only were known to Waagen as identical with upper carboniferous forms. By more recent investigations this number is now increased to seventeen. Among the Artinskian brachiopoda, as enumerated in Tschernyschew's tabular statement seventeen are identical with Salt Range types. Eight of them are ubiquitous ranging from upper carboniferous into permian strata. Of the remaining nine species two are restricted to the Amb beds, 1 to the Amb and Katta beds, 1 to the upper division of the middle Productus limestone. One extends from the Katta beds into permian strata, one from the Amb beds through the entire thickness of the middle Productus limestone, four are restricted to the upper division of the middle Productus limestone and to the upper Productus limestone. Excluding the ubiquitous species, which cannot be considered as leading, the distribution of identical species is as follows :—

$$
\begin{array}{rl}
4 & \left\{ \begin{array}{l} \text{Upper Productus Limestone : 4.} \\ \text{Virgal and Kálábagh beds : 6.} \end{array} \right. \\
7 & \left\{ \begin{array}{l} \text{Katta beds : 8.} \\ \text{Amb beds : 4.} \end{array} \right.
\end{array}
$$

It will be seen from this tabular statement, that there is no sufficient reason for correlating the Artinskian stage with the Katta beds only, but that equally strong faunistic affinities exist between the Artinskian brachiopoda and those of the upper division of the middle Productus limestone.

There is no doubt that this statistical method of correlation is open to grave objections, as the admixture of a few species only would suffice to influence the result considerably. It must be borne in mind that a comparison of the fauna of the middle Productus limestone with that of Artinsk is the more difficult, as no cephalopods are known from this Salt Range horizon, and we therefore must rely in our correlation almost entirely on brachiopoda alone, "which very generally do not keep strictly to an exact geological horizon and therefore can be used only in rare cases as reliable documents for the identification of a narrowly limited geological stage or zone." What I wanted to prove by this statistical method, is merely the fact, that there exist strongly marked affinities between the faunas of the Amb beds and of the Russian and Carnian upper carboniferous strata on one hand, and between the Artinskian and middle Productus limestone faunas on the other.

In the present state of our knowledge, the following modifications of Waagen's correlation of the Productus limestone series appear to be required. The lowest division of the uninterrupted sequence of marine beds, known as Productus limestone, is probably homotaxial with the topmost carboniferous strata in Europe. The equivalents of the Russian permocarboniferous horizon (Artinskian stage) must be looked for in the middle Productus limestone, but the question whether the boundary between the permocarboniferous horizon and the higher stages of the permian system ought to be drawn above the Kalabagh beds, cannot be answered, until cephalopod faunas have been discovered in one of the divisions of the middle Productus limestone.

The rich development of permocarboniferous or permian strata of a pelagic type in the Arctic region (Spitzbergen) is too little known to allow any exact correlation with the subdivisions of the anthracolithic system[1] in other countries.

The correlation of the permian deposits of a pelagic facies, according to the present state of our knowledge, is shown in the tabular statement on p. 103, it is, of course, only approximate, and in many instances yet open to grave doubts. Nevertheless I believe it to elucidate more clearly my views on this subject than would be possible in the text. In this tabular statement only such deposits have been admitted, whose correlation can be based on the presence of fossil remains. The Zechstein or magnesian limestone development of the permian system have been purposely excluded, because they do not represent a pelagic type of this system.

Having explained the present state of the permocarboniferous problem, I may enter now into a discussion of the relations existing between the fauna of the limestone crag of Chitichun No. I, and the faunas of homotaxial beds in other countries. This subject is so much more the important, as a determination of the geological age of the Chitichun limestone must be based on palæontological evidence only, the limestone crag itself exhibiting no stratigraphical connection with the surrounding beds.

It will be found convenient to treat each class of fossils separately, although

---

[1] I consider this term, which has been proposed by Waagen (loc. cit., p. 241) very fit for uniting the carboniferous and permian systems under one name.

the palæontological evidence must chiefly rely on the brachiopoda, which by far predominate over all the rest of organic remains.

The two pygidia of trilobites can be of but little service for the identification of the geological horizon of the Chitichun limestone. One of them, *Cheiropyge himalayensis*, belongs to a new genus. The second, *Phillipsia Middlemissi*, may be compared to types, which have been found in upper carboniferous and permocarboniferous beds of other countries. Karpinsky (*loc. cit.* p. 101) considers the Artinskian stage as a period in the history of development of organic life, characterised probably by the appearance of the last trilobites, which do not extend into higher permian strata. The geologically youngest trilobites have been discovered in the permian Fusulina limestone of Sicily, but their absence in the permian rocks of India and Armenia is perhaps only accidental. Nevertheless the presence of trilobites in the Chitichun limestone corroborates the evidence adduced by the character of the entire fauna, that it does certainly not correspond to a higher level than to the permian of Sicily.

The discovery of *Stacheoceras Trimurti* is of the highest geological importance. Up to now ammonites of this subgenus have never been found of an older than permocarboniferous age, whereas they are among the most common types in permocarboniferous and permian strata of a pelagic facies. The occurrence of this species in the Chitichun limestone, in correspondence with other evidence, clearly proves this limestone to be of a younger than upper carboniferous age.

Five sixths of the entire fauna of the limestone crag of Chitichun No. I is composed of brachiopoda. Among them the *Productidæ* and *Spiriferidæ* are about equally strongly represented, whereas the *Lyttoniidæ*, *Porambonitidæ*, *Rhynchonellidæ* and *Terebratulidæ* are represented by comparatively few forms.

The following eight species of brachiopods are peculiar to the present fauna and have not yet been found outside the crag of Chitichun No. I :—

> *Aulacegra tibeticus*, Diener.
> *Spirifer tibetanus*, Diener.
> *Martinia elegans*, Diener.
>   ,,   *acutomarginalis*, Diener.
> *Spirigerella pretiosula*, Diener.
> *Eatalstea Tschernyscheffi*, Diener.
> *Camarophoria gigantea*, Diener.
> *Notothyris triplicata*, Diener.

To this list *Productus boliviensis* var. *chitichunensis*, Diener, ought perhaps to be added, though its identity with a species from Timor is rather doubtful.

All the rest of species (31) are identical with forms that have been previously described from other countries and localities, although the exact age of the beds in which they occur has not everywhere been fixed with sufficient certainty.

The region, which is geographically least distant from Chitichun No. I, and in which the richest development of approximately homotaxial beds has hitherto been made known, is the Salt Range in the Punjab. The relations of the fauna of the Salt Range Productus limestone and of the Chitichun limestone are very close. Twenty-six species of brachiopoda are identical. The greatest number are such forms, as occur in the upper division of the middle Productus limestone (Virgal and Kálábagh beds). These species are the following :—

Productus lineatus, Waagen.	*Spirigerella Derbyi*, Waag.
,, *Cora*, d'Orb.	,, *grandis*, Waag.
,, cf. *subcostatus*, Waag.	*Athyris Royssii*, Lév.
,, *gratiosus*, Waag.	,, *subexpansa*, Waag.
,, *Abichi*, Waag.	,, *capillata*, Waag.
Marginifera typica, Waag.	*Uncinulus timorensis*, Beyr.
Lyttonia nobilis, Waag.	*Camarophoria Purdoni*, Dav.
Spiriferina octoplicata, Sow.	*Dielasma biplex*, Waag.
Spirifer musakheylensis, Dav.	*Hemiptychina sparsiplicata*, Waag.
,, *Wynnei*, Waag.	,, *himalayensis*, Dav.
Martinia semiplana, Waag.	,, *inflata*, Waag.
	Notothyris subvesicularis, Dav.

Of these 23 species, more than half of the entire brachiopod fauna of Chitichun No. 1, the following four do not extend into higher strata :—

Spirifer Wynnei, Waag.
Martinia semiplana, Waag.
Dielasma biplex, Waag.
Hemiptychina sparsiplicata, Waag.

All the rest occur also in the upper Productus limestone of the Salt Range but there is not a single species of the upper Productus limestone identical with a Chitichun form, which does not also occur in the middle Productus limestone. The affinity of the Chitichun fauna consequently appears to be more intimate with that of the middle than with the upper Productus limestone. This evidence is corroborated by the occurrence of a small number of species identical with such from the Amb beds and Katta beds, which in the Salt Range do not extend into higher divisions of the Productus limestone series. These species are the following :—

Productus semireticulatus, Mart.
Reticularia lineata, Mart.
Martinia cf. glabra, Mart.

On the other hand, there are among the Chitichun brachiopoda only ten identical with species from the Amb beds, and nine with those the Katta beds. Nor is there one of the three above quoted species, which could be considered as characteristic of a distinct geological horizon. This distribution of identical species indicates a homotaxis of the Chitichun limestone with the upper division of the middle Productus limestone (Virgal and Kálábagh beds,) in the Salt Range.

With upper carboniferous beds of Europe (Carnian Alps and Russia) the Chiti-chun limestone has ten species of brachiopoda in common.　These are :—

Productus lineatus, Waagen.
　　,,　　*Cora*, d'Orb.
　　,,　　*semireticulatus* Mart.
　　,,　　*cancriniformis*, Tschern.
Spiriferina octoplicata, Sow.
Spirifer musakheylensis, Dav.
Reticularia lineata, Mart.
Martinia glabra, Mart.
　　,,　　*semiplana*, Waag.
Athyris Royssii, Lév.

There is not a single form among them, which is restricted to carboniferous deposits and does not extend into higher strata.

The two species of brachiopods in the Chitichun limestone which have hitherto been found in carboniferous strata only, are *Productus mongolicus*, Diener, and *Martinia contracta*, Meek and Worthen.

Productus mongolicus is identical with a species from Loping in China, which has been described as *Prod. cf. cora* by Kayser, and has been compared to *P. compressus* by Waagen.　The fauna of Loping is distinguished by a rather mixed assemblage of forms, notwithstanding its well marked carboniferous aspect.　The question may therefore be raised, if *P. mongolicus* with its strong affinities to the permian *P. compressus* does not form part of the small fraction of permian types, which, like *Strophalosia horrescens*, Vern., are in this fauna mixed together with carboniferous species.

Martinia contracta has been quoted from the Chester group of the lower carboniferous series of Illinois, but it belongs to a group of forms, which are but very little adapted for the identification of geological horizons.　I need only allude to *Martinia planoconvexa*, Shumard, which extends from devonian into permian strata without any variation[1].　The presence of this species, therefore, cannot influence the evidence afforded by the general character of the fauna.

The relations of the Chitichun fauna to that of the Artinskian stage are more intimate than those with the faunas of any of the carboniferous stages of Russia.　The following twelve species can probably be quoted as identical with Artinskian ones :—

Productus lineatus, Waag.	*Spirifer musakheylensis*, Dav.
,,　 *Cora*, d'Orb.	,,　 *Wynnei*, Waag.
,,　 *semireticulatus*, Mart.	*Reticularia lineata*, Mart.
,,　 *cancriniformis* Tschern.	*Martinia semiplana*, Waag.
Marginifera typica, Waag.	*Athyris Royssii*, Lev.
Spiriferina octoplicata, Sow.	*Camarophoria Purdoni*, Dav.

The nearest region to the Punjab and to Hundes, in which beds corresponding exactly in age to the Artinskian stage have been found, is Darwas.　In

---

[1] *P. Frech*, in *K. Suess*, Beiträge zur Stratigraphie Central Asiens, Denkschr. K. Akad. Wiss. Wien, math. nat. Cl. LXI, p. 455.

a rock-specimen brought from this country, five species of ammonites were discovered by Karpinsky,[1] two of which are absolutely identical with Artinskian forms.

The permian rocks, collected by Stoliczka near Wosbjilga (Karakorum Pass), contain ammonites with ceratitic sutures recalling of *Xenodiscus*, Waagen, and must consequently be placed on a higher level than the Artinskian stage or the Fusulina limestone of Sicily.

There is yet a third rock-group in Central Asia, which is perhaps of permian age. This is a brachiopod-bearing limestone, which was discovered near the Gussass river (Western Kuen-Lun) by Bogdanowitsch. To the species, collected in this limestone by the Russian geologist and described by Frech (*loc. cit.*, p. 454) *Productus cancriniformis*, Tschern., must be added. The limestone unconformably overlaps all the geologically older beds including the upper carboniferous strata of the Moscovian stage with *P. semireticulatus*. Whether the Gussass fauna ought to be correlated with the Cora horizon of the Ural rather than with any permian stage cannot be asserted, as *P. cancriniformis*, the only characteristic fossil, ranges from the carnian Fusulina limestone of upper carboniferous age into permian strata.

The affinities of the Chitichun fauna to the hitherto known carboniferous and permian fauna of the main region of the Himálayas are considerably less distinctly marked than those with the Salt range or Artinskian faunas.

In my monograph of the Productus shales fossils (Pt. IV of the present volume) I have laid a special stress on the strongly marked difference between the brachiopod faunas of the Chitichun limestone and of the Productus shales of Painkhanda and Johár, which, on the evidence of their intimate stratigraphical connection with the triassic Otoceras beds, must be correlated with higher stages of the permian system rather than with the permocarboniferous horizon. The detailed examination of the brachiopoda in the collection from Chitichun No. I has confirmed my view. Five species only are identical in the two rock groups. These are:—

Productus cancriniformis, Tschern.
Spirifer musakheylensis, Davids.
Martinia cf. *glabra*, Mart.
Spirigerella Derbyi, Waag.
Athyris Royssii Lév.

There is none among them, with the sole exception perhaps of *Spirigerella Derbyi*, Waag., which is characteristic of a distinct geological horizon, and we might count these species just as well among the forms identical with upper carboniferous, permocarboniferous or truly permian strata. The presence of a large number of *Producti* of a distinctly permo-carboniferous type gives a more ancient aspect to the Chitichun fauna.

A comparison of the Chitichun fauna with that of the upper carboniferous rocks of Kashmir will not be possible until the latter has been studied in detail. It is, however, apparent, both from Davidson's descriptions and from a superficial ex-

[1] Verhandlgn. K. Russ. Mineralog. Gesellsch., St. Petersburg XVIII, p. 212.

amination of the fossil materials from Kashmir entrusted to me by the Director of
the Geological Survey of India, that the affinities between the two faunas are rather
distant, only less close probably than with the Amb beds or with the upper carboniferous strata of the Carnian Alps and of Russia. In the fauna of the Barus beds
of Kashmir there are distinct indications of Australian affinities. In the Chitichun-
fauna not the slightest trace of similar affinities to Australian carboniferous fauna
has been noticed.

I cannot point out any beds in the main region of the Himálayas which, with
any probability, could be placed on a level with the Chitichun limestone. The white
crinoidal limestone, which Mr. Hughes' discovered some twenty years ago, some-
where to the north of Milam in the vicinity of one of the frontier passes leading into
Hundes, may perhaps hold a similar geological position. Of the fossils collected by
Mr. Hughes the following have been determined by Prof. Waagen :—

> *Productus semireticulatus*, Mart.
> *Lyttonia sp. ind.*
> *Martinia cf. glabra*, Mart.
> *Athyris Royssii*, Lév.
> *Hemiptychina himalayensis*, Dav.
> *Notothyris subvesicularis*, Dav.

All these species are also represented in the fauna of Chitichun No. I, but among
them *Notothyris subvesicularis* alone is fit for an approximately exact determination
of the geological age, being restricted to the upper division of the middle Productus
limestone and to the upper Productus limestone in the Salt Range. Regarding the
small number and the rather indifferent character of the rest of species, a correlation
of these crinoidal limestones with the Chitichun limestone would not be advisable,
if only based on the evidence of this single fossil. On the other hand, it is not
at all impossible, that the fossils collected by Mr. Hughes during his shooting tour
from Milam to Laptal encamping ground, have been brought from the crag of Chitichun No. I itself or from one of the neighbouring crags. This explanation is the
more plausible, as Mr. Hughes probably went to Laptal by the Chitichun area, his
Milam Pass being certainly not identical with the Utadhura which he specially
mentions in his report as a different pass.

With the upper carboniferous beds of Loping the Chitichun limestone has but
very few species in common. These are the following :—

> *Productus semireticulatus*, Mart.
> "   *mongolicus*, Diener.
> *Reticularia lineata*, Mart.
> *Martinia cf. glabra*, Mart.

' T. W. Hughes and W. Waagen, Note on a trip over the Milam Pass, Kumaon, Records Geol. Survey of India,
XI, 1878, p. 182-187. The locality, where Mr. Hughes discovered these fossiliferous limestones, cannot be made out
exactly from his description. No similar beds have been noticed by Griesbach in the main sedimentary belt of the
Central Himalayas to the north of Milam.

There are however other beds in South-eastern China, which seem to approach more closely the Chitichun limestone in their geological age. These are the beds of Yar-ka-lo, discovered by Desgodins. From this locality the following species of brachiopoda have been quoted by Loczy[1]:

> *Productus semireticulatus*, Mart.
> „ cf. *scabriculus*, Mart.
> „ *kiangsiensis*, Kayser.
> *Marginifera tibetana*, nov. sp.
> *Strophalosia*, sp. ind.
> *Uncinulus timorensis*, Beyr.
> *Reticularia indica*, Waag.

This is a rather mixed assemblage of species. *Productus kiangsiensis* has hitherto been found in the upper carboniferous beds of Loping only, but has its nearest allies (*P. tumidus*, Waagen) in geologically younger strata. *P. semireticulatus* and *P. scabriculus* extend through the entire carboniferous system into permocarboniferous strata. *Reticularia indica* is in the Salt Range restricted to the upper division of the middle Productus limestone. *Uncinulus timorensis* does not descend into geologically older strata than the Virgal beds of the middle Productus limestone.

So far, as it is possible to judge from this fauna, the beds of Yar-ka-lo appear to be not more ancient than the middle Productus limestone of the Salt Range, and they should probably be placed about on a level with the Chitichun limestone.

Another group of rocks in the Chinese province of Kiangsu, which has been discovered by F. von Richthofen near Tschu-sz'-kang on the lower Yang-tze-kiang likewise corresponds in Froeh's[2] opinion to the middle Productus limestone of the Salt Range. By the same author a permian age is ascribed to the black shales of Ning-kwo-hsien (province of Nganhwei), but the determination of two unsatisfactorily preserved species of ammonites, on which this view has been based, remains somewhat doubtful.

There is yet the probability of one correlation to be discussed, and this is a correlation of the Chitichun limestone with the permian rocks of the island of Timor, the fauna of which has been described by Beyrich, Martin and Rothpletz. The number of species of brachiopoda identical in the two faunae, is eleven or perhaps even twelve. These species are the following:—

> *Productus semireticulatus*, Mart.
> „ ? *costatusaensis*, Diener (= *an* sp. ind., Rothpletz).
> „ *gratiosus*, Waag.
> „ *Abichi*, Waag.
> *Spiriferina octoplicata*, Sow.

[1] Die wissenschaftlichen Ergebnisse der Reise des Grafen Béla Szechenyi in Ost-Asien, I. Th., Wien. 1893, p. 723.

[2] F. Froeh.—Ueber palaeozoische Faunen aus Asien und Nordafrika, Neues Jahrbuch., 1895, p. 34.

Spirifer musakheylensis, Dav.
Reticularia lineata, Mart.
Martinia nucula, Rothpl.
Athyris Royssii, Lév.
   ,,   *capillata*, Waag.
Uncinulus timorensis, Beyr.
Hemiptychina sparsiplicata, Waag.

Among the permian brachiopoda of Timor sixteen species altogether—out of twenty-six—are identical with Salt Range forms, exhibiting about equally strong affinities with the upper division of the middle Productus limestone and with the upper Productus limestone. The occurrence of *Cyclolobus persulcatus*, Rothpl., is strongly in favour of a homotaxis of the permian rocks of Timor with the Fusulina limestone of Sicily or with one of the subdivisions of the upper Productus limestone. Similar ammonites have as yet never been found in permocarboniferous strata. I consequently think that the permian rocks of Timor ought to be placed on a slightly higher level than the Chitichun limestone.

The relations, which exist between the brachiopod fauna of the Chitichun limestone and those of beds in a similar geological position in America, Spitzbergen and Armenia, are nothing less than striking, while the number of forms identical with those found in the permian rocks of Julfa is remarkably small (2).

The names and distribution of all the species of Chitichun brachiopoda, which have been found in other parts of the world, are shown in the tabular statement on p. 103.

The most important fact, appearing on a first glance at this tabular view, is the intimate affinity of the Chitichun fauna to the Productus limestone fauna of the Salt Range and its remarkable difference from the permian fauna of the Himálayan Productus shales. A special stress must be laid on this faunistic difference between the limestone crag of Chitichun No. I and all the hitherto known beds of carboniferous or permian age in the main region of the Himálayas.

Of all the other classes of organic remains in the Chitichun limestone I need say but very little. They take a too insignificant part in the known fauna to materially affect the evidence adduced by the hitherto quoted forms of invertebrates. It is however worth mentioning that the only species of coral, which is very numerously represented in the Chitichun limestone, *Lonsdaleia indica* is identical with a form from the upper division of the middle Productus limestone in the Salt Range.

With regard to the general character of the fauna described in the preceding chapter, I may sum up my view in respect to the stratigraphical position of the Chitichun limestone in the following sentence.

The Chitichun limestone is approximately homotaxial with the upper division of the middle Productus limestone (Virgal and Kálábagh beds) in the Salt Range. It probably corresponds in age to the permocarboniferous horizon (Artinskian stage) in Russia, but the description of the brachiopods from the Fusulina limestone of Sicily must be awaited for, before it is possible to decide, whether it does not hold a slightly higher position in the stratigraphical sequence, than the Artinskian deposits.

Statement showing the distribution of the Chitichun permian fauna in other regions.

	Salt Range.										Kohma and Ceal Mts.			America.	
	Lower Productus (Amb) beds.	Middle Productus Limestone		Upper Productus Limestone	Productus shales of Pershkishak	Permian rocks of Timor.	Upper carboniferous at Leping.	Permian rocks of Jalla.	Fusulina carboniferous rocks of Spiti	Fusulina of the Cortina Alps (Upper carboniferous)	Upper Carboniferous.		Artinskian strata.	Carboniferous and Permian of North America.	Coal measures of Illinois and Nebraska.
		Katta beds.	Vurgal and Kalabagh								Lower division (Moscow stage)	Upper division (Omsk-line stage and Cora beds.)			
Productus lineatus, Wang	i	...	i	i	...	...	...	...	...	i	i	i	i	...	...
Productus Cora, d'Orb.	i	i	i	i	...	...	...	...	...	i	i	i	i	i	i
Productus semireticulatus, Mart.	i	...	...	...	i	i	i	...	i	i	i	i	i	i	i
Productus chitichunensis, Diener	...	...	...	...	...	i(?)	...	...	...	...	...	...	...	...	...
Productus cf. cubensis, Wang.	...	...	n(?)	n(?)	...	...	...	...	...	...	...	...	...	...	...
Productus gratiosus, Wang.	...	i	i	i	...	i	...	...	...	n(?)	...	...	...	i...	...
Productus cancriniformis, Tsch.	...	...	...	...	i	...	...	...	...	i	...	...	i	...	...
Productus Abichi, Wang.	...	...	i	i	...	i	...	i	...	...	...	...	...	...	...
Productus mongolicus, Diener	...	...	...	...	...	...	i	...	...	...	...	...	...	...	...
Marginifera typica, Wagg.	...	i	i	i	...	...	...	...	...	...	...	...	i	...	...
Lyttonia nobilis, Wang.	...	...	i	i	...	...	...	...	.	...	...	...	...	...	...
Spiriferina octoplicata, Sow.	i	i	i	i	...	i	...	...	i(?)	...	i	i	i	...	...
Spirifer musakheylensis, Dav.	i	i	i	i	i	i	...	...	i	i	...	...	n(?)	...	...
Spirifer Wynnei, Wang.	...	...	i	...	...	...	...	...	...	...	...	...	i(?)	i	...
Retinularia lineata, Mart.	i	...	...	...	...	i	i	i	...	i	i	i	i	...	...
Martinia cf. glabra, Mart.	n(?)	...	...	...	i	...	n(?)	...	...	i	i	...	...	...	...
Martinia ovonis, Rothpl.	...	...	...	...	...	i	...	...	...	...	...	...	...	...	...
Martinia semiplana, Wang.	...	...	i	...	...	...	...	...	...	i	...	...	n(?)	...	...
Martinia contracta, M. and W.	...	...	...	...	...	...	...	...	...	...	...	...	...	i	...
Spiriguella Derbyi, Wang.	...	...	i	i	i	...	...	...	...	...	...	...	...	...	i
Spiriguella grandis, Wang.	...	...	i	i	...	...	...	...	...	...	...	...	...	...	...
Athyris Roysii, Lêv.	i	i	i	i	i	i	...	...	i	...	i	i	...	...	...
Athyris subexpansa, Wang.	...	i	i	i	...	...	...	...	...	...	...	...	...	...	...
Athyris capillata, Wang.	...	...	i	i	...	i	...	...	...	...	...	...	...	...	...
Uncinulus timorensis, Beyr.	...	...	i	i	...	i	...	...	...	...	...	...	...	...	...

Statement showing the distribution of the Chitchan permian fauna in other regions—concld.

| | SALT RANGE. | | | | | | | | | | RUSSIA AND URAL MTS. | | | AMERICA. | |
| | | Middle Productus Limestone | | | | | | | | | Upper Carboniferous | | | | |
	Lower Productus Limestone (Amb beds)	Katta beds	Virgal and Kuttibugh beds	Upper Productus Limestone	Productus shales of Palækhenda (permian)	Permian rocks of Timor	Upper carboniferous of Lagang	Permian rocks of India	Permo carboniferous rocks of Spitzbergen	Fusulina of the Carnic Alps (upper carboniferous)	Lower division (Moscovian stage)	Upper division (Carbonian stage and Cora horizon)	Artinskian stage	Carboniferous and Permian of North America	Coal measures of Baghista and Turkestan
Camarophoria Purdoni, Davids.	...	...	i	i	...	...	...	...	...	..	...	i	i	...	...
Dielasma biplex, Waag.	...	..	i	...	...	...	...	...	...	...	...	...	...	...	...
Hemiptychina sparsiplicata, W.	i	i	i	...	i	...	...	...	...	...	...	...	...	...	...
Hemiptychina himalayensis, Dav.	i	i	i	i	...	...	...	...	...	...	...	...	...	...	...
Hemiptychina inflata, Waag.	...	...	i	i	...	...	...	...	...	...	...	...	...	...	...
Notothyris cf. subvesicularis, Dav.	...	...	i	...	...	...	...	...	...	...	...	...	...	...	...
	10	9	23	10	5	12	4	2	3	10	6	7	12	4	5

Statement of the approximate correlation of permian and upper carboniferous horizons of pelagic type.

		Mediterranean Region.	Russia and Ural Mts.	Salt Range.	Himalayas.	Central Asia.	Arctic-Pacific Region.	North America.
Permian.	Upper Division.	Bellerophon limestone of Tirol. Ozarppa beds of Jaffa.	Newkisk and Inland Sea Deposits.	Chidru beds.	Productus shales of Thmishsuda and Jabbar.	Cephalopoda beds of Washington.		Whichita beds of Texas.
							Fusulinalimestone rocks of Spitzbergen ?	Fusunulinaolinterous of Kansas and Nebraska.
	Middle Division.	Fusulina limestone of Sicily.	Artinskian stage.	Lala beds.		Cancrinellovich horizon of Gumsel ?	Permian rocks of Timor. Beds of Vardack.	
				Upper Productus Limestone: Kittilagh beds. Virgal beds. Katta beds.	Chidebhan limestone.	Cephalopoda beds of Darvas.		
	Lower Division. (Permocarboni-(serics).)							
Upper Carboniferous.	Upper Division.	Trogkofelkalk. Carnian Fusulina limestone of the Krone.	Schwagerina horizon. Over horizon (Ural). Gzhelan stage of Central Russia.	Amb beds.	Barus beds of Kashmir.	Carboniferous limestone of Fort Brisant.	Limestone of Lopina (China).	

PLATE I.

Pl. I.

CHITICHUN · FOSSILS (HIMALAYA)

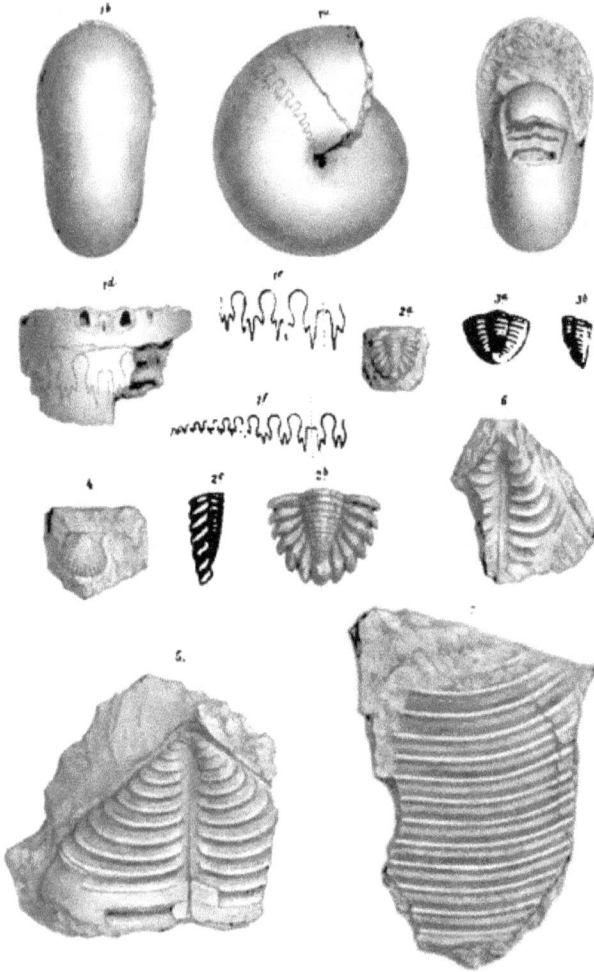

PLATE II.

Fig.   1, 3, 5.   PRODUCTUS SEMIRETICULATUS, Mart. Fig. 1, fairly complete specimen, representing the typical form of the species. 1a, ventral view ; 1b, dorsal view ; 1c, lateral view ; 1d, apical view. Fig. 3, largest specimen known to me. 3a, ventral view ; 3b, front view. Fig. 5, cast of the dorsal valve. 5b, lateral view.

  ,,   2, 4.   PRODUCTUS BOLIVIENSIS VAR. CHITICHONENSIS, two ventral valves. Figs. 2 and 4a, ventral views ; 4b, apical view ; 4c, lateral view.

  „   6.   PRODUCTUS CF. SUBCOSTATUS, Waagen. Cast of a dorsal valve. 6a, dorsal view ; 6b, lateral view ; 6c, part of the sculpture, exhibiting the deep grooves along the inner margins of the wings.

CHITICHUN - FOSSILS (HIMALAYA)

Pl. II.

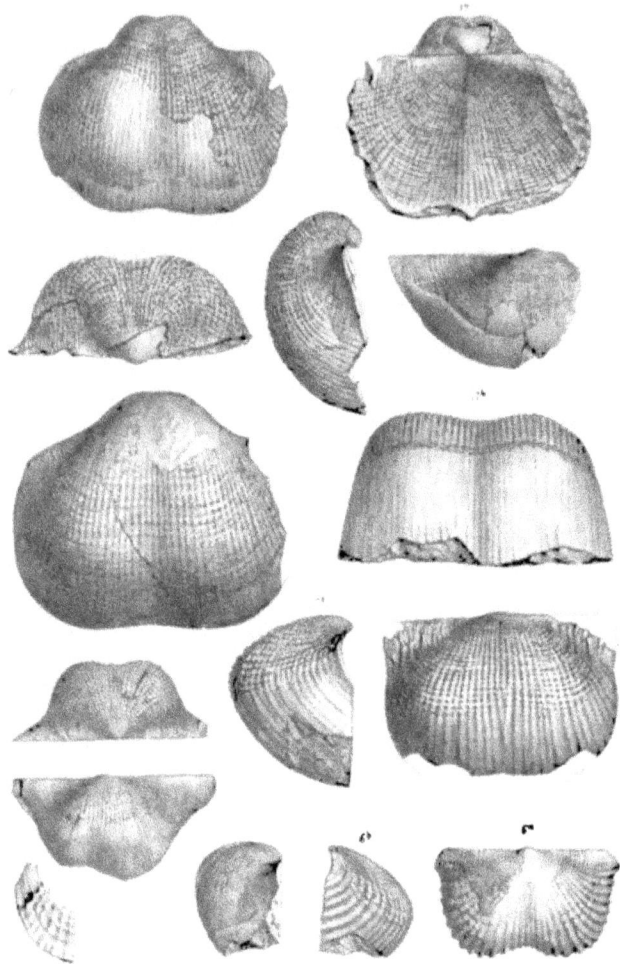

A. Swoboda del. et lith. Th. Bannwarth, Wien.

PLATE III.

CHITICHUN - FOSSILS (HIMALAYA)

Pl. III

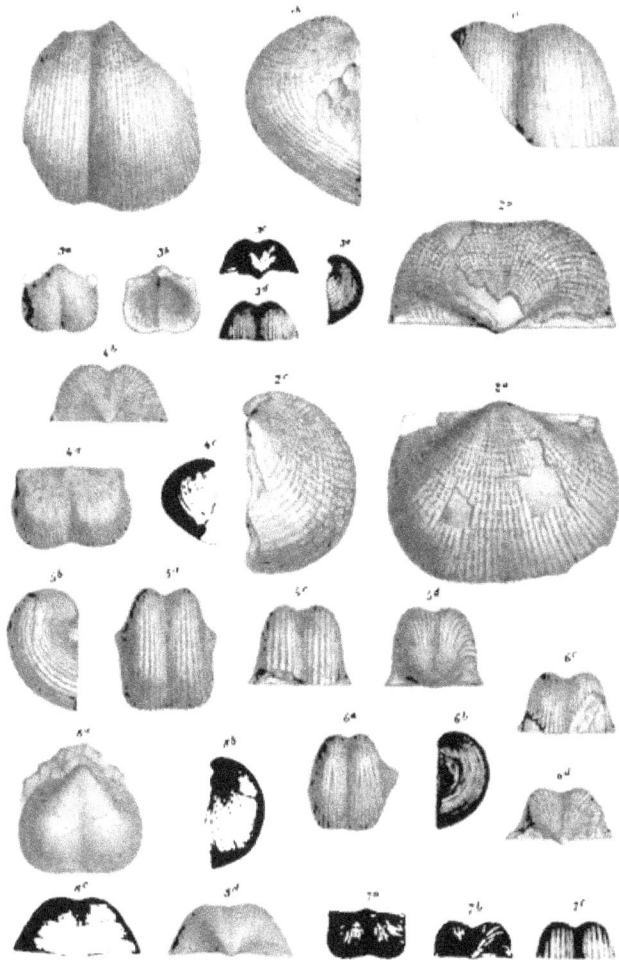

PLATE IV.

Figs.   1.   PRODUCTUS CORA, d'Orbigny. 1*a*, ventral view; 1*b*, lateral view; 1*c*, front view; 1*d*, apical view.

   „   2—5.   PRODUCTUS LINEATUS, Waagen. 2*a*, ventral view; 2*b*, lateral view; 2*c*, apical view. Fig. 3, ventral valve of a deeply sinuated specimen. Figs. 4, 5, largest specimens known to me. *a*, ventral view; *b*, lateral view; *c*, front view; *d*, apical view.

   „   6, 7.   PRODUCTUS CANCRINIFORMIS, Tschernyschew. Partially preserved ventral valves. *a*, ventral view; *b*, lateral view; *c*, front view; *d*, apical view.

   „   8, 9, 10.   PRODUCTUS MONGOLICUS, Diener. Fig. 8, ventral valve of a tolerably complete specimen. *a*, ventral view; *b*, lateral view; *c*, front view; *d*, apical view. Fig. 9, ventral valve with strongly geniculated trail. *a*, ventral view; *b*, lateral view; *c*, dorsal view. Fig. 10, fragment of a very large ventral valve.

   „   11, 12, 13.   MARGINIFERA TYPICA, Waagen. *a*, ventral view; *b*, dorsal view; *c*, lateral view; *d*, front view; *e*, apical view. Fig. 12, dorsal view of a specimen with wings broken off, thus showing the marginal ridges.

CHITICHUN - FOSSILS (HIMALAYA)

Pl. IV

Pl. VI.

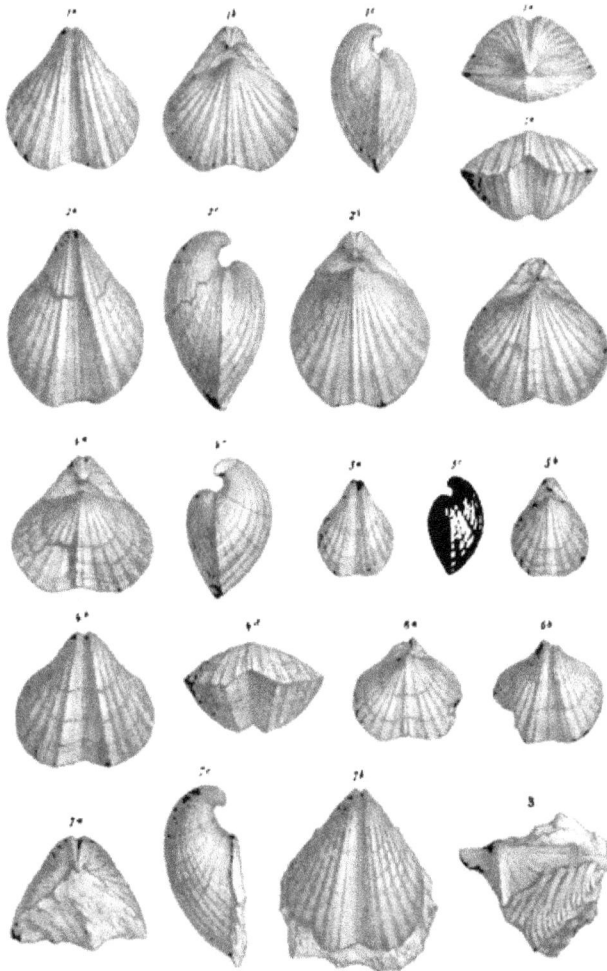

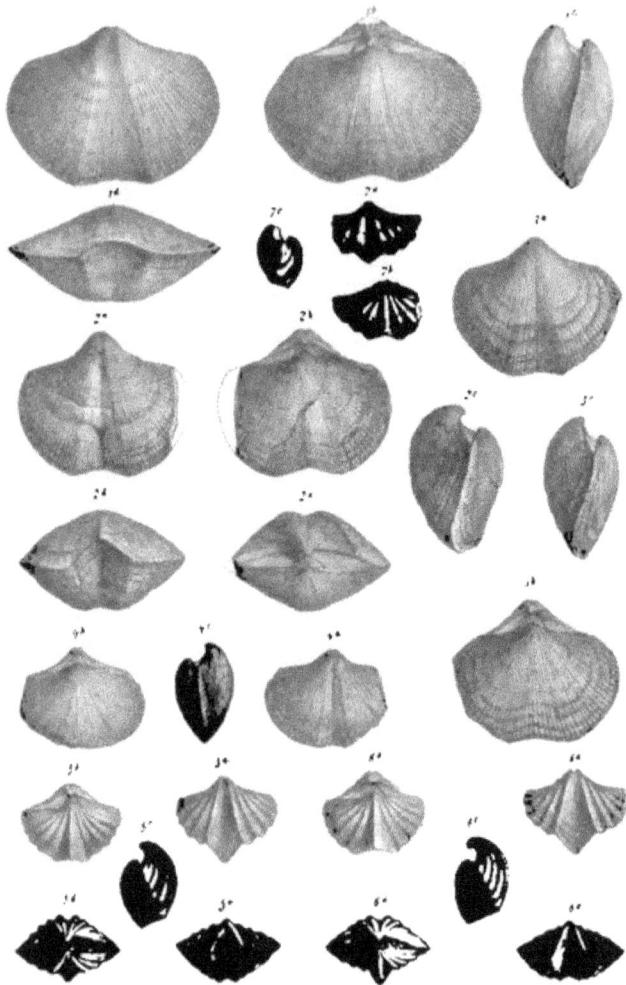

PLATE VIII.

Figs. 1, 2. MARTINIA ELEGANS, Diener. Fig. 1, type specimen. *a*, ventral view ; *b*, dorsal view ; *c*, lateral view ; *d*, front view ; *e*, apical view. Fig. 2, ventral valve with partly preserved shell. 2*b*, part of the shell enlarged, to show the punctation of the surface.

"   3, 4. MARTINIA ACUTOMARGINALIS, Diener. Fig. 3, type specimen. *a*, dorsal view ; *b*, ventral view ; *c*, lateral view ; *d*, front view. Fig. 4 *a*, ventral view ; *b*, dorsal view ; *c*, lateral view ; *d*, front view.

"   5, 6. MARTINIA NUCULA, Rothpletz. *a*, ventral view ; *b*, dorsal view ; *c*, lateral view ; *d*, front view.

"   7. MARTINIA SEMIPLANA, Waagen. *a*, dorsal view ; *b*, ventral view ; *c*, lateral view ; *d*, front view.

Pl. VIII

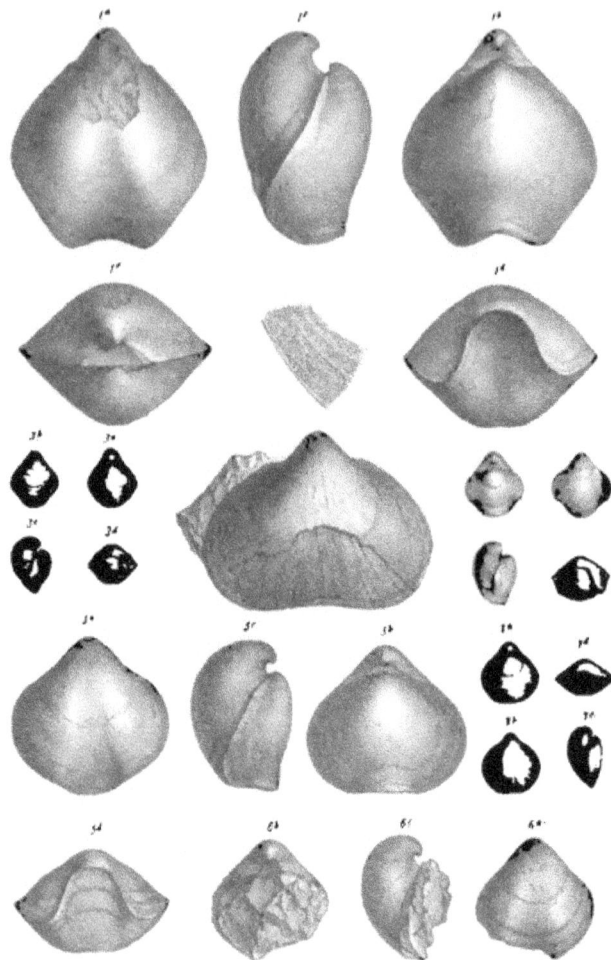

CHITICHUN- FOSSILS (HIMALAYA)

Pl. IX.

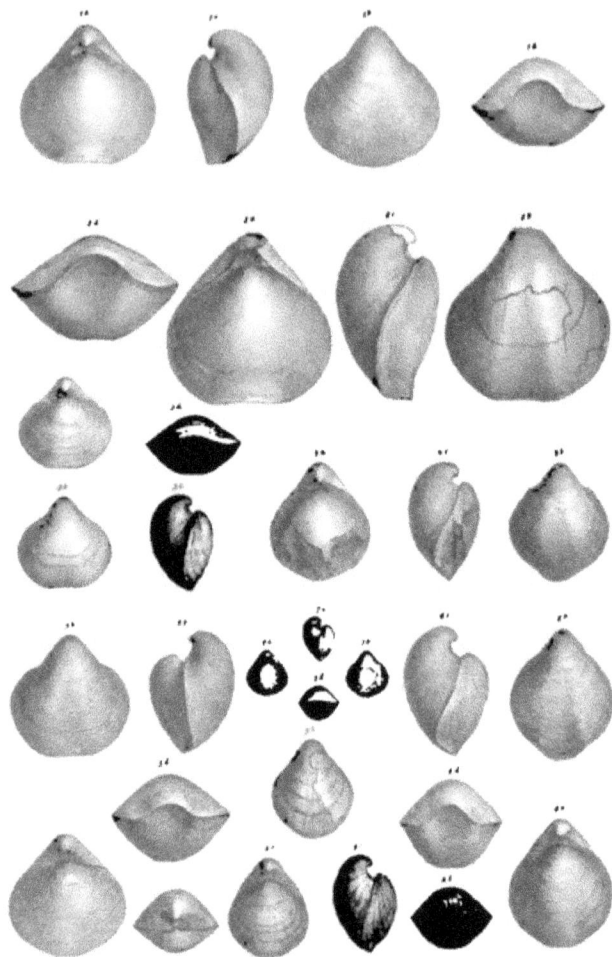

PLATE X.

Figs. 1, 2, 3, 6. ATHYRIS ROISSI, Léveillé. Fig. 6, specimen, with partly preserved internal characters, showing a portion of the spira. appendages.

,,   4. ATHYRIS SUBEXPANSA, Waagen.

,,   5. ATHYRIS CAPILLATA, Waagen. *a*, dorsal view ; *b*, ventral view ; *c*, lateral view ; *d*, front view.

,,   7—10. UNCINULUS TIMORENSIS, Beyrich. Fig. 7, very large specimen with a broad median fold. 7*a*, ventral view ; 7*b*, dorsal view ; 7*c*, lateral view ; 7*d*, front view. Fig. 8, specimen, agreeing with Waagen's *Uncinulus Theobaldi* from the Salt Range Productus limestone. 8*a*, dorsal view ; 8*b*, ventral view ; 8*c*, lateral view ; 8*d*, front view. Fig. 9, lateral view of a specimen with a strongly impressed ventral valve. Fig. 10, specimen with partly preserved shell and strongly developed sculpture ; ventral view.

CHITICHUN-FOSSILS (HIMALAYA)

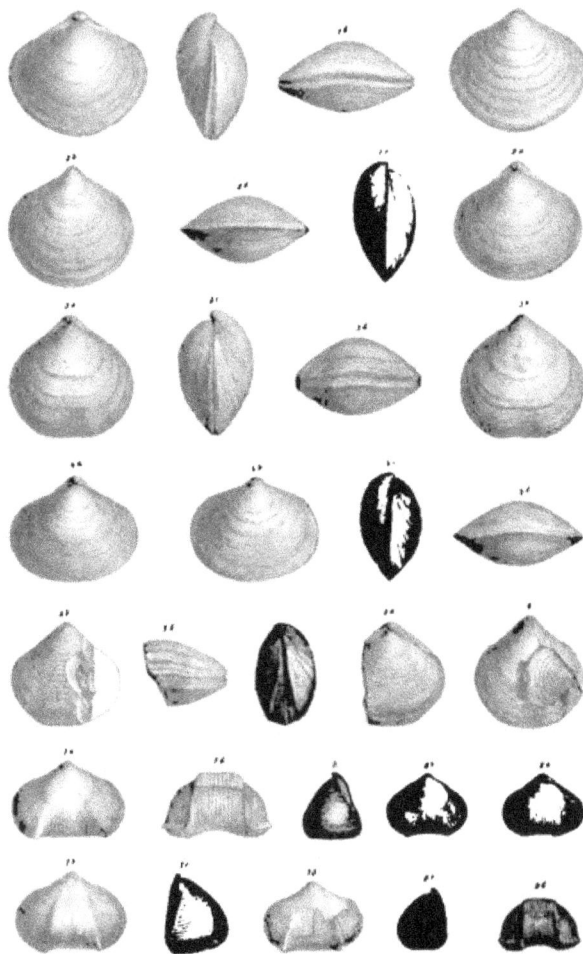

Geol. Surv. of India

CHITICHUN-FOSSILS (HIMALAYA)

Pl. XI.

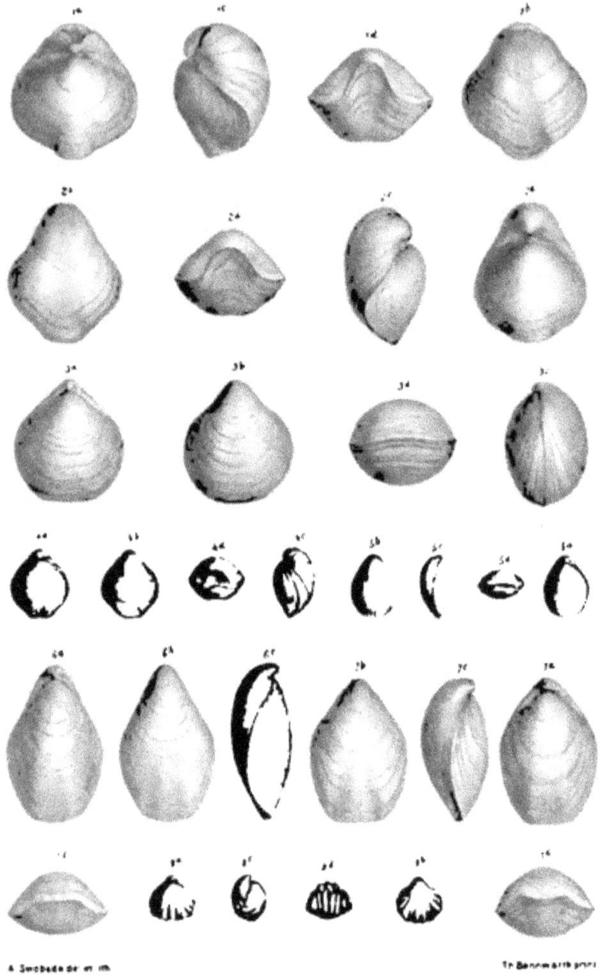

A. Swoboda del. et lith.

Th. Bannwarth print.

Pl XII.

PLATE XIII.

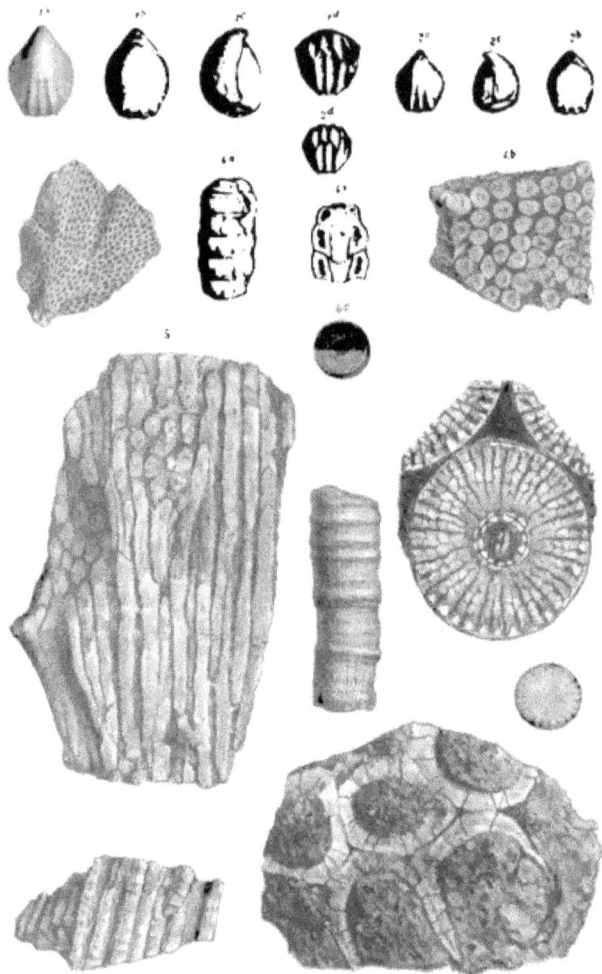

CHITICHUN-FOSSILS (HIMALAYA)

RECORDS OF THE GEOLOGICAL SURVEY OF INDIA.

VOLS. I TO XXX, 1868 TO 1897.

The RECORDS of the Geological Survey are issued quarterly,—in February, May, August, and November. They contain brief reports and papers; abstracts of more detailed work; notices of recent discoveries; donations to Museum; new additions to library, etc. They are of the same size as the 'Memoirs,' but are separately paged. Begun in June 1868.

The annual subscription for four numbers or parts is 2 Rs. (4s.). Postage additional; if for India, 4 As. for Great Britain, 8 As. (1s.).

MISCELLANEOUS PUBLICATIONS.

A Manual of the Geology of India. 4 Vols. With map. 1879-1887—
  Vol. 1. Peninsular Area. } By H. B. Medlicott and W. T. Blanford. Price 8 rupees (out of print).
  Vol. 2. Extra-Peninsular Area. }
  Vol. 3. Economic Geology. By V. Ball. Price 5 rupees (out of print).
  Vol. 4. Mineralogy. By F. R. Mallet. Price 2 rupees.
A Manual of the Geology of India, 2nd edition. By R. D. Oldham. (1893.) Price 8 rupees.
Popular guides to the geological collections in the Indian Museum, Calcutta—
  No. 1. Tertiary vertebrate animals. By R. Lydekker. 1879. Price 2 annas (out of print).
  No. 2. Minerals. By F. R. Mallet. (1879.) Price 2 annas.
  No. 3. Meteorites. By F. Fedden. (1880.) Price 2 annas.
  No. 4. Palæontological collections. By O. Feistmantel. (1881.) Price 2 annas.
  No. 5. Economic mineral products. By F. R. Mallet. (1883.) Price 2 annas.
Descriptive catalogue of the collection of Minerals in the Geological Museum, Calcutta. By F. R. Mallet. (1881.) Price 2 rupees.
An Introduction to the Chemical and Physical study of Indian Minerals. By T. H. Holland. (1895.) Price 8 annas.
Catalogue of the remains of Siwalik Vertebrata contained in the Geological Department of the Indian Museum. By R. Lydekker. Pt. 1. Mammalia. (1885.) Price 1 rupee. Pt. II. Aves, Reptilia, and Pisces. (1886.) Price 4 annas.
Catalogue of the remains of Pleistocene and Pre-Historic Vertebrata contained in the Geological Department of the Indian Museum. By R. Lydekker. (1886.) Price 4 annas.
Bibliography of Indian Geology. By R. D. Oldham. (1888.) Price 1 rupee 8 annas.
Report on the Inspection of Mines in India for 1893-94. By James Grundy. (1894.) Price 1 rupee.
Report on the Inspection of Mines in India for 1894-95. By James Grundy. (1895.) Price 2 rupees.
Report on the Inspection of Mines in India for 1895-96. By James Grundy. (1896.) Price 1 rupee.
Report on the Geological Structure and Stability of the hill slopes around Naini Tal. By T. H. Holland. (1897.) Price 3 rupees.
Geological Map of India. (1893.) Scale 1″=96 miles. Price 1 rupee per copy.

To be had on application to the Registrar, Geological Survey of India, Calcutta. London: Kegan Paul, Trench, Trübner & Co.

PALÆONTOLOGIA INDICA.

The price fixed for these publications is 4 annas (6 pence) per single plate.

To be had at the Geological Survey Office, Calcutta.—London: Kegan Paul, Trench, Trübner & Co.